Commercial Agriculture,
THE
Slave Trade
& Slavery
IN Atlantic
Africa

WESTERN AFRICA SERIES

The Economics of Ethnic Conflict:
The Case of Burkina Faso
Andreas Dafinger

Commercial Agriculture, the Slave Trade
& Slavery in Atlantic Africa
Edited by Robin Law, Suzanne Schwarz
& Silke Strickrodt

Afro-European Trade in the Atlantic World:
The Western Slave Coast c1550–c1885
Silke Strickrodt

Sects & Social Disorder:
Muslim Identities & Conflict in Northern Nigeria
Edited by Abdul Raufu Mustapha

*Creed & Grievance:
Muslims, Christians & Society in Northern Nigeria
Edited by Abdul Raufu Mustapha & David Ehrhardt

*forthcoming

Commercial Agriculture, THE Slave Trade & Slavery IN Atlantic Africa

Edited by
Robin Law, Suzanne Schwarz
& Silke Strickrodt

JAMES CURREY

James Currey
is an imprint of Boydell & Brewer Ltd
PO Box 9
Woodbridge, Suffolk IP12 3DF (GB)
www.jamescurrey.com

and of

Boydell & Brewer Inc.
668 Mt Hope Avenue
Rochester, NY 14620-2731 (US)
www.boydellandbrewer.com

© Contributors 2013

First published 2013
First published in paperback 2016

All Rights Reserved. Except as permitted under current legislation
no part of this work may be photocopied, stored in a retrieval system,
published, performed in public, adapted, broadcast,
transmitted, recorded or reproduced in any form or by any means,
without the prior permission of the copyright owner.

British Library Cataloguing in Publication Data
A catalogue record is available on request from the British Library

ISBN: 978-1-84701-136-7 (James Currey paperback)

The publisher has no responsibility for the continued existence or accuracy of URLs for
external or third-party internet websites referred to in this book, and does not guarantee that
any content on such websites is, or will remain, accurate or appropriate.

This publication is printed on acid-free paper.

Typeset in 12/13 Bembo with Albertus MT display by Avocet Typeset, Somerton, Somerset

To the memory of Yaw Bredwa-Mensah

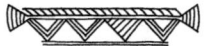

Contents

Acknowledgements	ix
Notes on Contributors	x
List of Maps, Tables and Figures	xii
List of Abbreviations	xiv

Introduction   1
ROBIN LAW, SUZANNE SCHWARZ
& SILKE STRICKRODT

1
The slave trade and commercial agriculture
in an African context   28
DAVID ELTIS

2
São Tomé and Príncipe: The first plantation economy
in the tropics   54
GERHARD SEIBERT

3
The export of rice and millet from Upper Guinea
into the sixteenth-century Atlantic trade   79
TOBY GREEN

4
'Our indico designe': Planting and processing indigo
for export, Upper Guinea Coast, 1684–1702   98
COLLEEN E. KRIGER

5
'There's nothing grows in the West Indies but will grow here': Dutch and English projects of plantation agriculture on the Gold Coast, 1650s–1780s   116
ROBIN LAW

6
The origins of 'legitimate commerce'   138
CHRISTOPHER LESLIE BROWN

7
A Danish experiment in commercial agriculture on the Gold Coast, 1788–93   158
PER HERNÆS

8
'The colony has made no progress in agriculture': Contested perceptions of agriculture in the colonies of Sierra Leone and Liberia   180
BRONWEN EVERILL

9
Church Missionary Society projects of agricultural improvement in nineteenth-century Sierra Leone and Yorubaland   203
KEHINDE OLABIMTAN

10
Agricultural enterprise and unfree labour in nineteenth-century Angola   225
ROQUINALDO FERREIRA

11
Commercial agriculture and the ending of slave-trading and slavery in West Africa, 1780s–1920s   243
GARETH AUSTIN

Index   266

Acknowledgements

The editors offer their profound thanks to the German Historical Institute London for its material and moral support for the conference in September 2010 from which this volume derives, and in the production of the volume itself. Thanks also to Angela Davies and Jane Rafferty, who assisted in the work of copy-editing, and Damien Bove, who prepared the maps. The volume is dedicated to the memory of Dr Yaw Bredwa-Mensah, who was a participant in the 2010 conference, and thus a potential contributor to this volume, but suffered a tragically premature death only a few months later, on 27 March 2011.

Notes on Contributors

GARETH AUSTIN is a professor in the Department of International History at the Graduate Institute of International and Development Studies, Geneva; author of *Labour, Land and Capital in Ghana: From Slavery to Free Labour in Asante, 1807–1956* (2005); and formerly an editor of the *Journal of African History*.

CHRISTOPHER LESLIE BROWN is Professor of History at Columbia University; author of *Moral Capital: Foundations of British Abolitionism* (2006), winner of the James A. Rawley Prize in Atlantic History, the Morris D. Forkosch Prize in British History, and the Frederick Douglass Book Prize of the Gilder Lehrman Center for the Study of Slavery, Resistance and Abolition; and co-editor of *Arming Slaves: From Classical Times to the Modern Age* (2006).

DAVID ELTIS is Professor Emeritus at Emory University, Atlanta; author of *Economic Growth and the Ending of the Transatlantic Slave Trade* (1987) and *The Rise of African Slavery in the Americas* (2000); co-compiler of *Atlas of the Transatlantic Slave Trade* (2010), and of *The Transatlantic Slave Trade: A Database on CD-ROM* (2011); and co-editor of vol. III of the *Cambridge World History of Slavery* (2011) and of *Extending the Frontiers: Essays on the New Transatlantic Slave Trade Database* (2008).

BRONWEN EVERILL is Lecturer in History at Gonville & Caius College, University of Cambridge; author of *Abolition and Empire in Sierra Leone and Liberia* (2013) and editor of *The History and Practice of Humanitarian Intervention and Aid in Africa* (2013).

ROQUINALDO FERREIRA is the Vasco da Gama Associate Professor of Early Modern Portuguese History in the History Department and the Department of Portuguese and Brazilian Studies at Brown University, Providence, Rhode Island; and author of *Cross Cultural Exchange in the Atlantic World: Angola and Brazil during the Era of the Slave Trade* (2012).

TOBY GREEN is Lecturer in Lusophone African History and Culture, King's College, University of London; author of *The Rise of the Trans-Atlantic Slave Trade in Western Africa, 1300–1589* (2012), and editor of *Brokers of Change: Atlantic Commerce and Cultures in Pre-Colonial Western Africa* (2012); and a member of the Council of the African Studies Association of the UK.

PER HERNÆS is Professor of African History at the Norwegian University of Science and Technology, Trondheim; author of *Modernizing Ghanaian Fisheries* (1991) and *Slaves, Danes, and African Coast Society* (1996); and Editor of the *Transactions of the Historical Society of Ghana*.

COLLEEN KRIGER is Professor of History at the University of North Carolina at Greensboro; author of *Pride of Men: Ironworking in Nineteenth-Century West Central Africa* (1999) and *Cloth in West African History* (2006), and a member of the editorial board of *African Economic History*. She is currently conducting research on Anglo-African trade along the Upper Guinea coast in the late seventeenth century.

ROBIN LAW is Emeritus Professor of African History, University of Stirling, and Honorary Senior Research Fellow in History, University of Liverpool; author of *The Slave Coast of West Africa 1550–1750* (1991) and *Ouidah: The Social History of a West African Slaving 'Port', 1727–1892* (2004); editor of *From Slave Trade to 'Legitimate' Commerce: The Commercial Transition in Nineteenth-Century West Africa* (1995); and formerly an editor of the *Journal of African History*.

KEHINDE OLABIMTAN teaches in the Department of Philosophy and Religious Studies, Bowen University, Iwo, Osun State, Nigeria. He is author of *Samuel Johnson of Yorubaland, 1846–1901* (2013).

SUZANNE SCHWARZ is Professor of History, University of Worcester; editor of *Slave Captain: The Career of James Irving in the Liverpool Slave Trade* (revised edn 2008) and co-editor of *Liverpool and Transatlantic Slavery* (2007). She is currently writing a book entitled *An Early African Colony: Contested Freedom, Identity and Authority in Sierra Leone*.

GERHARD SEIBERT is Associate Professor at the Universidade da Integração Internacional da Lusofonia Afro-Brasileira (UNILAB), São Francisco do Conde, Bahia, Brazil; and author of *Comrades, Clients and Cousins: Colonialism, Socialism and Democratization in São Tomé and Príncipe* (2006).

SILKE STRICKRODT is a Visiting Research Fellow in the Department of African Studies and Anthropology at the University of Birmingham; author of *'Those Wild Scenes': Africa in the Travel Writings of Sarah Lee (1791–1856)* (1998); and *Afro-European Trade in the Atlantic World: The Western Slave Coast, c. 1550–c. 1885* (2015); and editor of the first volume of *Women Writing Home, 1700–1920: Female Correspondence across the British Empire*, dealing with Africa (2006).

List of Maps, Tables & Figures

MAP

1 West Africa   xv
2 The Gold Coast   xv

TABLES

1.1 Estimated five-year annual average values of slaves and produce leaving Africa via the Atlantic in constant pounds Sterling, 1681–1807   33
1.2 Estimated five-year annual average values of provisions for slaves carried off from Africa and crews of vessels trading on the African coast, 1681–1807   39
1.3 Derivation of estimates of slave departures and produce exports from Africa into the Atlantic World, 1681–1807   49
2.1 Sugar mills in São Tomé, 1517–1736   71
2.2 Estimates of São Tomé's sugar production, 1517–1684   73
8.1 Liberian trade returns for 1859   199
11.1 Real prices of slaves in, and numbers of slaves shipped from, West Africa 1783–1807   247
11.2 Real prices of slaves in West Africa and Angola compared, 1783–1807   248
11.3 Slaves shipped from the seven 'coasts' of Western Africa, 1785–7 and 1804–6   252
11.4 Share of cowries in value of commodities shipped to the Bight of Benin from Britain, 'select years 1681–1724'   255
11.5 Cowrie imports to West Africa (lbs avoirdupois) by decade, 1700–1799   256

FIGURES

8.1	Sierra Leone imports and exports in Sterling	190
11.1	Slaves shipped from West Africa, 1784–1807: volume and price	247
11.2	Slaves embarked by region, 1783–1807	251

List of Abbreviations

ACS	American Colonization Society
AEH	*African Economic History*
AHA	Archivo Histórico de Angola
AHU	Archivo Histórico Ultramarino, Lisbon
AIS	African Improvement Society
CMS	Church Missionary Society
EHR	*Economic History Review*
GHIL	German Historical Institute London
HA	*History in Africa*
IJAHS	*International Journal of African Historical Studies*
JAH	*Journal of African History*
JEH	*Journal of Economic History*
JHSN	*Journal of the Historical Society of Nigeria*
JICH	*Journal of Imperial and Commonwealth History*
RAC	Royal African Company
RBGK	Royal Botanical Garden Kew
S&A	*Slavery and Abolition*
TASTD	Trans-Atlantic Slave Trade Database, ed. David Eltis, Martin Halbert et al. (Emory University, 2008)
THSG	*Transactions of the Historical Society of Ghana*
TNA	The National Archives, London
WIC	West India Company
WMQ	*William and Mary Quarterly*

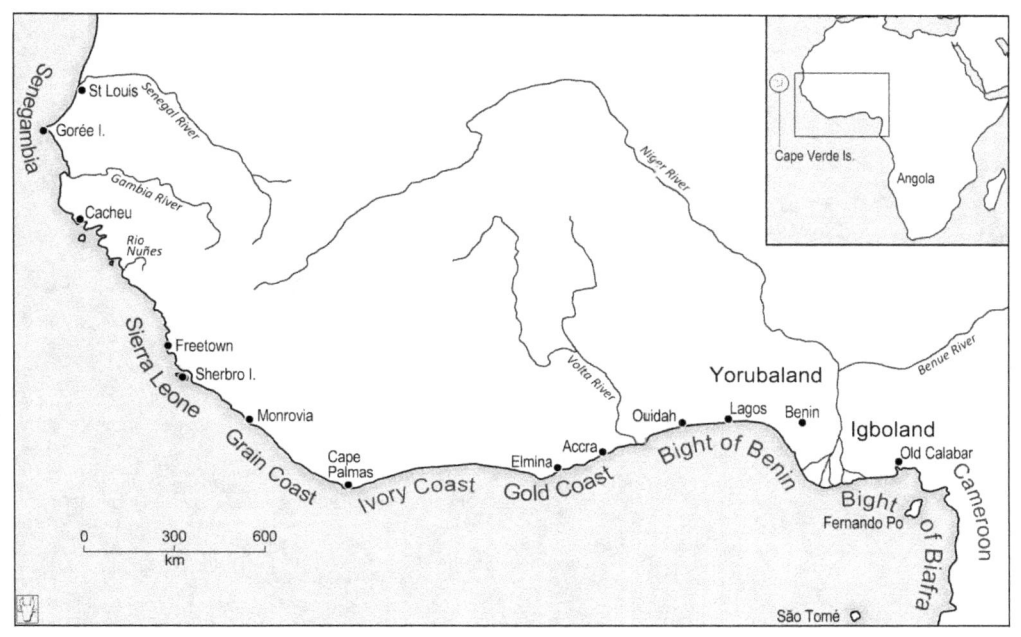

Map 1 West Africa

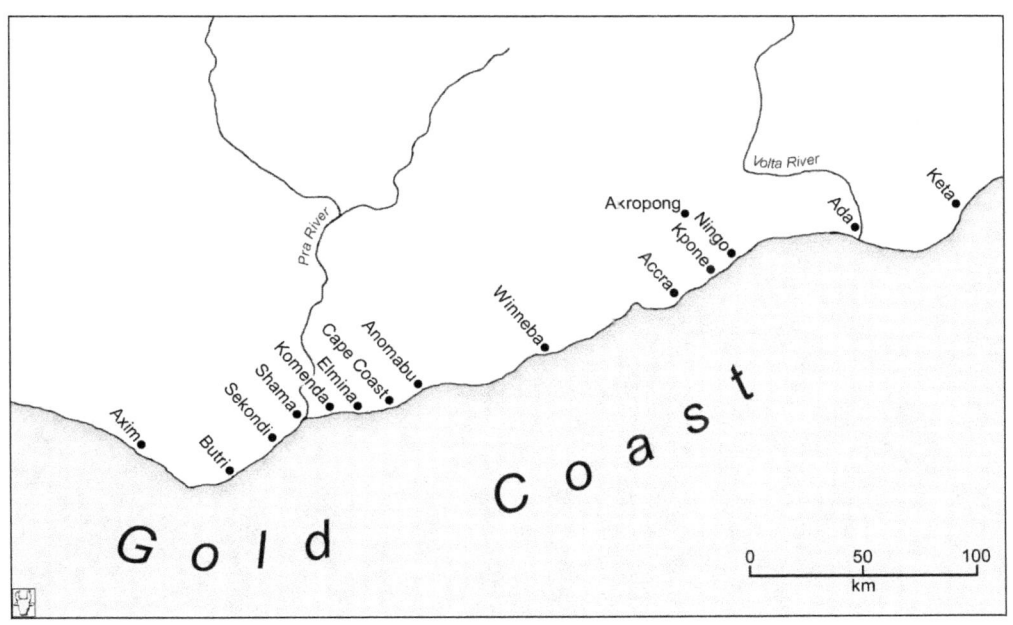

Map 2 The Gold Coast

Introduction

ROBIN LAW, SUZANNE SCHWARZ
& SILKE STRICKRODT

This volume presents a selection of papers from a conference held at the German Historical Institute London (GHIL), in September 2010, on the topic of 'Commercial Agriculture in Africa as an Alternative to the Slave Trade'. This Introduction begins by situating this topic in its context within the history and historiography of western Africa.[1]

The idea of 'legitimate commerce'

The idea of promoting the export of agricultural produce from Africa first became central to European thought in the context of the campaign to end the trans-Atlantic slave trade from the late eighteenth century onwards, with actual projects on the ground in West Africa beginning with Danish attempts to establish plantations on the Gold Coast (modern Ghana) from 1788, followed by the British[2] colony of Sierra Leone, after it was taken over by the Sierra Leone Company in 1791.[3] After the legal abolition of the slave trade in the early nineteenth century, proposed commercial alternatives to it became known in contradistinction as 'legitimate' (or 'legal' or 'lawful') commerce (or

---

[1] The editors thank the contributors to this volume, and also Tony Hopkins, for their comments on a draft of this Introduction.
[2] This volume covers a period before and after the Act of Union of 1707, which united England (and Wales) with Scotland to become the United Kingdom of Great Britain. As far as is practically possible, therefore, nomenclature is used appropriate to the historical context, though a post-1707 geographical mention may specifically indicate England. That the Royal African Company of England retained that name in full may, on occasion, create apparent inconsistencies.
[3] The Danish project is dealt with by Per Hernæs, this volume, chapter 7. For Sierra Leone, see Suzanne Schwarz, 'Commerce, Civilization and Christianity: The Development of the Sierra Leone Company', *Liverpool and Transatlantic Slavery*, ed. David Richardson, Suzanne Schwarz and Anthony Tibbles (Liverpool, 2007), 252–76.

trade).⁴ Strictly, the term 'legitimate commerce' designated trade in anything other than slaves, including non-agricultural commodities such as gold and ivory, but in practice interest was mainly concentrated on the promotion of commercial agriculture.

Rationales for this interest in agricultural exports were various. Most immediately, it was thought that West Africa could take the place of the Americas as a supplier of sugar and other tropical products to Europe, with African labour retained and employed locally rather than being transported across the Atlantic, thereby dispensing with the need for the trans-Atlantic slave trade: this was, for example, the thinking behind the Danish project of 1788. Beyond this, there was the desire to develop a form of trade which it was hoped would be more beneficial to Africa itself, given that the impact of the slave trade was commonly assumed to have been negative and destructive. William Wilberforce, for example, proposing abolition of the slave trade to the British Parliament in 1789, argued that the development of alternative forms of trade to replace it would represent a means of compensating for the harm which Europeans, through the slave trade, had done to Africa: 'Let us make reparation to Africa, as far as we can, by establishing a trade upon true commercial principles.'⁵

Further, it was recognized that African rulers and merchants could not be expected to give up the slave trade unless they were offered alternative means of obtaining the imported European and American goods which they had become accustomed to consume. Over time, indeed, the promotion of commercial alternatives came to be seen as a principal means of ending the slave trade, by stimulating a transformation of African societies, diverting their energies from warfare, which was the main source of slaves for sale into export, into productive labour. This idea can already be found in the plans of the Sierra Leone Company in the 1790s,⁶ but gained wider currency after the legal abolition of the slave trade in the early nineteenth century, in the context of the campaign to suppress the now illegal trade. Initially, efforts to end the slave trade, led by Britain, had concentrated on the demand side, in the Americas, by securing the legal prohibition of slave imports by all interested states, and employing naval patrols to police this ban by intercepting illegal slave-ships. But frustration at

---

⁴ The term was evidently coined only after the slave trade became illegal in 1807. 'Legitimate trade' was already used in *Report of the Committee of the African Institution: Read to the General Meeting on the 15th July 1807* (London, 1807). Even earlier, however, Abolitionists had employed similar terminology, referring e.g. to 'a fair and honourable commerce': Anthony Benezet, *Some Historical Account of Guinea* (2nd edn, London, 1789), 58, 123.

⁵ Speech in the House of Commons, 12 May 1789.

⁶ Schwarz, 'Commerce', 253.

the perceived failure of this approach led to increased interest in action on the supply side, in Africa, persuading Africans to cease to sell slaves by offering them what it was assumed would be a more attractive alternative. This approach achieved its classic formulation in Thomas Fowell Buxton's project of a 'Remedy' for the slave trade, published in 1840, which argued that 'Legitimate trade would put down the Slave Trade, by demonstrating [to Africans] the superior value of man as a labourer on the soil, to man as an object of merchandise.'[7]

From the perspective of European interests, it was also argued that alternative forms of trade might represent a more effective means of realizing the economic potential of Africa, including the possibility that developing export agriculture there might be more profitable than the existing trade in slaves, for Europeans as well as Africans. This argument played a critical role for the Abolitionist campaign in combating the widespread disposition to believe that abolition of the slave trade, even if agreed to be morally desirable, was not practically feasible, given its economic importance. As Chris Brown has persuasively argued, the project of African commercial agriculture served to underwrite the view that ending the slave trade need not be an economic disaster, but might even prove a benefit to Europe: it thus enabled the Abolitionists 'to present a moral case in the language of commercial and national interest', making Abolition 'not merely a humane wish but a viable political project'.[8]

The appeal of the idea was enhanced by the circumstance that it could also be connected to changing perceptions of the nature of British society and its interests, with focus in particular on the developing industrialization of the economy, and consequently on the raw materials and markets which British manufacturers required. In 1789, for example, Olaudah Equiano put forward an economic rationale for Abolition, by pointing to the growth of manufacturing industry: 'If I am not misinformed, the manufacturing interest is equal, if not superior, to the landed interest, as to the value … The manufacturing interest and the general interests are synonymous.' He argued that this interest would be better served by ending the slave trade, since 'a commercial intercourse with Africa [i.e. other than the slave trade] opens an inexhaustible source of wealth to the manufacturing interests of Great Britain, and to all which the slave-trade is an objection'.[9]

---

[7] Thomas Fowell Buxton, *The African Slave Trade and its Remedy* (London, 1840), 306.
[8] Christopher Leslie Brown, *Moral Capital: Foundations of British Abolitionism* (Chapel Hill NC, 2006), 322.
[9] Olaudah Equiano, *The Interesting Narrative of the Life of Olaudah Equiano, or Gustavus Vassa, the African, Written by Himself* (London, 1789), II, 250–2.

The Sierra Leone Company in 1791 likewise held out the prospect of 'a continually increasing market for the sale of British industry', as well as a supply of raw materials.[10] This perspective was reflected in the recurrent interest, despite repeated failures, in West Africa's assumed potential as a supplier specifically of cotton, for the British textile industry.[11]

The idea of commercial agriculture as a potential substitute for the slave trade can be traced in Britain already in the 1730s.[12] By the 1780s it had become widely accepted, in a variety of forms, within the Abolitionist movement. In the mid-nineteenth century, the view that 'legitimate' trade, especially the export of agricultural produce, would help to eradicate the Atlantic slave trade and bring mutual benefits to Britain and Africa was a central tenet of mainstream Abolitionist thought, as represented by Buxton. Buxton's ideas directly inspired the ambitious venture of the British government's expedition up the River Niger in 1841–2, with its project of the establishment of a 'model farm' in the interior (to cultivate cotton), which was intended to serve as an exemplary beacon for indigenous African enterprise.[13] Despite the catastrophic failure of that expedition, the ideas behind it continued to inform British policy thereafter.[14] When a British Consul was appointed for the Bights of Benin and Biafra (in terms of modern political geography, the West African coast from south-eastern Ghana to Gabon) in 1849, for example, his instructions included the promotion of 'legal commerce' as a means of driving out the slave trade, and more particularly, to 'endeavour to encourage the chiefs and people to till the soil and to produce available exports, so that they may obtain by barter the European commodities of which they may stand in need'.[15] The ambitious scale of the nineteenth-century project of the economic (and, by extension, also social and political) transformation

---

[10] Schwarz, 'Commerce', 255–6.

[11] See e.g. Barrie M. Ratcliffe, 'Cotton Imperialism: Manchester Merchants and Cotton Cultivation in West Africa in the Mid-nineteenth Century', *African Economic History (AEH)*, 11 (1982), 88–113.

[12] John Atkins, *A Voyage to Guinea, Brasil and the West Indies* (London, 1735), 121–2. Strictly, Atkins attributed this policy to an African ruler, King Agaja of Dahomey, who had sent a message to England proposing the establishment of European plantations in his country, which Atkins interpreted as implying a desire to end the slave trade. However, as is noted in Robin Law's contribution to this volume, chapter 5, this view of Agaja's motives was certainly fallacious.

[13] Howard Temperley, *White Dreams, Black Africa: The Antislavery Expedition to the River Niger, 1841–1842* (New Haven CT, 1991).

[14] Against the conventional view of the Niger Expedition as a failure of no long-term significance, see C.C. Ifemesia, 'The "Civilizing" Mission of 1841', *Journal of the Historical Society of Nigeria (JHSN)* 2/3 (1962), 291–310. On the long-term influence of Buxton's ideas, see also James B. Webster, 'The Bible and the Plough', *JHSN* 2/4 (1963), 418–34.

[15] House of Commons Parliamentary Papers, *Correspondence Relative to the Slave Trade*, 1849/50, Class B, no.11, Viscount Palmerston to Consul Beecroft, 30 June 1849.

of African societies has led to its being characterized as 'Britain's first development plan for Africa'.[16]

The idea was also taken up by Christian missionary organizations, which saw the promotion of agricultural exports as closely tied to the prospects of evangelization: they developed, in effect, a sort of proto-Marxist theory of religious conversion, in which change in the ideological sphere depended upon transformation of the economic base. This linkage between the promotion of 'legitimate' trade and Christian evangelization was, again, anticipated in the thinking of the Sierra Leone Company in the 1790s,[17] but became especially influential in the mid-nineteenth century, as reflected in the popular slogans linking 'Commerce and Christianity'[18] and 'the Bible and the Plough'.[19] Perhaps the best-known exponent of this approach was David Livingstone, who in 1857 advocated the promotion of cotton cultivation in East-Central Africa, as a concomitant of evangelization: 'We ought to encourage the Africans to cultivate for our markets, as the most effectual means, next to the Gospel, of their elevation.'[20] But such views were also commonplace among missionaries engaged in West Africa, taken up, for example, by Henry Venn, Secretary of the Church Missionary Society (CMS) from 1841 to 1872.[21] Attempts to put them into practice included the CMS project of cotton cultivation at Abeokuta (in Nigeria), pursued in 1850–63.[22] In the Gold Coast, likewise, the Basel Mission promoted agricultural innovation, experimenting with new crops such as coffee, tobacco and cocoa.[23]

The idea of commercial agriculture as a tool of Abolition persisted

---

[16] P.J. Cain and A.G. Hopkins, *British Imperialism: Innovation and Expansion 1688–1914* (London and New York, 1993), 353–62.

[17] Schwarz, 'Commerce', 265–6. A missionary who had served in Sierra Leone wrote that 'civilization and the arts are united in their plan with religion, and the commercial intercourse ... between [the interior] and Sierra Leone will facilitate the matter': Suzanne Schwarz, '"Apostolick Warfare": the Reverend Melvill Horne and the Development of Missions in the Late Eighteenth and Early Nineteenth Centuries', *Bulletin of the John Rylands Library of Manchester* 85/1 (2003), 65–94 (see 93).

[18] See Andrew Porter, '"Commerce and Christianity": the Rise and Fall of a Nineteenth-century Missionary Slogan', *Historical Journal* 28/3 (1985), 597–608.

[19] Commonly associated with Buxton, but the phrase was coined earlier, in 1817, by James Read, a missionary in South Africa: J.T. du Bruyn, 'James Read en die Tlhaping, 1816–1820', *Historia* (Journal of the Historical Association of South Africa) 35/1 (1990), 23–38. (Thanks to Phia Steyn for drawing our attention to this reference.) Buxton himself in fact cites it from Read: Buxton, *African Slave Trade*, 483.

[20] David Livingstone, *Missionary Travels and Researches in South Africa* (London, 1857), 720.

[21] J.F. Ade Ajayi, 'Henry Venn and the Policy of Development'. *JHSN* 1/4, (1959), 331–42.

[22] S.O. Biobaku, *The Egba and their Neighbours 1842–1872* (Oxford, 1957), 57–62; Jean Herskovits Kopytoff, *A Preface to Modern Nigeria: The 'Sierra Leonians' in Yoruba, 1830–1890* (Madison WI, 1965), 117–20.

[23] Edward Reynolds, 'Agricultural Adjustments on the Gold Coast after the End of the Slave Trade, 1807–1874', *Agricultural History* 47/4 (1973), 308–18 (see 315).

even beyond the final ending of the trans-Atlantic slave trade in the 1860s. In the later nineteenth century the focus of Abolitionist concern shifted to the slave trade within Africa, to whose suppression the major powers formally committed themselves at the Brussels Congress of 1890. The suppression of the slave trade became a principal element in the rhetorical justification of the extension of European colonial rule over Africa. The term 'legitimate trade', in distinction from slave-trading, thus retained some currency into the colonial period.[24] Moreover, the belief that the development of commercial agriculture and the ending of slave-trading were necessarily linked persisted in relation to the internal trade. When the British mounted a military expedition against the Aro, the principal slave-trading community in south-eastern Nigeria, in 1901, for example, its proclaimed objectives included both 'to put a stop to slave dealing' and 'to induce the natives to engage in legitimate trade'; and following the conquest, policy was defined as 'to make the Aro interest himself in the produce trade … instead of as hitherto slaves'.[25]

The historiography of the 'commercial transition'

The implications of this commercial transition for the African societies involved have been the subject of a considerable amount of scholarly analysis and debate, beginning with the very first academic monograph in the field of African history, published in 1956, K.O. Dike's study of the trading states of the Niger Delta in the nineteenth century.[26] The subject has attracted not only detailed case-studies, but also ambitious attempts at general synthesis, of which the most influential has undoubtedly been that by Tony Hopkins, published in 1973.[27] Hopkins' central argument was that the nineteenth-century transition represented a major discontinuity in West African economic structures: whereas the slave trade was by its nature a large-scale business, dominated by a small elite of wealthy entrepreneurs, and more

---

[24] For an instance of its use by a British official as late as 1935, see A.E. Afigbo, *The Abolition of the Slave Trade in Southeastern Nigeria, 1885–1950* (Rochester NY, 2006), 91.
[25] Ibid., 44, 57.
[26] K. Onwuka Dike, *Trade and Politics in the Niger Delta 1830–1885* (Oxford, 1956).
[27] A.G. Hopkins, *An Economic History of West Africa* (London, 1973), chapter 4. Alternative perspectives, stressing continuity rather than revolutionary transformation, include Ralph A. Austen, *African Economic History: Internal Development and External Dependency* (London, 1987), chapters 4–5. See also Robin Law, 'The Historiography of the Commercial Transition in Nineteenth-century West Africa', *African Historiography: Essays in Honor of Jacob Ade Ajayi*, ed. Toyin Falola (Harlow, 1993), 91–115.

particularly by rulers and military leaders,[28] the production and marketing of agricultural produce were open to small-scale participation, and thus to the mass of ordinary farmers and petty traders. Since such small-scale agricultural production for the export market has continued to characterize West African economies down to the present, Hopkins suggested that the nineteenth-century transition represented 'the start of the modern economic history of West Africa'.[29] He also argued that since existing rulers found it difficult to maintain their control over the new trade, and therefore suffered a decline in their incomes, which tended to undermine their effective authority, the transition posed a 'crisis of adaptation' for West African ruling elites.[30] Further, the strains of the transition caused conflicts between African suppliers and European merchants, as well as within African societies, which were held to have contributed to the causation of the European Partition of Africa at the end of the nineteenth century. Thus, according to Hopkins, the problems encountered in establishing the new economy of 'legitimate commerce' formed part of 'the economic basis of imperialism'.[31]

The nineteenth-century commercial transition and, more particularly, Hopkins' analysis of it were the subject of a conference held at the University of Stirling, Scotland, in 1993, organized by one of the co-editors of the present volume, papers from which were published in 1995.[32] This focused on the period of the legal abolition of the slave trade and its immediate aftermath, with a primary concern in evaluating Hopkins' concept of the 'crisis of adaptation', including its suggested links to the establishment of European colonial rule later in the nineteenth century. The new conference held at the German Historical Institute London in 2010 sought to follow on this earlier discussion, but also to extend and diversify treatment of the subject, both chronologically and thematically. Papers were invited on any aspect of agricultural development and trade in African commodities from the fifteenth to nineteenth centuries (and, in practice, the

---

[28] This and similar formulations (by Hopkins and others) were not of course intended to imply (though they have sometimes been misunderstood in this sense) that only states were involved in the supply of slaves. The mechanisms of slave supply in 'stateless' societies evidently differed, but in them also enslavement was dominated by a small number of wealthy and powerful 'big men': see e.g. Walter Hawthorne, *Planting Rice and Harvesting Slaves: Transformations along the Guinea-Bissau Coast, 1400–1900* (Portsmouth NH, 2003), chapters 3–4.

[29] Hopkins, *Economic History*, 124.

[30] In *Economic History*, 142, Hopkins refers to 'acute problems of adaptation'. The term 'crisis of adaptation', which has gained wider currency, was used in an earlier publication: A.G. Hopkins, 'Economic Imperialism in West Africa: Lagos, 1880–92', *EHR* 21/3 (1968), 580–600.

[31] This is the title of chapter 4 of the *Economic History*.

[32] Robin Law, ed., *From Slave Trade to 'Legitimate' Commerce: The Commercial Transition in Nineteenth-century West Africa* (Cambridge, 1995).

chronological scope of some of those presented extended into the period of European colonial rule in the twentieth century). Eleven (of 24) of these papers are included in this volume, the selection being determined partly by the authors' willingness to make them available, and partly by the editors' judgement of thematic relevance.

The pre-Abolitionist period

One of the priorities of the 2010 conference was to extend consideration of the history of commercial agriculture in West Africa back beyond the late eighteenth century, into the pre-Abolitionist era.[33] Five of the papers in this volume relate to this earlier period: David Eltis offers a general overview, while Gerhard Seibert, Toby Green, Colleen Kriger and Robin Law deal with local case-studies.

The European traders who opened up maritime trade with West Africa from the mid-fifteenth century onwards were not interested in purchasing slaves only, but any commodities which might yield a profit. Indeed, initially it was gold, rather than slaves, which formed the main focus of European interest: exports of gold continued to exceed those of slaves in value until the later seventeenth century.[34] In addition to gold, a wide range of other commodities was purchased, such as ivory, hides and bees-wax. These included at least one agricultural product of some significance – pepper, of the indigenous African species, 'malagueta' or 'grains of paradise' (*aframomum melegueta*), purchased mainly on what was called the 'Grain Coast' (modern Liberia):[35] some of this pepper may have been gathered from wild plants, but it was also cultivated.[36] A distinct species, 'Guinea pepper' (*piper guineense*), was obtained from the kingdom of Benin (in Nigeria).[37] There were also items of silvan produce, i.e. items gathered from wild plants, rather than cultivated, such as dyewoods –

---

[33] There was of course 'commercial agriculture' in West Africa oriented towards internal markets, which was largely independent of, and whose origins predated, the European Atlantic trade. In this Introduction, unless otherwise indicated, 'commercial agriculture' refers specifically to production for sale into overseas markets.

[34] David Eltis, 'The Relative Importance of Slaves and Commodities in the Atlantic Trade of Seventeenth-century Africa', *Journal of African History (JAH)* 35/2 (1994), 237–49.

[35] Malagueta had reached Europe via the trans-Saharan trade even before the opening of direct maritime contact in the 15th century: Raymond Mauny, *Tableau géographique de l'Ouest africain au moyen âge* (Dakar, 1961), 249–50.

[36] It was reportedly 'sown like corn': Pieter de Marees, *Description and Historical Account of the Gold Kingdom of Guinea (1602)*, trans. and ed. Albert van Dantzig and Adam Jones (Oxford and NewYork, 1987), 160. See also J.M. Lock, J.B. Hall and D.K. Abbiw, 'The Cultivation of Melegueta Pepper (*Aframomum melegueta*) in Ghana', *Economic Botany* 31/3 (1977), 321–30.

[37] Alan Ryder, *Benin and the Europeans 1485–1897* (London, 1969), 31, etc.

purchased mainly in the Sierra Leone area,[38] and gum arabic (secreted from acacia plants, and used in Europe to stiffen cloth) – from the valley of the River Senegal.[39] Palm-oil (extracted from the fruit of the oil-palm) was also sometimes imported into Europe, where it was used for medicinal purposes, 'applied warm, as a treatment for flatulence, chills of the shoulders and dislocation of the limbs',[40] but evidently in only trivial quantities, until it became an industrial raw material (for the manufacture of soap and candles) in the second half of the eighteenth century. Europeans also sought African commodities for re-sale elsewhere on the coast, and these also included an important item of agricultural produce, kola nuts, which were purchased mainly in Sierra Leone, and taken for sale at Cacheu and other places in what is today Guinea-Bissau.[41]

The importance of these other commodities relative to slaves declined over time, with slaves accounting for over 90 per cent of the value of West Africa's trans-Atlantic exports by the end of the eighteenth century.[42] Some historians, including notably Walter Rodney and Joe Inikori, have inferred from these changing relativities that the growth of the slave trade tended to undermine other forms of commerce, both by diverting energies and labour from them and by fomenting disorder which was disruptive of peaceful economic activities.[43] In this volume Eltis offers an estimate of the importance of commodities other than slaves in European trade with western Africa, insisting that, despite the overall dominance of slaves, 'the non-slave component was always significant'. Although slaves exceeded non-human commodities exported in value from the second half of the seventeenth century onwards, the latter still accounted for 16 per cent of total exports from 1681 to 1807, and probably more than a quarter over the entire period from the mid-fifteenth century to the beginning of the nineteenth century. Moreover, although the non-slave share of produce in total trade declined, in absolute terms it actually

---

[38] Walter Rodney, *A History of the Upper Guinea Coast, 1545–1800* (Oxford, 1970), 158–60.
[39] James L.A. Webb, 'The Trade in Gum Arabic: Prelude to French Conquest in Senegal', *JAH* 26 (1985), 149–68.
[40] Jean Barbot, *Barbot on Guinea: The Writings of Jean Barbot on West Africa, 1678–1712*, ed. P.E.H. Hair, Adam Jones and Robin Law (London, 1992), II, 461.
[41] Rodney, *History*, 206–7. Like pepper, kola nuts were collected from wild plants, but the trees were also cultivated, at least in recent times: Edmund Abaka, *'Kola is God's Gift': Agricultural Production, Export Initiatives and the Kola Industry in Asante and the Gold Coast, c.1820–1950* (Oxford, 2005), 38–40.
[42] David Eltis, 'Precolonial Western Africa and the Atlantic Economy', *Slavery and the Rise of the Atlantic System*, ed. Barbara L. Solow (Cambridge, 1991), 91–119.
[43] Walter Rodney, 'Gold and Slaves on the Gold Coast', *Transactions of the Historical Society of Ghana (THSG)* 10 (1969), 13–28; Joseph E. Inikori, *Africans and the Industrial Revolution in England: A Study in International Trade and Economic Development* (Cambridge, 2002), 381–93.

grew (albeit much more slowly than the slave trade), at least to the 1770s: the decline of gold exports was offset by the expansion of trade in gum, dyewoods and hides. This finding casts doubt on the argument that the rise of the slave trade had the effect of undermining that in other commodities. However, as Eltis observes, this had little to do with 'commercial agriculture', strictly defined. Even apart from non-vegetable commodities such as gold and ivory, most of the vegetable produce exported was gathered, rather than cultivated: the principal exception, pepper, amounted to less than 1 per cent of the value of non-slave exports prior to the nineteenth century.

The demonstrable willingness of Europeans to seek and of Africans to supply commodities other than slaves serves, nevertheless, to raise the issue of why it was that in the long run the dominant tendency was to transport labour from Africa to cultivate crops in America, rather than to employ that labour in Africa to produce directly for European markets. Why did Europeans not establish plantations to produce sugar and other tropical crops in Africa, or alternatively purchase these commodities from African producers? Part of the answer is, of course, that to some extent they did. The earliest European plantation colonies, producing sugar, were in fact established on islands in the eastern Atlantic, including those off the African coast, before the system migrated to the Americas. The most successful instance was São Tomé, which is studied by Seibert in this volume. He charts the rise and decline of the island's sugar industry from its establishment in the 1510s to its end in the eighteenth century. Its origins derived from the transfer of personnel, skills and sugar plants from Madeira. With production on Madeira in decline, São Tomé briefly became the world's largest sugar producer, with output reaching a peak in the 1570s, followed by a long decline. Its initial success was due to a combination of fertile soil and the availability of cheap supplies of enslaved African labour from the adjacent mainland. Its decline is attributed mainly to competition from Brazil, which developed as a sugar producer from the 1530s onwards. This was compounded by ecological factors, in the form of drought, pest infestation and impoverishment of the soil by intensive exploitation. There were also destructive raids on the island by other European powers, notably by the Dutch in 1599. Control of the labour force also presented difficulty, the island's mountainous and densely forested interior affording conditions favourable to the maintenance of runaway slave communities. There were several slave rebellions, the most important in 1595, which resulted in the destruction of many of the sugar-mills. In fact, Seibert observes that the sugar industry 'never completely recovered' from the destruction in the 1590s.

There was also recurrent interest in establishing plantations, to produce crops such as sugar, cotton and indigo (used for dyeing cloth), on the West African mainland. Experiments with commercial plantations, employing enslaved Africans locally, were undertaken, for example, by the English Royal African Company (RAC) and the Dutch West India Company (WIC) in the seventeenth and eighteenth centuries. The motivation for this was a desire to find alternative or supplementary sources of profit, at a time when conditions in the existing branches of the African trade (in gold, as well as slaves) were becoming increasingly difficult for the established monopoly companies, in competition with independent traders (or 'interlopers'). None of these projects, however, achieved long-term success. Their failure has been attributed by some historians to the political influence of established European planters in the Americas, who feared their competition;[44] but an alternative interpretation might be that the African plantations were in any case uncompetitive. After all, the trans-Atlantic slave trade – assuming that it was an economically rational enterprise – only makes sense on the assumption that the productivity of labour was greater in America than in Africa.[45] The causes of this differential in productivity might relate to ecological, social or technical factors, or very likely some combination of the three.

In this volume, these early European projects of West African plantations are studied by Kriger and Law, the former with reference to attempts by the English RAC to cultivate indigo in Sierra Leone from the 1680s onwards, while the latter focuses on Dutch and English attempts to cultivate indigo, cotton and sugar on the Gold Coast between the 1650s and 1780s. Kriger explores the problems encountered in the Sierra Leone indigo project, in unsuitable soils, European lack of expertise, and difficulties in enforcing labour discipline – the last compounded by the more onerous and unpleasant conditions of work which the large-scale processing of indigo involved, by comparison with the indigenous African techniques, with which at least some of the workers were familiar. However, the decisive factor in the abandonment of the project is identified as the 'chronic insecurities' caused by war with France and attacks on the RAC's factories by English interlopers. Law likewise details problems of incompetent management, inconsistent metropolitan support, shortage of labour, and lack of secure tenure of land which dogged the plantation projects on the Gold Coast. He also examines opposition to the African projects by established planters in the West Indies, which played some role at least in the English case, but concludes

[44] Most recently by Inikori, ibid., 389.
[45] Hopkins, *Economic History*, 105.

that the basic reason for failure was the unsuitability of local soils for the particular crops upon which European interest centred at this period – as opposed to those which did become successful export crops later, such as palm-oil and cocoa.

Law notes that both Dutch and English companies expressed interest in the purchase of agricultural commodities from Africans, as well as their production on European plantations, but such projects also enjoyed little success. Since they are less well-documented, the reasons for their failure are unclear. In the specific cases of cotton and indigo, however, a relevant factor may have been the existence of a considerable internal demand for these products. The manufacture of cotton textiles was long-established, and cotton cloth widely traded within West Africa. This indigenous cotton textile industry not only survived the impact of the early Atlantic trade (despite the competition it faced from imported European and Indian cloth), but was even stimulated by it, since West African cloth was purchased by Europeans, mainly for re-sale elsewhere on the African coast, although small amounts were also taken to Europe and the Americas.[46] The project of turning West Africa into a supplier of raw cotton for European industry implied the diversion of local cotton production from the internal market into export (and the substitution of imported for locally made cloth in African consumption), but in practice Europeans encountered great difficulty in achieving this, even later in the colonial period, when reductions in transport costs enhanced their competitiveness.[47]

The provisions trade

Another aspect of the pre-Abolition history of exports of agricultural produce is that the slave trade itself had the effect of stimulating commercial agriculture in parts of western Africa to supply food crops for the provisioning of slave-ships in the Middle Passage to America. Recent research has documented especially the purchase of rice, on the Upper Guinea coast, that is, Senegambia and Sierra Leone.[48] But other crops were also purchased by European slavers, notably grain

[46] Colleen E. Kriger, '"Guinea Cloth": Production and Consumption of Cotton Textiles in West Africa before and during the Atlantic Slave Trade', *The Spinning World: A Global History of Cotton Textiles, 1200–1850*, ed. Giorgio Riello and Prasannan Parthasarathi (Oxford and New York, 2009), 105–26.

[47] See Marion Johnson, 'Cotton Imperialism in West Africa', *African Affairs* 73/291 (1974), 178–87; and for a recent case-study, Richard L. Roberts, *Two Worlds of Cotton: Colonialism and the Regional Economy in the French Soudan, 1800–1946* (Stanford CA, 1996).

[48] Hawthorne, *Planting Rice*, chapter 5.

(millet and maize) and yams (and later, cassava), and also palm-oil, which was bought for use in cooking slaves' food, before it became a raw material for European industry. Malagueta pepper was also purchased for slaves' consumption, partly because it was thought to have medicinal properties, against 'the flux [i.e. dysentery] and dry belly-ach[e]'.[49] In addition to supplying slave-ships, there was also a significant trade in provisions to the urban centres on the coast which served as commercial entrepôts for the Atlantic trade, which were not normally self-sufficient in food, but obtained it by purchase from the adjacent hinterland: this has been documented, for example, for the cases of the French island colonies of Gorée and Saint-Louis in Senegal, the port of Ouidah in the modern Republic of Bénin, and Luanda in Angola.[50] Food was also produced for sale for the provisioning of slaves in transit to the coast, and indeed for trading caravans more generally.

In this volume, Eltis offers an estimate of the value of African provisions supplied for slaves' subsistence, based on the assumption that this represented around 25 per cent of the purchase price of the slaves themselves. This yields the arresting conclusion that this provisions trade may have averaged £1 million annually by the end of the eighteenth century, which was substantially greater than the value of the non-human commodities then shipped to Europe. This demonstrates, as Eltis observes, that 'the roots of commercial [i.e. export] agriculture in [western Africa] were well established long before 1800'. It also tends to reduce the significance conventionally attributed to the 'commercial transition' of the first half of the nineteenth century: exports of palm-oil to Britain did not attain a comparable value until after 1850, so that given the decline of the slave trade, and hence of the provisions trade which supported it, it might even be that 'commercial agriculture' contracted rather than expanded in this period.

This general survey is complemented by Green, who studies the export of rice and grain from the Upper Guinea coast in the sixteenth century. He stresses the market for provisions in the Portuguese settlements on the Cabo Verde islands, which were too arid to produce their own food requirements, as well as the provisioning of slave-ships. He argues that the production and sale of agricultural surplus were

---

[49] Thomas Phillips, 'Journal of a Voyage in the Hannibal of London, Ann. 1693, 1694', *A Collection of Voyages and Travels*, ed. Awnsham and John Churchill (London, 1732), VI, 173–239 (see 195).
[50] James F. Searing, *West African Slavery and Atlantic Commerce: The Senegal River Valley, 1700–1860* (Cambridge, 1993), 84–5; Robin Law, *Ouidah: The Social History of a West African Slaving 'Port', 1727–1892* (Oxford, 2004), 84; Joseph C. Miller, *Way of Death: Merchant Capitalism and the Angolan Slave Trade, 1730–1830* (Madison WI, 1988), 270, 297. This was also true, of course, for ports not involved in slave-trading: see e.g. Philip Misevich, 'The Sierra Leone Hinterland and the Provisioning of Early Freetown, 1792–1803', *Journal of Colonialism and Colonial History* 9/3 (2008).

nothing new for these communities, as they had probably been supplying provisions for the trans-Saharan trade before. What was new was the scale of the demand. This led to significant changes in agricultural practice, both in techniques (greater use of iron tools used in clearing ground for planting) and in the social organization of production. Changes in the latter sphere included, in different areas, the development of the co-operative exchange of labour through a system of age-grades (as among the 'stateless' Balanta of modern Guinea-Bissau) and of a more authoritarian form of political structure (such as the Kassanké kingdom of the Casamance area of Senegal).

Early commercial agriculture as an 'alternative' to the slave trade

It might be objected – as indeed it was by some of the participants in the 2010 conference – that these cases of the sale of agricultural produce did not, in any meaningful sense, represent an 'alternative' to the trans-Atlantic slave trade. Clearly, early European projects of African plantations were not intended as a substitute for the slave trade, in the sense of replacing it altogether. São Tomé, as Seibert shows, functioned as an entrepôt for the trans-Atlantic slave trade at the same time as it produced sugar. Likewise, those Europeans who experimented with plantations on the mainland in the pre-Abolition era explicitly assumed that the slave trade would continue. To this extent, commercial agriculture was evidently conceived as a supplement to the slave trade, rather than a replacement for it. However, the choice of the term 'alternative', rather than 'substitute', in the title of the 2010 conference was deliberate; logically, 'alternatives' may co-exist. From the point of view of individual European and African entrepreneurs, trade in agricultural produce was, in a real sense, potentially an alternative to the slave trade. European traders could opt to seek other commodities, including agricultural produce, rather than slaves – these were, to some extent, two separate trades. Although slave-ships could and did also carry other commodities, some ships carried only produce for European markets, and these latter could and did sail directly home to Europe, rather than via the Americas, as slave-ships had necessarily to do.[51] Likewise, for Africans non-slave commodities were an alternative to the slave trade, insofar as those who wished for access to imported European goods could obtain them by supplying produce, rather than slaves. In particular, trade in commodities other than slaves potentially offered opportunities, as Hopkins suggested for the nine-

[51] See K.G. Davies, *The Royal African Company* (London, 1957), 185–6. Also David Eltis, this volume, chapter 1.

teenth century, to persons outside the elite who normally dominated the slave trade.⁵²

This option also operated, to some degree, at the level of whole communities or regions as well as individuals. As is noted by Eltis, there was a degree of regional specialization within Africa as regards what was supplied to the Atlantic trade. This was reflected in the designations of different sections of the West African coast which had become conventional by the seventeenth century: from west to east, the 'Grain [i.e. Pepper] Coast', already mentioned, 'Ivory Coast' (still today called Côte d'Ivoire), and 'Gold Coast', in distinction from the 'Slave Coast' further east (roughly, the Bight of Benin). The Gold Coast supplied gold, and few slaves, for the first two hundred years of its participation in the Atlantic trade; only from around the end of the seventeenth century did traders there opt to supply slaves instead, provoking the Director-General of the Dutch WIC to observe in 1705 that 'this Gold Coast has turned into a real Slave Coast.'⁵³ Likewise Benin, although it supplied some slaves from the beginning of its contacts with Europeans in the 1480s, sold mainly other commodities – pepper, ivory, gum, and locally made beads and cloth – until the early eighteenth century.⁵⁴ In the Senegal valley, in contrast to other areas of West Africa, slave exports stagnated and even declined in the eighteenth century, and the trade in gum was more important.⁵⁵ The large-scale export of palm-oil was pioneered by Old Calabar (in south-eastern Nigeria) from the 1770s, and a contemporary British observer attributed this initiative to the port's failure in competition for the slave trade with the rival port of Bonny, in the Niger Delta to the west.⁵⁶

The supply of provisions for slave-ships might seem to be a different matter, since it was evidently integrated with and dependent upon the slave trade, rather than distinct from it, and thus, it might be argued, could not in any sense represent an 'alternative' to it. But again, for individual African (as distinct from European) entrepreneurs it could

---

⁵² As noted (though without citing Hopkins) for bees-wax: Michael W. Tuck, 'Everyday Commodities, the Rivers of Guinea, and the Atlantic World: The Beeswax Export Trade, c.1450–c.1800', *Brokers of Change: Atlantic Commerce and Cultures in Precolonial West Africa*, ed. Toby Green (Cambridge, 2012), 285–303 (see 286).
⁵³ Albert van Dantzig, *Les Hollandais sur la Côte de Guinée à l' Époque de l'Essor de l'Ashanti et du Dahomey 1680–1740* (Paris, 1980), 146.
⁵⁴ Ryder, *Benin*, chapters, 2, 3, 5–6.
⁵⁵ Searing, *West African Slavery*, 129–30, 150–4.
⁵⁶ John Adams, *Remarks on the Country Extending from Cape Palmas to the River Congo* (London, 1966 [1823]), 143. This has been disputed, on the grounds that Calabar remained an important slaving port after the beginning of oil exports, but even a relative decline might have inspired a search for alternatives: A.J.H. Latham, *Old Calabar 1600–1891: The Impact of the International Economy upon a Traditional Society* (Oxford, 1973), 55–6.

nevertheless represent an alternative, in that they could obtain access to desired European goods by selling provisions to slavers, rather than slaves. Admittedly, more research is needed on how far African suppliers of provisions and of slaves were in practice different persons, or whether the same entrepreneurs usually supplied both sorts of commodities. But here too, again as noted by Eltis, there was a measure of regional specialization – that is, some African communities at some times opted to supply provisions *rather than* slaves. In the case of the Gold Coast in the seventeenth century, for example, some coastal communities had become heavily involved in supplying provisions to slave-ships, even though they as yet supplied few slaves. It became common practice for European slave-ships to take in grain on the Gold Coast, before proceeding further east to purchase their cargoes of slaves. The town of Anomabu, on the central Gold Coast, in particular, was prominent in this trade, being described in 1689 as 'the principal granary' from which all English slave-ships drew their supplies.[57] Likewise in Upper Guinea, Green notes that some communities (in the Rio Nuñes, for example) supplied provisions without being directly involved in slave-trading, though it is implied that this was not common.

Linkages from the pre-Abolitionist era to 'legitimate commerce'

Although not motivated by any Abolitionist intention, this earlier history of commercial agriculture in West Africa was not entirely without links to the subsequent development of 'legitimate' commerce. On the African side, it has been suggested that the earlier experience of the provisions trade to some degree, at least in some particular areas, formed a basis for the subsequent development of the export trade in other crops. In the case of the Bight of Biafra (southeastern Nigeria), for example, David Northrup argued that the supply of provisions (mainly yams) for slave-ships played a critical role in 'preparing' inland farmers and traders for participation in the later export of palm-oil: 'It is no coincidence that the same areas which were most involved in this trade in provisions were the ones which responded first to the demand for increased exports of palm oil.'[58] Likewise, in the Senegal valley, James Searing suggested that the provisions trade (here in grain, and directed to the trans-Saharan as well as the trans-Atlantic trade) 'laid the groundwork' for the later develop-

[57] Van Dantzig, *Les hollandais*, 116.
[58] David Northrup, 'The Compatibility of the Slave and Palm Oil Trades in the Bight of Biafra', *JAH* 17/3 (1976), 353–64 (see 362–3).

ment of an export trade in groundnuts.[59] On the Upper Guinea coast, where rice continued to be the principal export even after the ending of the slave trade, the continuity is even more evident.[60]

On the European side also, one link to later Abolitionist thought was in establishing the idea that African cultivation was a viable alternative to the slave trade. Despite the general pattern of failure of early European attempts to establish plantations in West Africa, recollection of them played some role in encouraging Abolitionists in the belief that the crops produced on American plantations could be cultivated in Africa. Some reference to these earlier ventures was made, for example, in evidence given to the official Privy Council enquiry into the slave trade conducted in Britain in 1788.[61] Even the supply of provisions for slave-ships was cited as evidence for the fertility of African soils, and thereby of their potential for agricultural improvement.[62] In this volume, Chris Brown addresses this issue. He identifies as a critical link the British political economist Malachy Postlethwayt, who in the 1740s was a leading propagandist for the RAC, and thereby also for British involvement in the trans-Atlantic slave trade, but who had moved by the 1750s to advocating alternative forms of commerce, as offering potentially a more profitable market for British industry, and whose works were frequently cited by Abolitionist writers in the 1770s and '80s. Brown also stresses the significance of the British colony of Senegambia, the first formal British colony in Africa (1765–83), which, although not an Abolitionist venture, was primarily focused on trade in other commodities than slaves, especially gum. Although short-lived, the Senegambia venture also helped to popularize the idea of a commercial alternative to the slave trade.

Abolitionism and after

Projects of commercial agriculture in the Abolitionist era have been more extensively studied, but there still remain questions to be investigated. In this volume, five papers deal with this later period. Four present particular case-studies, all relating to developments within European

---

[59] Searing, *West African Slavery*, 198.
[60] Hawthorne, *Planting Rice*, chapter 6.
[61] *Report of the Committee of Council appointed for consideration of all matters relating to Trade and Foreign Plantations ... [on] the State of the Trade to Africa* (House of Commons, London, 1789), section on 'Produce', evidence of John Matthews, Robert Norris, J.B. Weuves, Archibald Dalzel. Witnesses referred to former 'indigo works' in Sierra Leone, and Dutch and English attempts to cultivate cotton on the Gold Coast.
[62] For example, C.B. Wadström, *Observations on the Slave Trade and a Description of the Coast of Guinea* (London, 1789), 42.

and American colonial enclaves: Per Hernæs deals with the Danish plantation project of 1788; Bronwen Everill with the British colony of Sierra Leone and the settlement of free Blacks from the USA in what became the independent Republic of Liberia; Kehinde Olabimtan with missionary projects of agricultural improvement in Sierra Leone; and Roquinaldo Ferreira with the Portuguese colony of Angola. Gareth Austin offers a general overview, which focuses on independent African initiatives, in both the pre-colonial and colonial periods.

One question arising is, in fact, the shift in focus which occurred in Abolitionist thought from European to African initiative. The earliest projects for the promotion of commercial agriculture in West Africa, in the late eighteenth century, were generally conceived within an explicitly colonial framework.[63] This was true, for example, of the pioneering Danish project, which Hernæs shows envisaged no less than 'a Danish settler colony in Africa'. Its motivation, indeed, derived not only from its Abolitionist purpose, but also from a desire to reconfigure the relationship between Denmark and its African possessions, involving the consolidation and territorial extension of the latter, although this proved unfeasible in the circumstances of the time, and was not the most appropriate method for developing agricultural exports. The contemporaneous Sierra Leone project was likewise based on the establishment of a territorial colony (initially in the form of a chartered company, but taken over as formal Crown Colony from 1808), though the Sierra Leone Company also emphasized trading for 'legitimate' commodities with neighbouring African societies beyond colonial jurisdiction.[64] Both these projects also involved the establishment of European-managed plantations. In the Danish case, as Hernæs expresses it, what was envisaged was 'not to generate "African production", but European production in Africa'. In Sierra Leone, although there was some experimentation with plantations, greater emphasis was placed on small-scale farming, but this was also a form of 'settler' agriculture, though the settlers were black (initially from Britain and America, but from 1808 also African-born 'recaptives' liberated by the British navy from illegal slave-ships) rather than white. By 1819, however, partly in response to the perceived failure of the Sierra Leone experiment, it was being argued that realization of Africa's commercial potential would depend on 'promoting and encouraging the growth of merchantable commodities by *the natives themselves*', rather than by European plantations.[65] Likewise, Hernæs

---

[63] Philip D. Curtin, *The Image of Africa: British Ideas and Action, 1780–1850* (Madison WI, 1964), chapter 4.
[64] Schwarz, 'Commerce', 257–8, 260–1.
[65] G.A. Robertson, *Notes on Africa* (London, 1819), 23 (emphasis in original).

cites the comment of a Danish observer on the Gold Coast in 1831 that for commercial agriculture to succeed 'it must probably be carried out by the natives themselves'. The attempt of the British Niger Expedition of 1841 to establish its model farm in the interior was a belated venture still predicated on the assumption of the need for external stimulus, and its failure helped to consolidate the perception that agricultural initiatives were better left to the Africans themselves. Experience showed, indeed, that Africans were more successful than Europeans in identifying and developing the best prospects for commercial agriculture – a pattern which continued into the colonial period, when it was again indigenous African farmers, rather than European planters or colonial officials, who pioneered the development of cocoa-farming.[66]

A challenging new perspective on the saliency of African initiative is offered in Austin's contribution to this volume, which in part argues for the 'endogenous' origins of the decline of the trans-Atlantic slave trade. His starting point is the fact that the volume of the slave trade began to decline from the 1780s, even before its legal abolition by any European or American state. This initial decline has commonly been attributed to developments outside Africa which diminished the effective demand for slaves, specifically the slave insurrection in the French colony of Saint-Domingue (modern Haïti) from 1791, which eliminated what had lately been the principal market for slaves, and the European wars of 1793–1815, which disrupted shipping. Austin argues, however, that this is difficult to reconcile with the trend of slave prices in this period, which in West (as opposed to West-Central) Africa was actually upwards, suggesting constraints on the supply side in Africa. This might have been due to increasing costs in the procurement and delivery of slaves, as communities subject to slave-raiding organized more effective self-defence, and slaves were brought to the coast from further in the interior. In the case of Senegal, however, Searing argued that the decline in slave exports there in the second half of the eighteenth century reflected a growing internal demand for slaves, for the provisions and gum trades.[67] Although Searing himself implied that these conditions were specific to Senegal, Austin suggests that this analysis might in fact be more generally applicable. This implies a more general growth of production for the market, and again, tends to contradict the thesis that the slave trade was overall destructive of market activity in Africa.

[66] A.G. Hopkins, 'Innovation in a Colonial Context: African Origins of the Nigerian Cocoa-farming Industry, 1880–1920', *The Imperial Impact: Studies in the Economic History of Africa and India*, ed. Clive Dewey and A.G. Hopkins (London, 1978), 83–96; see also R.H. Green and S.H. Hymer, 'Cocoa in the Gold Coast: a Study in the Relations between African Farmers and Agricultural Experts', *Journal of Economic History (JEH)* 26 (1966), 299–319.
[67] Searing, *West African Slavery*, 130.

Seymour Drescher has recently argued, against some earlier analyses, that in Britain Abolitionism had no close connection to 'imperialism' in Africa, partly because the major 'imperialist' initiatives which Abolitionists sponsored – the Sierra Leone colony and the 1841 Niger Expedition – had proved embarrassing failures.[68] This argument rests on a narrow definition of 'imperialism' as formal territorial expansion: a different picture might be painted if the analysis was extended to include various forms of 'informal imperialism', i.e. intervention short of annexation.[69] But even on the narrower conception of 'imperialism' in terms of formal sovereignty, there is arguably more to be said of its connection with Abolition. The colony of Sierra Leone certainly continued to play an important role in Abolitionist policy towards Africa. In the shift of emphasis towards 'African' initiative, a key role was commonly assigned to persons of African birth or descent who had been assimilated into European-Christian culture, especially those from the community of 'liberated Africans' settled in Sierra Leone.[70]

In this volume, Everill reappraises the early history of settler agriculture in Sierra Leone and Liberia. She contests contemporary notions of failure in both cases, arguing that many settlers did engage in farming, including production for export as well as subsistence agriculture. While, however, this study shows the continuing importance attached to the promotion of cultivation, it also highlights the disparity between metropolitan expectations and the practical realities in West Africa. Hostile press reporting of both settlements promoted stereotypical views of Africa and Africans, and typically failed to recognize the specific difficulties associated with cultivation, particularly in Sierra Leone. Agricultural reform was seen as a panacea for Africa, which would suppress the slave trade, transform the moral character of former slaves, and serve metropolitan commercial interests in the supply of raw materials and markets for exports of manufactured goods. This, however, incorporated a number of different visions – projects of large-scale export agriculture, ideals of self-sufficient 'yeoman' farmers, and the priority of suppression of the slave trade (which pointed logically to a focus on the interior, rather than the coastal settler colonies) – which in practice proved contradictory. These multiple and unrealistic expectations, it is argued, inevitably generated disappointment among external observers in Europe and America.

Olabimtan studies efforts by a CMS missionary in Sierra Leone,

---

[68] Seymour Drescher, 'Emperors of the World: British Abolitionism and Imperialism', *Abolitionism and Imperialism in Britain, Africa and the Atlantic*, ed. Derek R. Peterson (Athens OH, 2008), 129–49.
[69] See Robert Gavin, 'Palmerston and Africa', *JHSN* 6/1 (1971), 93–99.
[70] This was, for example, Buxton's conclusion, after the failure of the Niger expedition: Temperley, *White Dreams*, 163.

John Ulrich Graf, with the support of the CMS secretary Henry Venn in London, to promote agricultural improvement through the African Improvement Society, formed at Freetown in the 1849. This illustrates how Sierra Leone was still viewed as a potential base for the development of 'legitimate trade' in the 1850s, and moreover that, a decade after the failure of the Niger Expedition, Venn and Graf remained explicitly committed to Buxton's policies of 'Commerce, Civilization and Christianity'. An innovative feature of the scheme was the aim to develop African capacity for agricultural and commercial improvement by sending a small number of liberated Africans from Sierra Leone to Britain for training. Olabimtan explores the case of Henry Johnson, who was sent to the Royal Botanical Gardens at Kew in 1853 for training in plant science and methods of cultivation. Although the scheme was short-lived and ineffective, it exemplifies similar principles to those which linked missionary endeavour and economic reform in the contemporary CMS project of cotton cultivation at Abeokuta, mentioned earlier. Moreover, although both these projects failed, in the longer run it can be argued that the vision of Graf and Venn was vindicated, in the key role which was subsequently played by Christian Africans in the development of cocoa-farming in southern Nigeria.

In any case, the disconnection of Abolition from imperialism, even if thought valid for the case of Britain, does not fit that of Portugal, as shown in Ferreira's contribution to this volume. Portuguese efforts to promote commercial agriculture were focused on its existing colony of Angola, and were part of an attempt to shape a new role for that territory, in the context of the secession of Brazil from Portuguese rule (in 1822), as well as of the decline of the slave trade. The Angolan projects also mostly employed Portuguese (or Brazilian) management, though there was also some involvement of independent African cultivators.

Another issue is the shift which occurred over time as regards which particular crops were thought to offer the best prospects of a viable trade. Initially, the emphasis was on the cultivation in Africa of crops which were currently grown by the labour of enslaved Africans in the Americas. The Danes on the Gold Coast in 1788, for example, hoped to cultivate cotton; while the Sierra Leone Company in the 1790s was initially mainly interested in sugar, though it also tried to cultivate coffee, and to obtain cotton from neighbouring African societies.[71] In Liberia, the early emphasis was on cotton and coffee. The Portuguese in Angola tried sugar, cotton, tobacco and (most successfully) coffee.

---

[71] Schwarz, 'Commerce', 257–9, 261.

Over time, however, interest spread to other crops. This was partly a question of finding out by experiment which crops were viable in West African conditions, but partly also of responding to changes in patterns of demand in the world economy. Although a wide range of crops was contemplated or experimented with, in practice the one that achieved most success, prior to the establishment of colonial rule at the end of the nineteenth century, was palm-oil, used mainly in the manufacture of soap.[72] Palm-kernels – the hard core, as opposed to the fleshy pericarp, of the fruit of the oil-palm (initially used to make soap, and later also margarine) – were also exported from the 1850s. Other exports which were successfully developed during the nineteenth century included groundnuts (also used for soap), from the 1830s; and later, rubber, which was harvested from wild plants (used in Europe for waterproof clothing and footwear, electrical insulation, and bicycle and automobile tyres), and cocoa, introduced from America, which was initially cultivated on islands off the West African coast (São Tomé and Fernando Po), and transferred to the mainland (Ghana and Nigeria) from the 1880s.

It is noteworthy that Buxton in 1840, although acknowledging that the palm-oil trade represented 'the most important branch of our commerce with Africa', was dismissive of its significance (along with that in other commodities such as timber, gum, bees-wax and ivory) on the grounds that it was not the product of 'cultivation', and so was supposedly 'obtained at comparatively little labour and cost'.[73] This judgement has been echoed in some modern scholarship, which argued that since palm produce (like slaves) was gathered rather than cultivated, it required 'only a relatively marginal involvement by agriculturalists', and consequently could not have transformed the domestic economy in the manner suggested by Hopkins.[74] This, however, is misconceived: although the palm-fruits were generally harvested from wild trees,[75] it was not the fruits which were exported but the oil processed from them, by pressing and boiling, which was in fact a quite laborious process.[76]

---

[72] Martin Lynn, *Commerce and Economic Change in West Africa: The Palm Oil Trade in the Nineteenth Century* (Cambridge, 1997).
[73] Buxton, *African Slave Trade*, 308–9, 321.
[74] Ralph A. Austen, 'The Abolition of the Overseas Slave Trade: a Distorted Theme in West African History', *JHSN* 5/2 (1970), 257–74 (see 268–71).
[75] Strictly, oil-palms were 'semi-cultivated', in that they were preserved and protected when land was cleared for farming: David Northrup, *Trade without Rulers: Pre-colonial Economic Development in South-eastern Nigeria* (Oxford, 1978), 185. Moreover, in some cases oil-palms were in fact planted, e.g. in Dahomey: Law, *Ouidah*, 211.
[76] See e.g. Northrup, *Trade without Rulers*, 184–6.

Commercial agriculture, slavery and 'modernity'

Hopkins' thesis that the transition from the slave trade to commercial agriculture posed a 'crisis of adaptation' for African ruling elites has commanded declining attention in recent years, with only limited discussion of it in recent published studies,[77] although a reappraisal of Hopkins' original case-study (in Yorubaland) broadly endorses his analysis.[78] However, his broader argument that it was the rise of 'legitimate commerce' in the nineteenth century, rather than the subsequent establishment of European colonial rule, which marked the beginnings of economic 'modernity' in West Africa remains salient. The question of the relative importance of large-scale and small-scale production in West African commercial agriculture, which was central to Hopkins' analysis, remains contested: although subsequent research has demonstrated that both existed, the balance between them is difficult to determine, and very probably varied among different areas, and perhaps also over time.[79] There is also the related, though strictly distinct, question of the relative importance of enslaved and free labour.

European projects of cultivation in West Africa began with large-scale enterprises employing slaves, representing in effect the transfer of the existing plantation system from America, as in the Danish project of 1788. The Sierra Leone Company also initially envisaged the establishment of large-scale plantations, albeit using free labour, in the explicit but unrealized hope that they would prove the superior efficiency of free over enslaved labour.[80] In Angola, as Ferreira shows, commercial agriculture depended largely on slave labour, as was reflected in a doubling of the number of slaves in the colony during the 1850s.

[77] Chima J. Korieh and Femi J. Kolapo, eds, *The Aftermath of Slavery: Transitions and Transformations in Southeastern Nigeria* (Trenton NJ, 2007), although its editorial introduction suggests a focus on the concept, in fact includes only one contribution which deals centrally with the organization of production and marketing: Chima J. Korieh, 'Gender and the Political Economy of the Post-Abolition Era: the Bight of Biafra (Nigeria) and its Hinterland', 41–58. Slight reference to the 'crisis of adaptation' is made by Hawthorne, *Planting Rice*, 177–8; Law, *Ouidah*, 261–2, the former dismissive but the latter expressing qualified support. G. Ugo Nwokeji, *The Slave Trade and Culture in the Bight of Benin* (Cambridge, 2010), chapter 7, although arguing that the transition exacerbated internal conflicts, does not explicitly link this to Hopkins' analysis.
[78] Robin Law, 'The "Crisis of Adaptation" Revisited: The Yoruba War of 1877–1893', *Africa, Empire and Globalization: Essays in Honor of A.G. Hopkins*, ed. Toyin Falola and Emily Brownell (Durham NC, 2011), 125–44.
[79] Law, *From Slave Trade*, 'Introduction', 11–16; A.G. Hopkins, 'The "New International Economic Order" in the Nineteenth Century: Britain's First Development Plan for Africa', *From Slave Trade to 'Legitimate' Commerce*, ed. Robin Law (Cambridge, 1995), 240–62 (see 254).
[80] Schwarz, 'Commerce', 257.

The use of slave labour in theory ceased to be an option in European colonies with the legal abolition of slavery, effective in British colonies from 1834 and French colonies in 1848 (though not completely in Portuguese colonies until 1868). But even after legal abolition, owners could often retain effective control over their former slaves, by redefining them as wives, clients or tenants.[81] In Angola, according to Ferreira, the status of former slaves (*libertos*) was in practice 'barely distinguishable from that of slaves'. In the case of the Gold Coast, the situation was complicated by uncertainty about the basis and extent of British jurisdiction: although British subjects were prohibited from owning slaves, their African wives could still legally do so, so that the ban could be evaded by attributing ownership of the slaves to the latter.[82] Considerable use was also made of other forms of unfree labour. In Sierra Leone, for example, until 1847, many of the recaptives supposedly 'liberated' from slave-ships by the British authorities were then assigned to European or settler employers as 'apprentices', in a system which critics considered no more than a disguised form of slavery.[83] In Liberia also, Everill notes that workers from the interior were recruited under a similar system of 'apprenticeship', giving rise to allegations that the settler farmers were reproducing in Africa the slave plantation system of the USA.

On the African side of the frontier, slavery remained legal, until the extension of European colonial rule over the interior. Slavery was a widespread and important institution in pre-colonial African societies: this reflected (as discussed by Austin in this volume) the high cost of free labour in a situation of land surplus. Although slaves were employed in a wide range of roles, many were engaged in agricultural production, including the cultivation of export crops, and also in their transport to the coast. European Abolitionists who were aware of this had to face the prospect that African production for 'legitimate' trade might be based upon slave labour.[84] Some of them hoped that the new trade would undermine slavery in Africa, as well as the overseas export of slaves, by giving slaves opportunities to participate in export production on their own account, and thereby accumulate the means to purchase their freedom.[85] Most modern scholarship, however, has

[81] As, for example, in Lagos after its annexation by Britain: Kristin Mann, 'Owners, Slaves and the Struggle for Labour in the Commercial Transition at Lagos', *From Slave Trade to 'Legitimate' Commerce*, ed. Robin Law (Cambridge and New York, 1995), 144–71.
[82] Silke Strickrodt, 'British Abolitionist Policy on the Ground in West Africa in the Mid-nineteenth Century', *The Changing Worlds of Atlantic Africa: Essays in Honor of Robin Law*, ed. Toyin Falola and Matt D. Childs (Durham NC, 2009), 183–200.
[83] Christopher Fyfe, *A History of Sierra Leone* (London, 1962), 106–7, 182–3, 251.
[84] Curtin, *Image of Africa*, 284–5, 451–3.
[85] Law, 'Historiography', 93–5.

held that the growth of 'legitimate' trade in the nineteenth century in fact strengthened the institution of slavery in West Africa, since the wealthy and powerful were able to participate in it on a large scale, employing slave labour. There was thus an expansion of the demand for slaves within local economies, which to some extent offset the loss of overseas markets.[86] This is sometimes presented as evidence against Hopkins' analysis, which emphasized small-scale production by independent free farmers, although Hopkins himself had in fact noted the use of slaves in 'legitimate' production, and made the inference that the rise of 'legitimate' trade led to an increased scale of slavery in West Africa.[87] The issue for Hopkins was to understand the 'relative importance' of large-scale (slave-based) and small-scale (free household) production, and 'the extent to which they could (or could not) coexist'.[88]

The use of slave labour in export production continued into the colonial period. Despite the Abolitionist rhetoric of the European colonizing powers, slavery initially persisted in the newly acquired territories. In the case of Britain, the Abolition Act of 1833 in principle automatically applied to newly acquired colonies, as for example Lagos and the Gold Coast, after their formal annexation, respectively in 1861 and 1874.[89] But it did not apply to 'protectorates', which was the status favoured for territories acquired from the 1880s onwards: the adoption of the 'protectorate' system allowed colonial rulers to evade any immediate requirement to confront the issue of domestic slavery (as distinct from slave-trading). The legal abolition of slavery in these new territories was a protracted process: slavery was not entirely prohibited, for example, in Sierra Leone (i.e. the hinterland Protectorate, as distinct from the original Colony) until 1928, and in northern Nigeria until 1936.[90] In consequence, slave labour played a significant role in the expansion of commercial agriculture under colonial rule, as research has indicated, for example, for the cultivation of cocoa in south-western Nigeria and southern Ghana,[91]

---

[86] Paul E. Lovejoy, *Transformations in Slavery: A History of Slavery in Africa* (Cambridge, 1983), chapter 8.
[87] Hopkins, *Economic History*, 143.
[88] Hopkins, 'New International Economic Order', 261, n.14.
[89] For the Gold Coast, see most recently, Kwabena Akurang-Parry, 'The Administration of the Abolition Laws, African Responses and Post-Proclamation slavery in the Gold Coast, 1874–1940', *Slavery and Abolition (S&A)* 19/2 (1998), 149–66.
[90] John Grace, *Domestic Slavery in West Africa, with Particular Reference to the Sierra Leone Protectorate, 1896–1927* (London, 1975); Paul E. Lovejoy and Jan S. Hogendorn, *Slow Death for Slavery: The Course of Abolition in Northern Nigeria, 1897–1936* (Cambridge, 1993).
[91] Sara S. Berry, *Cocoa, Custom and Socio-economic Change in Rural Western Nigeria* (Oxford, 1975), 129–30; Gareth Austin, *Labour, Land and Capital in Ghana: From Slavery to Free Labour in Asante, 1807–1956* (Rochester NY, 2005), chapter 13.

and of groundnuts in northern Nigeria.[92]

However, generalization about the importance of slavery tends to obscure the variety of forms of social relations of production in which slaves could be involved.[93] Slaves might be employed in large-scale enterprises comparable to American plantations, involving the co-operative organization of production and the direct appropriation of the product by the owner;[94] or they might operate as independent producers or traders, working on their own account, while paying a duty to their owner.[95] The optimistic view of the liberating effects of 'legitimate' trade put forward by nineteenth-century Abolitionists, noted earlier, implicitly (but unrealistically) assumed a version of the latter system as the norm. But often, slaves were incorporated (in smaller numbers) into individual households, working alongside their free members. The acquisition of slaves by smaller households evidently complicates Hopkins' analysis, though he himself has maintained in response that 'the basic question remains unchanged … it concerns the relative advantages of large-scale and small-scale producers'; the nature of the labour force is 'an element in understanding this problem, but it is not the problem itself'.[96]

In this volume, this issue is addressed by Austin, who stresses the importance of enslaved labour in the development of commercial export agriculture in West Africa, in the nineteenth century and on into the early colonial period. While acknowledging that the palm-oil and groundnut trades brought small-scale farmers into export production, as Hopkins argued, Austin suggests that 'the basic social units' of commercial agriculture in the nineteenth century were 'large slave estates combined with "ordinary" households expanded by the incorporation of captives'. This pattern, moreover, initially persisted in the early colonial period, until slavery was undermined by the combination of colonial legislation and the growth of opportunities for employment of free labour created by colonial economic developments. On this view, the 'modern' social form of commercial agriculture characterized by small households employing only free labour emerged not, as Hopkins suggested, in the nineteenth century but

---

[92] Mohammed Bashir Salau, 'The Role of Slave Labor in Groundnut Production in Early Colonial Kano', *JAH* 51/2 (2010), 147–63.

[93] Cf. Law, 'Historiography', 109; *From Slave Trade*, 'Introduction', 17.

[94] E.g. in Dahomey: Law, *Ouidah*, 211. See also Paul E. Lovejoy, 'Plantations in the Economy of the Sokoto Caliphate', *JAH* 19/3 (1978), 341–68.

[95] See e.g. Robin Law, '"Legitimate" Trade and Gender Relations in Yorubaland and Dahomey', *From Slave Trade to 'Legitimate' Commerce*, ed. Robin Law (Cambridge, 1995), 195–214 (see 198, 200); Paul E. Lovejoy, '*Murgu*: the Wages of Slavery in the Sokoto Caliphate', *S&A* 24/1 (1992), 168–85.

[96] Hopkins, 'New International Economic Order', 255.

only in the early twentieth. This generalization, as Austin acknowledges, is subject to local exceptions, since research has indicated that slavery played little role in commercial agriculture in some areas.[97] This reinforces the argument for a disaggregated approach, recognizing variation across space and time. As Hopkins has observed, with reference to the nineteenth-century 'transition', it is possible to envisage 'different groups of conclusions', rather than only one.[98]

The contributions to this volume show that commercial agriculture did form a potential or actual alternative to the slave trade, even at the height of the latter in the eighteenth century, though the degree of involvement in it differed from region to region, and even within regions. Moreover, this earlier history is critical for understanding the nature and significance of the subsequent development of 'legitimate' trade. Insofar as the idea of a 'commercial transition' remains viable, it was evidently a protracted and uneven process. On the one hand, its origins in some areas actually predated the rise of the Abolitionist campaign, as with gum arabic in Senegal and palm-oil in the Bight of Biafra. On the other, the persisting importance of enslaved labour in commercial agriculture meant that its 'civilizing' (or in recent parlance, 'modernizing') effects were initially more limited than Abolitionists had hoped. As one contemporary critic of the 1841 Niger Expedition suggested, the implication of 'legitimate' trade might be merely to transform the African entrepreneur 'from the slave-hunter to the slave-driver'.[99] It is important, nevertheless, not to minimize the significance of this transformation, not least from the perspective of slaves themselves. The employment of slaves within Africa was clearly, in a straightforward sense, an 'alternative', and surely a preferable alternative, to exporting them.

[97] E.g. in Igboland (south-eastern Nigeria): Susan Martin, 'Slaves, Igbo Women and Palm Oil in the Nineteenth Century', *From Slave Trade to 'Legitimate' Commerce*, ed. Robin Law (Cambridge, 1995), 172–94.
[98] Hopkins, 'New International Economic Order', 243.
[99] *The Eclectic Review*, quoted in Curtin, *Image of Africa*, 453.

1

The slave trade & commercial agriculture in an African context

DAVID ELTIS[1]

What is commercial agriculture? For present purposes two types come to mind: one that produces for local markets (where the output is sold within say 50 km of the point of production) and the other that is able to supply markets further afield. We can assume that before low-cost sea or inland waterway transportation developed, the populations of large cities everywhere survived on the basis of local commercial agriculture. In that sense, commercial agriculture in Africa, as in the rest of the world, must go back almost to the point where human beings began to draw most of their sustenance from agriculture as opposed to hunting and gathering. But the thrust of the present collection is, I believe, on commercial output for faraway markets. For this we have to start with the recognition that distance and lack of transportation infrastructure were the major barriers to trade in produce almost everywhere around the globe until quite recently.

As a consequence agriculture that produced food, drink and the raw material for clothing (including dyes) that could be traded over long distances was quite rare before the early modern era and where it existed it contributed little to the basic food, clothing and shelter requirements of either the sellers of the produce or its buyers. Kola nuts and spices, to take two random examples, were not central to the well-being of consumers nor did they provide a livelihood for many producers. Where agricultural products *were* carried long distances before the nineteenth century, it was often through piggybacking on, or as an accompaniment of high-value, low-volume, lightly weighted luxury goods. Thus slaves taken from the interior to the African coast

[1] Thanks to Steve Behrendt and Daniel Domingues da Silva for advice as this essay evolved, as well as to participants at the GHIL conference in Sept. 2010, especially Robin Law, for their comments on the original paper, which have been incorporated into this, its published successor.

provided cheap transportation for commodities, including food. In the Atlantic, the first palm-oil, dyewoods (and probably cloves) left Africa on slave vessels long before it was economical to dispatch a vessel specializing in such products. From the sixteenth-century Caribbean, hides and some sugar were sent to Europe on ships whose primary purpose was the transportation of precious metals. Nowadays it is usual to think that the slave and the palm-oil trades were complementary rather than competing activities, but this is not such a great insight: the same could be said of many pairings of commodities that left Africa in the pre-colonial era. It is true that in the eighteenth-century Atlantic a small specialized wheat trade developed between Philadelphia and Great Britain, and the export of rice from South Carolina to (eventually) southern Europe is well documented, but generally the long-distance bulk shipping of semi-processed agricultural commodities as a stand-alone commercial activity was very much a nineteenth-century phenomenon.

The late development of commercial agriculture was thus not confined to Africa, but there was an additional consideration specific to much of the continent – rainfall distribution and whether navigable rivers (and lagoons) connect areas with differing precipitation levels. This is relevant because trade in agricultural produce is more likely to evolve where regions that specialize in different products have feasible transportation links. In Africa isohyetal lines predominantly run parallel to the major rivers and coastal lagoons – with the significant exception of the Nile, where large-scale long-distance trade in foodstuffs, and of course major urban areas, developed early. In West Africa the isohyetal/riverine axis is east-west, and in West-central Africa, or at least the region east of Luanda, the axis is predominantly north-south.[2] The impact of such a pattern is that the waterways that did offer the possibility of cheaper transportation of bulkier commodities tended to tie together areas capable of producing similar products – a situation that would inhibit shipping agricultural commodities large distances.[3] The real significance of, first, Arab and then European contact with Africa was that both introduced a small but, by the nineteenth century, rapidly growing north-south trade which increased the possibilities for exchange of produce, notwithstanding the fact that for several centuries non-Africans were interested chiefly in slaves.

[2] The relatively small exception to this assessment is the Niger after it swings south.
[3] This is not to ignore the very active trade between pastoral nomadic economies on the one hand and the grain-producing areas on the other – some of it coastal. Curtin estimated that half the caloric intake of some areas of the steppe was supplied from trade: Philip D. Curtin, *Economic Change in pre-Colonial Africa: Senegambia in the Era of the Slave Trade* (Madison WI, 1975), 218. This was not, however, a result of a regional or long-distance trade in foodstuffs, but rather of the movement of cattle on the hoof.

Thus, in the early days of oceanic contact between Europe and Africa, foodstuffs began to be shipped from millet- and then rice-growing regions of Upper Guinea north to the Cabo Verde Islands, and Phil Misevich's recent study of Sierra Leone in the eighteenth and early nineteenth century provides a further demonstration of this re-orientation away from what might be termed the isohyetal-riverine straightjacket. Misevich shows a large-scale north-south trade in rice carried on in coastal vessels and canoes developing in the aftermath of the establishment of Freetown.[4]

If the additional north-south orientation of trade introduced by non-Africans came to be dominated by slaves the non-slave component was always significant. Sub-Saharan African commodities reached European (and to a lesser extent, American) markets via two routes: on vessels carrying slaves that sailed to Europe via the Americas, and on what have come to be called 'produce ships' that sailed out and back to Africa without carrying slaves.[5] Produce ships, it should be noted, did trade in slaves, but they did not carry slaves as part of their return cargo.[6] Which of these two routes carried the greater share of produce is an interesting question. In the pre-1641 slave trade to the Spanish Americas vessels carried off significant quantities of bees-wax from Senegambia, as well as large numbers of hides and 'elephants' teeth', even though Spanish regulations allowed only slaves.[7] Produce exports from Upper Guinea on Portuguese vessels continued at high levels even when the slave trade reached its acme after 1750.[8] Further south the picture was different. One vessel arriving in Santo Domingo with 40 slaves carried the bulk of its cargo in the form of sugar from São Tomé. Small quantities of cloth from the Cabo Verde Islands show up in the early seventeenth century, but by the 1630s more slave vessels arriving in the Americas were sailing from Angola, rather than West Africa, and no produce is mentioned in the records of the Angolan

---

[4] Philip Misevich, 'On the Frontier of "Freedom": Abolition and the Transformation of Atlantic Commerce in Southern Sierra Leone, 1790s to 1860' (PhD thesis, Emory University, 2009, available at https://etd.library.emory.edu/view/record/pid/emory:199w5). For early 17th-century Upper Guinea see Linda A. Newson and Susie Minchin, *From Capture to Sale: The Portuguese Slave Trade to Spanish South America in the Early Seventeenth Century* (Leiden, 2007), 77–85.

[5] Some of the vessels returning direct to Europe from Africa also carried slaves, particularly in the first quarter of the 16th century – before the African trans-Atlantic slave trade was established. Occasional shipments of slaves to Europe continued after 1525 to at least 1756, but compared to either the slave trade to the Americas or the number of vessels carrying produce only to Europe, such voyages were numerically insignificant and disregarded here.

[6] Among the many log and account books that establish this point see National Maritime Museum, London, Log M/64, logbook of the *African Queen*, a Bristol produce trader.

[7] David Wheat, 'The First Great Waves: African Provenance Zones for the Transatlantic Slave Trade to Cartagena de Indias, c.1570–1640', *Journal of African History (JAH)* 52/1 (2011), 1–22.

[8] Walter Hawthorne, *Planting Rice and Harvesting Slaves: Transformations Along the Guinea-Bissau Coast, 1400–1900* (Portsmouth NH, 2003), 40–50.

vessels.⁹ The greater availability of commodities as well as slaves in Upper Guinea probably explains that region's ability to continue to compete with Angola after 1600 as a slave-supply region, given that Upper Guinea was always a marginal source of slaves in the broad spectrum of trans-Atlantic slaving. Without non-slave commodities, fewer slaves would have been sold and *vice versa*.¹⁰

But in addition to slave-trading vessels numerous produce vessels left northwestern European ports before 1640 returning direct to Europe with gold from Africa west of Accra and sugar from São Tomé. We now know that more than twice as many Dutch vessels went to Africa for produce than for slaves in this period.¹¹ The direct trade in commodities between Africa and Europe of course predates the trans-Atlantic slave trade by many decades, and down to 1640 and for all nations except the Spanish there were probably more vessels that traded for commodities alone than sought slaves.¹² After 1650, the English, French and Dutch sent many vessels to the coast for African produce alone, even as their respective slave trades rapidly expanded. While working on the trans-Atlantic slave trade database (TASTD) over the last two decades, researchers have brought to light records of no less than 1,600 voyages (and 300 more probables) that sailed from Europe to Africa and back again without bringing slaves between 1672 and 1807.¹³ This figure constitutes only 15 per cent of the total number of slave-ships sent out over the same period, and moreover produce vessels were smaller than the average slave-ship. Nevertheless, any reasonable estimate of the total value of outbound cargoes on produce ships, added to the share of the cargo on slave vessels that was designated for non-human commodities as reflected in the above discussion, suggests that the value of non-human African commodi-

---

⁹ Personal communication from David Wheat, 10 July 2010; Rudy Bauss, 'Rio de Janeiro: The Rise of Late Colonial Brazil's Dominant Emporium, 1777–1808' (unpublished PhD thesis, Tulane University, 1977), 105.

¹⁰ For a fuller exposition of this argument see David Eltis, Frank D. Lewis, and Kimberly McIntyre, 'Accounting for the Traffic in Africans: Transport Costs on Slaving Voyages', *Journal of Economic History (JEH)* 70 (2010), 940–63.

¹¹ A comparison of the Binder database (note cards held at the Wilberforce Institute of the University of Hull and the Rijksarchief, The Hague, Netherlands) and www.slavevoyages.org shows 42 Dutch vessels that cannot be identified as slave-ships compared to 19 slave voyages. Even in the two decades after 1640 when the Dutch were supposedly supplying slaves to the whole Atlantic world, slavevoyages.org shows 159 Dutch slave voyages compared to 30 probable produce vessels in the Binder database. In other words about one in six Dutch African voyages at this time were in the produce trade. For early English involvement in the African produce trade see Hilary Jenkinson, 'The Records of the English African Companies', *Transactions of the Royal Historical Society* 6 (1912), 185–210.

¹² David Eltis, *The Rise of African Slavery in the Americas* (Cambridge, 2000), 24–8.

¹³ Stephen Behrendt, David Eltis, David Richardson, and Jelmer Vos, 'Produce Vessels Sailing from Europe to Africa, 1672 to 1807', unpublished database.

ties carried off probably exceeded the value of slaves prior to the second half of the seventeenth century, despite the 26,000 slaves a year estimated to have been carried off to the Americas between 1600 and 1650.[14] Further, and as shown below, produce constituted 17 per cent of total exports for the period 1681–1807. And overall, from first oceanic contact to the beginning of the nineteenth century, produce likely constituted one quarter of all Atlantic trade from Africa. This constitutes a substantial and largely neglected trade.

The most systematic data for such assessments are available from the beginning of the fourth quarter of the seventeenth century as records of European chartered companies trading with Africa come on stream. Table 1.1 summarizes annual series of the value and relative importance of the traffics in both slaves and produce from Atlantic Africa between 1681 and 1807 (the full data are available in the Appendix to this chapter), and presents them in the form of five-year averages. It draws on three databases: the first of trans-Atlantic slave voyages, the second of slave prices and the third, the aforementioned data set of produce voyages. The base of the Table is the estimate of five-year annual averages of the total value of slaves leaving Africa for the Atlantic World shown in column 1 and is obtained by multiplying slave prices and slave departures. The resulting values are in terms of the trade goods as they left Europe adjusted for inflation. Given the high cost of transporting these goods from Europe these values would have to be doubled to reflect their c.i.f. (cost, insurance, freight) value on the African coast. As already noted, produce left the African coast via two routes. The direct produce trade is estimated in column 2. It derives from an annual series of produce voyages shown in the appendix. The values shown are the product of the ratio of produce vessels to slave vessels (adjusted for their different sizes) for each year multiplied by the total volume of slaves shown in column 1.

The second and somewhat more important route by which African produce reached the Atlantic world was on slave-ships. The key questions here are what proportion of the total value of the cargo of a slave vessel was exchanged for African produce and how did this change over time. Sources and additional databases discussed in the Appendix suggest that in the early 1680s 45 per cent of the value of goods and people on board slave vessels leaving Africa comprised produce, a ratio that had declined to 5 per cent by the beginning of the nineteenth century. Applying this diminishing ratio to the total value of slave exports yields an annual series for produce on slave-ships shown in column 3. Adding column 3 (estimated values of produce on slave

[14] http://slavevoyages.org/tast/assessment/estimates.faces?yearFrom=1601&yearTo=1650

Table 1.1 Estimated five-year annual values of slaves and produce leaving Africa via the Atlantic in constant pounds Sterling, 1681–1807 (valued in terms of the goods exchanged for slaves and produce as those goods left Europe and the Americas)

	Average annual value of slaves leaving Africa for America	Average annual value of produce leaving Africa on produce ships	Average annual value of produce leaving Africa on slave-ships	Average annual value of all produce leaving Africa (col 2 + 3)	Average annual value of slaves and produce (col 1+ 4)	Produce as per cent of all African trade (%)	Produce on slavers as per cent of all produce (%)
	1	2	3	4	5	6	7
1681–1685	107,587	7,092	88,026	95,118	202,705	46.9	92.5
1686–1690	86,610	8,324	66,528	74,852	161,462	46.4	88.9
1691–1695	110,731	6,772	75,736	82,508	193,239	42.7	91.8
1695–1700	202,723	19,359	122,160	141,519	344,241	41.0	86.3
1701–1705	196,478	32,408	106,153	138,561	335,039	41.4	76.6
1706–1710	196,241	11,056	93,686	104,742	300,983	34.8	89.4
1711–1715	248,689	13,171	103,108	116,279	364,967	31.9	88.7
1716–1720	361,423	37,550	132,260	169,810	531,233	32.0	77.9
1721–1725	512,750	92,908	160,117	253,025	765,775	33.0	63.3
1726–1730	392,838	64,175	104,461	168,636	561,474	30.0	61.9
1731–1735	262,800	63,118	60,077	123,194	385,995	31.9	48.8
1736–1740	385,436	79,663	74,008	153,671	539,107	28.5	48.2
1741–1745	405,049	39,406	63,300	102,706	507,756	20.2	61.6
1746–1750	449,280	53,819	55,212	109,031	558,310	19.5	50.6
1751–1755	537,147	80,983	58,008	138,992	676,139	20.6	41.7
1756–1760	396,019	55,520	40,515	96,035	492,054	19.5	42.2
1761–1765	796,786	95,921	77,173	173,094	969,880	17.8	44.6
1766–1770	1,246,728	159,029	114,593	273,622	1,520,350	18.0	41.9
1771–1775	1,503,464	223,674	130,358	354,032	1,857,496	19.1	36.8
1776–1780	698,969	68,695	57,006	125,701	824,670	15.2	45.4
1781–1785	1,317,247	34,926	100,483	135,409	1,452,656	9.3	74.2
1786–1790	2,381,871	131,156	169,740	300,896	2,682,768	11.2	56.4
1791–1796	1,717,428	98,142	114,798	212,940	1,930,368	11.0	53.9
1796–1800	1,812,184	30,120	111,280	141,401	1,953,584	7.2	78.7
1801–1807	2,165,957	77,050	119,938	196,988	2,362,945	8.3	60.9
1681–1807	**762,158**	**63,577**	**96,327**	**159,904**	**922,062**	**17.3**	**64.7**

(Source: Appendix)

vessels) to column 2 (estimated values of the direct produce traffic) allows us to compute annual values for all produce leaving Africa via the Atlantic Ocean shown in column 4. The remainder of Table 1.1 shows the total value of Africa's Atlantic trade derived from adding slaves and produce together (column 5) and a couple of interesting ratios, one of which is the produce share of all African trade (column 6), and finally, the share of all produce carried off on slave-ships (column 7).

The size and fluctuations in the slave trade shown in Table 1.1 and the Appendix are well known and require little comment, though the tenfold increase in the average annual value of slave departures over the course of the eighteenth century is a new finding worthy of note. What interests us here is the trend in produce exports. Column 4 shows that starting in the 1680s at just below the value of the slave trade these clearly grew much more slowly than did the slave trade over the whole period.[15] In addition, produce trade, or at least that part of it carried by produce ships, was disproportionately affected by war, probably because produce vessels were smaller and carried fewer crew per ton than slavers. But the more interesting pattern is that over the long term they peaked in the early 1770s and thereafter declined, and on their way to this peak they underwent two distinct growth periods: one before 1725, and the second in the third quarter of the eighteenth century. Produce exports from Africa are traditionally associated with economic growth and, eventually, industrialization in Europe but the nearly 60 per cent fall in the value of produce exports in the final quarter of the eighteenth century is scarcely consistent with such an interpretation. Moreover both growth phases, before 1725 and between 1751 and 1775, were driven by the rapid expansion of the direct trade between Europe and Africa shown in column 3. In the first period the increase was distributed between the Upper Guinea coast and the Gold Coast. In the second period expansion was almost entirely accounted for by Upper Guinea, more specifically Senegambia and to a lesser extent Sierra Leone. Indeed, there is no doubt that in the eighteenth century a hugely disproportionate share of the produce trade between Europe and Africa was really between Europe and Upper Guinea. At least some of the pre-1725 growth was in the dyewood trade with Sherbro and immediately adjacent regions

---

[15] These new series suggest a small revision to my previous contributions to the debate on the relative importance gold and slaves in Africa's Atlantic trade. In the 17th century, the value of African produce, but particularly gold, carried off from Africa exceeded that of slaves, but the reversal of this situation now seems to have taken place around 1680 rather than 1700. See Ernst van den Boogaart, 'The Trade Between Western Africa and the Atlantic World, 1600–90', *JAH* 33/3 (1992), 369–85; David Eltis, 'The Relative Importance of Slaves in the Atlantic Trade of Seventeenth Century Africa', *JAH* 35/2 (1994), 237–49.

(later spreading to Gabon and Loango, north of the Congo) and this subsequently declined. The second growth surge was a direct result of the British occupation of Senegal (1758–79) during which the trade in gum arabic and a range of other products, including ostrich feathers and mules surged, the mules being dispatched to the plantations of the Caribbean.

But what does this have to do with commercial agriculture? A close examination of those non-human commodities reveals little that was actually cultivated. Gold, ivory and sugar from São Tomé would have accounted for almost all the produce exported (by value) from sub-Saharan Africa prior to 1650. Gum arabic, dyewoods, and hides became more important later (though hide exports remained flat in the eighteenth century). The African produce that at one stage exceeded in value the slave trade and continued to be significant thereafter was nearly all collected rather than grown. Or as the *Liverpool General Advertiser* put it in 1768, 'the trade carried on to Africa by the Europeans consists in … slaves, teeth, and gold … [B]ut all this while, there is not the least use made of the land.'[16] The output of commercial agriculture appears to have been limited to some pepper, both malagueta and Guinea pepper, and those cloths that were not made from unravelled imported cloth. Pepper was the more important of the two – we have records of produce ships returning to England with fifteen tons of pepper in their holds – but pepper must have amounted to less than one per cent of the value of non-slave exports before the nineteenth century.[17] Like other peoples of the early modern era, Africans fed, clothed and sheltered themselves without drawing on the trans-oceanic and trans-desert worlds, or indeed, to any great extent, on other regions on their own continent. They quickly incorporated non-African commodities into their lives but did so by growing and processing them themselves rather than by trading. At the beginning of the eighteenth century, Sir Dalby Thomas, one of the longest-surviving Europeans in West Africa, wrote that while 'there has never been a sugar cane in Comp^y [i.e. Royal African Company] grounds … the blacks have many years had them grow in Plenty.'[18] If they did not import such commodities, neither did they export them. This was despite the fact that, as with the sixteenth-century Caribbean,

---

[16] 8 April 1768.
[17] For vessels carrying mainly pepper, see the *Guinea Frigott* arriving in England with 15 tons of pepper in 1689 (The National Archives, London (TNA), T70/11, 34, Samuel Humphreys, Wright & John Boylston, 12 Feb. 1689); and the *Broughton*, with 17 tons in 1706 (T70/14, fol. 128, Josias Collins, Spithead, 16 July 1706). Malagueta and Guinea pepper appears to have been traded throughout West Africa and the Atlantic islands. Every European port trading with Africa must have imported the spice.
[18] TNA, T70/5, fol. 63, Dalby Thomas to Royal African Company (RAC), 22 Oct. 1709.

the first African long-distance commodity exports could get to Europe for relatively low cost – given that produce vessels had relatively abundant return cargo space created by the exchange of relatively bulky manufactured goods for relatively tiny quantities of gold.[19]

This did not inhibit the expectations of Europeans that a trans-oceanic agricultural export economy could be established. In 1768 the leading newspaper of the leading slave-trading port in the world advocated the development of coffee, sugar, tea and spice plantations in Africa – a strategy that would have destroyed the trans-Atlantic slave trade. The potential of Africa fascinated European promoters, merchants, and colonial officials from the fifteenth century to the onset of the colonial period but came to nothing in the face of African resistance to European occupying forces. Ninety per cent of all European 'plantation' projects in the seventeenth century sense of that word, anywhere in the world, were unsuccessful. But for African ventures the ratio was close to 100 per cent. Portuguese Angola might appear as an exception, but not when we compare Angola with Brazil where the Portuguese established a presence at about the same time. If there was an exception it was São Tomé, which for several decades in the sixteenth century became the major source of sugar for Europeans and continued as a significant producer to the middle of the following century.[20] Yet São Tomé provides some underpinning for the resistance argument presented here. The decline of its plantation sector was of course brought about by cheaper and better sugar from Brazil, but an important auxiliary factor in this decline was the inability of Europeans to maintain control over the whole island. Given the trajectory of scholarly interest in slavery and resistance during the course of my own career, it is hard to understand why the English-language literature has paid little attention to the São Tomé slave rebellion of 1595.[21] This was the second most successful slave revolt of the Atlantic world and accelerated the decline of one of the two major sugar export economies of the time. Its importance to my own argument is that if Europeans could not maintain a plantation economy on an island off the African coast, within easy reach of European naval vessels, what were their prospects on the mainland?[22]

---

[19] Philip D. Curtin, 'Africa and the Wider Monetary World, 1250–1850', *Precious Metals in the Later Medieval and Early Modern World*, ed. J.F. Richards, (Durham NC, 1983), 231–68.

[20] See Gerhard Seibert, this volume, chapter 2. Many Dutch vessels made return voyages to Africa in the 1620s and 1630s carrying sugar to Europe from São Tomé. See Binder database.

[21] But see Gerhard Seibert, 'São Tomé's Great Slave Revolt of 1595: Background, Consequences and Misperceptions of One of the Largest Slave Uprisings in Atlantic History', *Portuguese Studies Review* 18/2 (2010), 29–50.

[22] 'Remarks relative to the Extension and Improvement of the African Trade', *Liverpool General Advertiser*, 8 April 1768. For the resistance argument see Eltis, *Rise of African Slavery*, 137–63.

So is the search for commercial agriculture in an African context prior to the nineteenth century a futile exercise? I think not. There is one relatively unexplored activity which merits attention. In the eighteenth century over six million people departed from a 3,000-mile range of the western African coast via the Atlantic Ocean alone. The average was somewhat greater than 60,000 a year, though the outflow peaked in the 1780s at close to 100,000 departures annually. They left with provisions for a three-month journey (even though the voyage itself averaged only 67 days), only a small portion of which was carried to the African coast on the outbound voyage of the slave vessels.[23] Moreover after allowing for mortality among captives en route to the coast and while awaiting embarkation we can assume that more than this number would have had to be fed while they were in transit from the point of enslavement or from the point at which they were sold to an owner who had the trans-Atlantic slave trade in mind for his human property. Few captives grew their own food so that what they consumed was the output of commercial agriculture that would not have existed without the slave trade. In some cases captives carried the food themselves. Bartholomew Stibbs saw '30 or 40 [captives] strung together with thongs with a bundle of corn or elephants tooth on their heads' while up the Gambia River in 1723.[24] In addition to this we can see from Table 1.3 in the Appendix that by the second half of the eighteenth century, between three and four hundred trans-oceanic sailing vessels a year were spending several months on the African coast with on average 32 crew on board (amounting to about 10,000 men).[25] While crew provisions were for the most part carried from the home port of the vessel, captains were instructed '[w]hile Shipes are in ye country to Ketch fish and ...[buy] Rice and Corne to Lengthen out ye Sea provisions at Least 2 or 3 months'.[26]

A rough estimate of the value of the provision trade is possible. Curtin claimed that of the total cost of placing a slave on board a slave ship, only 80 per cent constituted the price of the slave (that is the price paid by the factor on the coast to obtain a slave), with the

---

[23] Agents of slave-traders in Jamaica complained that factors on the African coast 'send more provisions than are necessary' (TNA, T70/26, p. 40, John Stewart and John Wright, 29 May 1714), and Georg Nørregård, who wrote the basic study of the Danish slave trade, commented that '[o]nly one case of a ship having been insufficiently provisioned has come down to us': Georg Nørregård, *Danish Settlements in West Africa, 1658–1850* (Boston MA, 1966) 88.

[24] See the journal of Bartholomew Stibbs, in Francis Moore, *Travels into the Inland Parts of Africa*, 2nd edn (London, 1755), 175–214.

[25] Crew average from the TASTD downloaded from www.slavevoyages.org.

[26] TNA, T70/1222, p. 28. This standing order was written in 1663 for Company of Royal Adventurers' captains.

remainder comprising provisions.²⁷ Provisions thus had a value equivalent to 25 per cent of the goods sent out to Africa from the home port (column 1 of Table 1.1). Such a figure would include provisions for slaves that originated in Europe or the Americas as well as provisions consumed by slaves in either the forts prior to embarkation or on the slave vessel itself prior to departure. Provisions from Europe did not comprise a major item, amounting to between one and two per cent of the price of a slave on the coast. By contrast, provisions consumed prior to the middle passage were significant given that over the whole period of the slave trade a slave ship spent 137 days on average between the embarkation of the first slave and the vessel's departure for the Americas.²⁸ Obviously, not all slaves would have spent this amount of time awaiting departure, but four and a half months might be used as the average time on the coast for all slaves whether in European forts (as on the Gold Coast and Senegambia) or in the hands of African traders. Slaves would thus require food for seven and a half months (four and a half waiting for departure and three for the voyage). Prices for slaves recorded in the documents implicitly incorporated the African-sourced provisions that were consumed by the slaves prior to embarkation, and in addition sometimes included African-sourced provisions that were consumed on board.²⁹ Table 1.2 provides an estimate of those African-sourced provisions based on the estimate that 25 per cent of column 1 of Table 1.1 is accounted for by provisions.³⁰ But Table 1.1 is valued in terms of prices for the goods exchanged for slaves and produce in Africa as those goods left Europe and the Americas. Valuation in Africa itself, by contrast, is obviously more appropriate and column 2 of Table 1.2 provides an approximation of such a valuation by doubling column 1 and thereby allowing for the cost of transporting the trade goods to the African coast (a procedure that is well established in the literature). Column 3 adds an estimate of provisions consumed by crews of vessels visiting the African coast –

---

[27] Curtin, *Economic Change*, 230; Newson and Minchin, *From Capture to Sale*, 71, have a smaller ratio, but the food listed in their Table 2.5 presumably includes the cost of food only after the slave was acquired on the coast by the shipper or his agent.

[28] Computed from 3,854 cases in the TASTD available on the download page at www.slavevoyages.org.

[29] Eltis, Lewis and Mcintyre, 'Accounting for the Traffic'.

[30] The fixed ratio means that prices of provisions are assumed to have changed in lockstep with the prices of slaves. If they increased at a slower rate then there will be upward bias in the series of provision values. Joseph C. Miller, *Way of Death: Merchant Capitalism and the Angolan Slave Trade 1730–1830* (Madison WI, 1987), 397, suggests a ninefold rise in the price of manioc from 'before 1710' to the 1810s in Luanda – the largest single slave embarkation point in Africa. Prices of slaves on the African coast, shown in column 1 of Table 1.3, increased about six-fold over the same period (1681–1709 compared to 1801–1807).

Table 1.2 Estimated five-year annual average values of provisions for slaves carried off from Africa and crews of vessels trading on the African Coast, 1681–1807, in constant pounds sterling (1700=100)

	Total Value of African-sourced provisions for slaves (fob Europe/ Americas)	Total Value of African-sourced provisions for slaves (cif Africa)	Total Value of provisions for crews of ships trading in Africa	Total Value of African-sourced provisions for all Atlantic trade
	1	2	3	4
1681–1685	26,897	53,794	1,614	55,407
1686–1690	21,653	43,305	1,299	44,604
1691–1695	27,683	55,365	1,661	57,026
1695–1700	50,681	101,361	3,041	104,402
1701–1705	49,119	98,239	2,947	101,186
1706–1710	49,060	98,120	2,944	101,064
1711–1715	62,172	124,344	3,730	128,075
1716–1720	90,356	180,712	5,421	186,133
1721–1725	128,187	256,375	7,691	264,066
1726–1730	98,209	196,419	5,893	202,311
1731–1735	65,700	131,400	3,942	135,342
1736–1740	96,359	192,718	5,782	198,500
1741–1745	101,262	202,525	6,076	208,600
1746–1750	112,320	224,640	6,739	231,379
1751–1755	134,287	268,573	8,057	276,631
1756–1760	99,005	198,009	5,940	203,950
1761–1765	199,197	398,393	11,952	410,345
1766–1770	311,682	623,364	18,701	642,065
1771–1775	375,866	751,732	22,552	774,284
1776–1780	174,742	349,485	10,485	359,969
1781–1785	329,312	658,624	19,759	678,382
1786–1790	595,468	1,190,936	35,728	1,226,664
1791–1796	429,357	858,714	25,761	884,475
1796–1800	453,046	906,092	27,183	933,275
1801–1807	541,489	1,082,979	32,489	1,115,468
1681–1807	**186,494**	**372,989**	**11,190**	**384,179**

(Source: Table 1.1 and text)

set at 3 per cent of column 2.³¹ Column 4 is the sum of columns 2 and 3.

How significant was the fact that African-sourced provisions averaging £1,000,000 a year were being sold to those participating in Atlantic commerce in the two decades from 1786 (inflation adjusted prices, 1701=100)? Given the focus on African commodity production during the nineteenth century it is logical that we draw first on later comparisons. The African export commodity that has received the greatest attention is palm-oil. Strikingly, not until after 1850 did imports into Britain breach the £1,000,000 sterling level in constant prices.³² At the other end of the period for which systematic data are available – the 1680s – African gold carried off annually by Royal African and Dutch West India Company vessels certainly exceeded the annual average value of all African-sourced provisions, but by the late eighteenth century the value of provisions was many times that of gold even when gold exports were at their peak.³³ The production of basic foodstuffs around the early modern Atlantic World has attracted only a tiny fraction of the scholarly attention paid to the major plantation crops. Yet the above comparisons should not be surprising.³⁴ Pre-industrial peoples everywhere spent well over half their incomes on food, and high wheat prices in Europe often presaged social upheaval. Growing and supplying provisions to sustain the long-distance movement of 60,000 people per year (the annual average dispatched during the eighteenth century) would not have required a significant redistribution of resources from a sub-continental economy comprising perhaps tens of millions of people. Nevertheless the total value of such produce would still be large relative to other African commodities exported to Atlantic markets which, before the nineteenth century, were largely the result of collection in the wild, rather than cultivation, and chiefly used in the production of small quantities of luxury goods. We can be sure that in any society that had passed as far through the agricultural revolution as most of Africa had, silvan produce would inevitably be of minor importance. Gold production certainly had the potential to be a high-value activity, but we should keep in mind that in global terms African gold was trivial.

---

³¹ Assuming 32 crew members per vessel, a stay of 4.5 months, and an African-sourced provision ration of half that of slaves.

³² Current pound values are from Martin Lynn, *Commerce and Economic Change in West Africa: the Palm Oil Trade in the Nineteenth Century* (Cambridge, 1997), 31, converted into inflation-adjusted values via the Schumpeter/Gilboy and Rostow/Gayer/Schwartz indexes.

³³ Curtin, 'Africa and the Wider Monetary World', 245–46.

³⁴ Commercial agriculture would have sustained long-distance trade in the sense of providing provisions for traders, and when the commodities in which those traders dealt were human, then the size and importance of the activity would have greatly increased.

It would be easy to extend the above analysis to other phases of the system that generated slaves for Atlantic markets and for other slave trades out of Africa (the Atlantic slave trade being only one of the conduits for removing people from the continent) in each case allowing for ever-larger numbers of slaves as mortality took its toll. However, the further we move away from the Atlantic the more speculative assessments become. Suffice it to say that hard evidence on the distance from point of origin to embarkation point does however exist for three regions (one of which is Angola) for the nineteenth century and these suggest the vast majority of slaves may be identified with places within two hundred miles of their point of shipment, and most within one hundred miles.[35] The same may be said about the location of most wars that historians have been able to link with slave departures. Allowing for food production to sustain the movement of people to the coast as well as those crossing the Sahara might easily double the above estimates. But despite the apparent and largely spurious appearance of precision in Tables 1.1 and 1.2, the point of this exercise is to generate only orders of magnitude, rather than exact estimates. There is certainly enough evidence here to suggest scale.

Where did the provisions for the slave trade that were grown in Africa come from? What were the crops involved? And how were these crops produced? We cannot answer all these questions at the present time, but they are certainly worth asking. As already mentioned, all slave vessels carried some food to the coast for their captives. Those setting out from the Americas, where provisions were cheaper than in Europe, carried the greater quantities. North American slave-ships are invariably referred to as rum ships, but in fact the provisions they sold on the coast to slave-traders and slave-trading communities were also very important. For the one half of the North American slave trade originating outside Rhode Island, provisions may have been more important than rum, and in any event some of the rum was given to captives on the voyage.[36] Slave-ships leaving Rio de Janeiro for Angola carried some manioc flour as well as the counterpart of rum in the South Atlantic, cachaça.[37] Vessels from Europe carried beans (horsebeans or *vicia faba*) and for those obtaining slaves south of Senegambia, some animal protein. The English factor at Ouidah requested that captains bring from Europe, 840 lbs of beans,

---

[35] Misevich, 'On the Frontier of "Freedom"'; Daniel B. Domingues da Silva, 'Crossroads: Slave Frontiers of Angola, c.1780–1867' (unpublished PhD dissertation, Emory University, 2011), 88–115; G. Ugo Nwokeji and David Eltis, 'Characteristics of Captives Leaving the Cameroons for the Americas, 1822–37', *JAH* 43/2 (2002), 191–210.
[36] The early New York trade to Madagascar was based on provisions.
[37] Bauss, 'Rio de Janeiro', chapter 6.

1,100 lbs of beef and 224 lbs of cheese for every 100 slaves they intended to board, although by the 1720s beans for consumption by slaves were being grown in the vicinity of Ouidah.[38] 'Desire Beans by every Ship which the Slaves like better than Corn,' wrote the factors at James Fort in the Gambia in 1722.[39] Stockfish or 'stinkfish' – salted fish from a variety of northern seas, including the North Atlantic – might be substituted for beef.

The greater share of food consumed by slaves was nevertheless grown in Africa, and by Africans. Until the second quarter of the eighteenth century, four crops dominated – millet, maize, yams and the cassava root, which yielded manioc flour. Millet and maize are often hard to distinguish from each other in the documents given the English predilection for describing both as 'corn', but millet at least appears to have been grown in varying quantities throughout western Africa outside the rainforest. It was the predominant grain purchased by slave-traders in both Upper Guinea and Angola in the seventeenth century, though it lost ground later to crops introduced from the Americas.[40] Millet was probably the first foodstuff exported from the African mainland by vessels other than slave-ships as both the Cabo Verde Islands and São Tomé drew provisions from the mainland. Yams were grown in commercial quantities from the western Gold Coast to the Cameroon, with smaller quantities in Angola. By contrast, manioc production was mainly confined to Angola, though, as with 'corn' in the English case, the Portuguese sources use the term '*farinha*' without being clear about the crop from which the *farinha* was made. Some rice was grown in the southern rivers of Senegambia and along the Gambia River in the early seventeenth century and this crop increased rapidly in importance in Upper Guinea after 1720, becoming indeed *the* staple for the region. But Upper Guinea dispatched far fewer slaves than did other regions and thus rice was always a minor source of provisions compared to the other four. Palm-oil was grown in most of coastal western Africa and formed a part of all provisions purchased by slave captains on the coast, but it was always cheaper in the Bight of Biafra and was often among the provisions purchased at São Tomé by captains who had embarked

---

[38] TNA, T70/28, fol. 62, Captain Willis, 1 May 1705, 'A List of Negro Provisions Supposed to be Best to be sent for 100 negroes' specified 'beans – 30 quarters, grout – 3 quarters, bread – 5 cwt, cheese – 2 cwt, beef – 10 cwt, flower (sic) – 3/4 cwt, vinegar – 1/2 barrel or less, tobacco – 1 cwt, [tobacco] pipes – 4 gross.' There are other similar lists in this volume. For African-grown beans see T70/958, 19–21, 72. The *Bladen* (Voyageid 75181) shipped one ton of beans at Ouidah in 1723 (for 400 slaves), and beans were just as common in the traffic from Luanda to Brazil (Miller, *Way of Death*, 393).

[39] TNA, T70, fol. 23, Glynn and Ramsey, Gambia, to RAC, 2 Jan. 1722.

[40] See the literature surveyed in Newson and Minchin, *From Capture to Sale*, 78–93.

slaves on the Gold and Slave Coasts.[41] Malagueta pepper was mainly grown west of the Gold Coast, but was traded widely and formed a part of slave provisions on ships leaving from most parts of West Africa. In Luanda the animal protein additive was dried *savelha*, a fish relatively abundant in the South Atlantic off Angola.[42]

The above discussion suggests a wide range and a considerable volume of foodstuffs produced for commercial distribution on the African coast. Writing of Senegambia Curtin suggested that slaves needed a kilogram of millet per day.[43] Using the seven and a half months figure from the earlier discussion we can estimate that landing a slave in the Americas required the production and sale of 225 kilograms of millet. The calorific composition of 100 grams of millet is similar to that of 100 grams of yam. The other staples of the slave trade have slightly different calorie yields. The same quantity of manioc provides about one third more calories, rice about 15 per cent more, and maize about one third less.[44] During the eighteenth century, then, we can estimate that the transportation of 60,000 slaves per year from Africa during the eighteenth century required the annual production of 13,500 metric tons of some combination of millet, maize, yams, rice and manioc. But other produce was also required. For every 100 slaves taken on board, John Willis at Ouidah recommended 24 gallons of palm-oil, 12 gallons of lime juice, 112 lbs of malagueta pepper, and 224 lbs of salt ('which may be had in Africa').[45] To the grains and flour we can therefore add over 20,000 gallons of palm-oil and lime juice, 60 metric tons of pepper and 120 metric tons of salt. But these quantities of ancillary produce were for the voyage alone. They need to be more than doubled to allow for time spent on the African coast both before and after embarkation, and of course increased again if we are to allow for other phases of the enslavement process and other slave trades out of Africa. To reiterate, the concern here is not to generate exact estimates, but rather to suggest orders of magnitude and to make the argument that a significant export-oriented commercial agriculture was established in sub-Saharan Africa long before the nineteenth century.

If commercial agriculture was dispersed along the full range of the coast from which slave-ships departed, then its import would be less.

---

[41] Awnsham Churchill and John Churchill, *A Collection of Voyages and Travels*, 6 vols. (London, 1744–46), II, 225.
[42] Miller, *Way of Death*, 393. For cassava as provisions for captives leaving Ouidah in the mid-19th century see Robin Law, *Ouidah: The Social History of a West African Slaving 'port', 1727–1892* (Athens OH, 2004), 192.
[43] Curtin, *Economic Change*, 169; Newson and Minchin concur (*From Capture to Sale*, 82).
[44] www.nal.usda.gov/fnic/cgi-bin/nut_search.pl. Obviously the nutritional composition of modern crops might be different from those of 250 years ago.
[45] See note 38.

But in fact there was considerable regional specialization. The domination of the western Gold Coast in the supply of millet and maize to vessels embarking slaves from a very wide range of the coast (from the Gold Coast to places just north of the Congo River) is particularly striking. 'Corn' appears to have been as important as gold in Anomabu long before that port began to embark more slaves than any other place west of Ouidah. While Loango, Cabinda and other Vili ports came to supply their own provisions, before 1730 it was common for slave vessels to obtain at least some of their millet and maize on the Gold Coast before sailing on to what became northern Angola.[46] The availability of malagueta pepper, of course, ensured that part of the African coast west of the Gold Coast was called the Grain Coast. Yams, by contrast, were widely available, but were always cheaper in the Bight of Biafra. Slave vessels leaving from Gold Coast and Bight of Benin would buy supplies of them at islands in the Gulf of Guinea.[47] The entry of small-scale farmers into yam production for provisions generated some regional specialization even before the nineteenth century with large canoes carrying yams to both Bonny and Old Calabar.[48] Cassava production was also regionally specialized and there are no known instances of a slave-ship leaving West Africa provisioned with this staple. In Angola, the Jesuits, prior to their expulsion in 1760, grew cassava, maize and beans along the banks of the Bengo and Kwanza rivers and sold them to slave-ship captains leaving Luanda. A state granary was established in Luanda after their expulsion but attempts to concentrate the supply from the interior and to regulate the price were soon abandoned. But both before and after these institutional interventions, Africans had a major role in cassava and maize production.[49]

[46] The *Prosperous* (Voyage id 9886) embarked 575 slaves from Cabinda but previously obtained 130 chests of 'corn' at Anomabu (TNA, T70/3, fol. 94). The *John and Thomas* (Voyage id 9866) took on board 200 chests at Cape Coast Castle before going on to Angola for slaves (T70/16, fol. 53). In 1722 the *Helden* (Voyage id 75628) went to buy its 'corn' at São Tomé (no doubt grown on the Gold Coast) before embarking slaves at Cabinda (T70/7, fol. 37). Ouidah supplied its own provisions, but perhaps because of the enormous number of slaves dispatched, many of the slave vessels leaving the port first obtained millet and maize at Anomabu.

[47] Stephen D. Behrendt, 'Ecology, Seasonality and the Transatlantic Slave Trade', *Soundings in Atlantic History: Latent Structures and Intellectual Currents, 1500–1830*, eds. Bernard Bailyn and Patricia L. Denault (Cambridge MA, 2009), 44–85. Slave-ship captain Francis Buttram indicated that he would prefer a part cargo of yams instead of all corn (to the factors at Cape Coast Castle, 9 Feb. 1688, Bodleian Library, Oxford, Rawlinson collection, C.747, fol. 180). For the lower prices in the Gulf of Guinea see Captain Phillip's journal in Churchill, *Collection*, II, 224–5: 'Here [Cape Coast Castle] is some palm oil, but it is cheaper at Whidow [but] the island of St Thomas is the cheapest place.'

[48] David Northrup, *Trade Without Rulers: Pre-Colonial Economic Development in South-Eastern Nigeria* (Oxford, 1977), 177–81, 214–20; Stephen D. Behrendt, A.J.H. Latham, and David Northrup, *The Diary of Antera Duke, an Eighteenth-century African Slave Trader* (New York, 2010), 102–13, 140.

[49] A major argument in Mário José Maestri Filho, *A Agricultura Africana nos Séculos XVI e XVII no Litoral Angolano* (Porto Alegre, 1978). I thank Daniel Domingues da Silva for this reference. For the later period see Miller, *Way of Death*, 270–71, 392–400.

As in Europe and the Americas, almost all of staple crops were grown without a plantation system. Indeed, this is probably why historians have largely ignored the subject of provisions. Unlike sugar there appear to have been no economies of scale in the production and marketing of grain and root crops anywhere in the Atlantic world. In those parts of Africa with European enclaves, Europeans organized some production – usually adjacent to forts, but these activities accounted for a very small share of total provisions, with no European production at all at 'the principal granary' of Anomabu. At Cape Coast Castle Africans were able to undersell the European factors.[50] One requirement was certainly access to riverine transportation. On the banks of the Senegal, Gambia, Niger, Cross, Bengo and Kwanza rivers small farmers could get their crops to the coast with relative ease. They sent large quantities of yams on the Cross River from sixty miles or more upstream from Old Calabar, and in the Gambia could even sell directly to ocean-going vessels for a few months each year over a hundred miles upriver. We can explain the prominence of the Fante people in the supply of millet and maize by soil conditions and the pattern of two crops a year made possible by the relatively even monthly distribution of the precipitation in their region. But how did the Fante and their neighbours on the central and western Gold Coast manage to bring such enormous quantities of millet to the coast? We simply do not yet know. The commercial production of foodstuffs for oceanic markets developed much more slowly in southeast Africa. Before 1740 almost all the captives entering the Atlantic from the region left from Madagascar. They were rice-eaters and there were relatively few of them.[51] The big upsurge in the southeast African component of the trans-Atlantic trade occurred after 1780 (mostly after 1810) and came from the mainland. We have no information on what these captives ate, but it was probably cassava.

The aim of this very preliminary and rather superficial survey is no more than to ask scholars to pay attention to the issue of provisions, and, more generally, to remember just how important food production could be in any pre-industrial society. If, indeed, growing and marketing provisions in eighteenth-century sub-Saharan Africa was more valuable than say gold exports in the seventeenth century and

[50] Of Cape Coast Castle, which did have extensive gardens, Captain Phillips of the *Hannibal* wrote: 'We took imports of the Indian corn ordered us for the provision of our negroes to Barbados, the allowance being about a chest, which contains about 4 bushels, for every negro. It is charged to the company at two accies per chest, and bare measure but we could buy better off the blacks at an achy and ½ and heaped.' (Churchill, *Collection*, II, 224).
[51] Jane Hooper, 'An Empire in the Indian Ocean: The Sakalava Empire of Madagascar' (Unpublished PhD thesis, Emory University, 2010, available at https://vmch-etd.library.emory.edu/file/view/pid/emory:7tcps/hooper_dissertation.pdf).

palm-oil for the first half of the nineteenth century, then several broad conclusions follow. First, despite the fact that almost all the produce dispatched from sub-Saharan African into the Atlantic world was collected rather than cultivated, the roots of commercial agriculture in the sub-continent were well established long before 1800. Second, the scale of the provisioning trade was greater than historians have typically recognized. Indeed, the transition from the slave trade to produce trade – what used to be called the emergence of 'legitimate commerce' –is perhaps not a transition at all, or at least not a seamless transition. Exports of palm-oil and other products may have increased in the nineteenth century, but it is probable that as the trans-Atlantic slave trade declined (and the demand for provisions along with it) in the aftermath of Brazilian abolition in 1830, commercial agriculture contracted rather than expanded, at least outside the palm-oil producing heartlands in the Bight of Biafra. Third, the slave trade not only encouraged this particular form of commercial agriculture; it was self-evidently essential to its existence. Moreover, provisions generally came from small-scale producers, not plantations. This complicates the picture of the economic impact of the slave trade painted by Joseph Inikori, Walter Rodney and others where the slave trade favours elites and inhibits the emergence of non-slave trade activities.[52]

Both the provisions trade and the later produce export trends do show that African farmers could respond as well as farmers anywhere else in the Atlantic world to increased demand for what they could produce whether this required expanding output of an existing crop or switching to new crops. But it is not likely that nineteenth-century produce exports were dependent on (or would not have happened without) either the preceding slave trade or the provisions trade that accompanied it. It also seems unlikely that the slave trade prevented the development of commercial agriculture as Rodney and Inikori have argued. Nevertheless, scholars need to pay more attention to the unglamorous business of growing staples before we can properly assess the emergence of commercial agriculture in an African context. Thanks to Walter Hawthorne and Judith Carney, we have a good grasp on how, where and when rice was produced in Africa, but in comparison we know almost nothing about the millet, maize, yams and manioc that formed the bulk of the estimated 13,500 metric tons of grains and root crops that enabled the slave trade in the eighteenth century, and rice could have made up much less than ten per cent of this total.

---

[52] Joseph E. Inikori, *Africans and the Industrial Revolution in England: A Study of International Trade and Economic Development* (Cambridge, 2002), 392–94; Walter Rodney, *How Europe Underdeveloped Africa* (Washington DC, 1981).

Appendix

DERIVATION OF ESTIMATES OF THE VALUE OF AFRICAN COMMODITIES PRODUCED FOR ATLANTIC TRADE, 1681–1807

The foundation of Table 1.3 is the estimates page of www.slavevoyages.org where annual estimates of slave departures are readily available. The value of these slaves (in other words, the price on the African coast) is derived from two separate databases; one is described in Eltis, *Rise of African Slavery in the Americas* (2000), 293–97, which provides average prices of slaves from 1681 to 1698; the other, for 1699–1807 is from David Richardson's 'Prices of Slaves in West and West-Central Africa: Toward an Annual Series, 1698–1807,' *Bulletin of Economic Research* 43 (1991), 21–56. Column 4, which estimates the number of vessels in the slave trade is simply column 2 – the total number of slaves – divided by the average number of slaves carried off per vessel from www.slavevoyages.org. Column 5 is taken directly from yet another database – of produce vessels – referenced in note 13, and column 6 is simply column 5 multiplied by 1.5 to allow for produce vessels that sailed without leaving a trace in the documents. Column 7 estimates a value for the African commodities carried on the produce vessels shown in column 6. The value is set according to the ratio of produce ships to slave-ships for each year with the resulting ratio applied to the value of the slave trade shown in column 3. But African produce was also carried on slave-ships and to allow for this we need to know what proportion of the cargo of a slave-ship was traded for commodities and what proportion was traded for slaves.

For the beginning of the period, a detailed analysis of the accounts of 22 Royal African Company slave-ships that left London for Africa, then Barbados, Jamaica, or Nevis between 1683 and 1685 is possible.[53] This indicates that 45 per cent of the goods that the Company took to the coast on slave vessels were intended for African produce (including gold).[54] Column 8 assumes that the produce share of the slave-ship's cargo changed slowly over time over time. When slave prices were relatively high, as for example in 1700, the Company instructed its agent that 'since the procuring of Negroes is so difficult And the prizes [sic] so high Wee shall not so much covet them but rather advise you to use y^r utmost dilligence to Improve our Trade in

---

[53] Eltis, Lewis and Mcintyre, 'Accounting for the Traffic'.
[54] Calculated from Eltis, Lewis and Mcintyre, 'Accounting for the Traffic', and the database summarizing the cargo breakdowns that underpins that essay.

all other Commodities and Employ all our Vessels in Search thereof both up ye rivers and without ye same.'[55] For the last few years of its trading activities on the coast – the early 1730s – the Company carried very few slaves and eventually abandoned the slave trade altogether, carrying only African commodities back to London.

For the second half of the eighteenth century there are even more detailed accounts for some of the major British slave-traders.[56] There is general agreement that the produce share of what slave-ships carried away from Africa (slaves and produce combined) declined steeply after the 1720s to about 5 per cent of the British trade at the beginning of the nineteenth century.[57] Most of the annual ratios in column 8 are interpolated from Royal African Company (RAC) data from the early 1680s to the 5 per cent ratio adopted for 1807. Column 9 uses the ratio in column 8 to derive an estimate of annual values of produce carried on slave-ships. As with other columns, the series leans heavily on estimates of the value of the slave trade in column 3 (detailed spreadsheet available from the author). Finally, column 10 adds together the valuations of commodities on slave-ships to the values of cargoes carried off on vessels plying the direct produce trade. Column 11 simply provides an idea of the relative importance of the produce carried on slave-ships as a share of African produce exported from sub-Saharan Africa.

[55] TNA, T70/51, fol. 75, RAC to Thomas Gresham, Nathan Pile & Thomas Rayner, 10 Oct. 1700.
[56] Nicholas James Radburn, 'William Davenport, the Slave Trade, and Merchant Enterprise in Eighteenth-Century Liverpool' (unpublished MA thesis, Victoria University of Wellington, 2009, available at http://researcharchive.vuw.ac.nz/bitstream/handle/10063/1187/thesis.pdf). See also discussion of the issue in David Richardson, 'Prices of Slaves in West and West-Central Africa: Toward an Annual Series, 1698–1807', *Bulletin of Economic Research* 43/1, (1991): 21–56.
[57] But can English trade with Africa be taken as representative of the African trades of other European nations? The Dutch carried off more produce than the English – at least before 1720, while the Portuguese took somewhat less than other nationalities.

Table 1.3 Derivation of estimates of slave departures and produce exports from Africa into the Atlantic World, 1681–1807

	Average price of slaves in constant pounds Sterling	Total slaves carried off from sub-Saharan Africa	Total value of slaves (Col 1 x col 2)	Estimated number of slave-ships leaving Africa (Col 2/300.6)	Number of ships sent to carry only African produce	Estimated number of produce ships (Col 5 x 1.5)	Total value of produce on produce ships ((Col 6/col 4) x col 3))	Ratio of TV of produce on slave ships to TV of all cargo on slave-ships	Total value of produce on slave-ships (Col 3 x (1-col 8) x (col 8))	Total value of produce carried from Africa (Col 7 + col 9)	Produce on slave-ships as share of all produce carried from Africa (Col 9/col 10)
	1	2	3	4	5	6	7	8	9	10	11
1681	3.2	34,223	110,282	111	2	3	2,970	0.45	90,231	93,200	0.97
1682	3.5	27,925	97,888	91	1	2	1,615	0.45	80,090	81,705	0.98
1683	4.0	32,149	129,770	105	4	6	7,440	0.45	106,175	113,615	0.93
1684	4.0	22,751	90,089	74	7	11	12,773	0.45	73,709	86,482	0.85
1685	3.9	28,499	109,907	93	6	9	10,663	0.45	89,924	100,587	0.89
1686	3.8	23,051	88,629	75	6	9	10,630	0.44	70,903	81,533	0.87
1687	3.8	28,959	108,608	94	8	12	13,825	0.44	84,951	98,776	0.86
1688	3.6	26,078	94,025	85	6	9	9,969	0.43	71,901	81,870	0.88
1689	3.7	20,220	75,632	66	1	2	1,724	0.43	56,540	58,264	0.97
1690	4.0	16,710	66,157	54	3	5	5,473	0.42	48,346	53,819	0.90
1691	4.5	25,752	116,893	84	2	3	4,183	0.42	83,495	87,678	0.95
1692	4.4	28,448	124,780	93	6	9	12,127	0.41	87,110	99,237	0.88
1693	4.2	25,983	108,895	85	3	5	5,794	0.41	74,293	80,087	0.93
1694	4.1	23,880	97,948	78	2	3	3,780	0.40	65,299	69,079	0.95
1695	4.3	24,293	105,137	79	4	6	7,977	0.39	68,484	76,461	0.90
1696	4.5	30,008	134,731	98	2	3	4,138	0.39	85,738	89,876	0.95
1697	4.1	37,104	151,135	121	6	9	11,262	0.38	93,949	105,211	0.89
1698	5.1	47,541	244,370	155	8	12	18,949	0.38	148,367	167,316	0.89
1699	4.8	42,970	207,248	140	11	17	24,447	0.37	122,881	147,329	0.83
1700	5.2	53,578	276,129	174	16	24	37,998	0.37	159,864	197,862	0.81
1701	4.2	55,754	235,251	181	14	21	27,220	0.36	132,968	160,188	0.83
1702	2.8	45,713	129,156	149	23	35	29,944	0.36	71,258	101,203	0.70

Year	Average price of slaves in constant pounds Sterling (1)	Total slaves carried off from sub-Saharan Africa (2)	Total value of slaves (Col 1 x col 2) (3)	Estimated number of slave-ships leaving Africa (Col 2/300.6) (4)	Number of ships sent to carry only African produce (5)	Estimated number of produce ships (Col 5 x 1.5) (6)	Total value of produce on produce ships ((Col 6/col 4) x col 3)) (7)	Ratio of TV of produce on slave ships to TV of all cargo on slave-ships (8)	Total value of produce on slave-ships (Col 3 x (1-col 8) x (col 8)) (9)	Total value of produce carried from Africa (Col 7 + col 9) (10)	Produce on slave-ships as share of all produce carried from Africa (Col 9/col 10) (11)
1703	8.5	34,476	294,141	112	17	26	66,834	0.35	158,384	225,218	0.70
1704	4.5	37,195	167,898	121	12	18	24,961	0.34	88,218	113,178	0.78
1705	4.1	38,452	155,943	125	7	11	13,081	0.34	79,937	93,018	0.86
1706	3.7	31,546	117,614	103	11	17	18,898	0.33	58,807	77,705	0.76
1707	10.4	39,852	413,443	130	4	6	19,122	0.33	201,596	220,718	0.91
1708	2.7	39,927	106,730	130	8	12	9,854	0.32	50,740	60,594	0.84
1709	5.8	32,118	185,862	105	0	0	0	0.32	86,131	86,131	1.00
1710	4.0	39,207	157,556	128	4	6	7,407	0.31	71,154	78,561	0.91
1711	4.2	36,283	152,019	118	1	2	1,931	0.31	66,889	68,819	0.97
1712	2.3	34,247	79,044	111	2	3	2,127	0.30	33,876	36,003	0.94
1713	12.1	41,913	508,033	136	3	5	16,756	0.29	212,014	228,770	0.93
1714	6.2	49,716	308,956	162	8	12	22,909	0.29	125,513	148,422	0.85
1715	4.0	48,818	195,391	159	12	18	22,132	0.28	77,247	99,379	0.78
1716	10.6	46,922	499,228	153	11	17	53,930	0.28	192,011	245,940	0.78
1717	8.2	47,813	391,268	156	10	15	37,709	0.27	146,352	184,061	0.80
1718	5.2	52,274	271,465	170	11	17	26,323	0.27	98,715	125,037	0.79
1719	5.6	48,442	272,268	158	10	15	25,899	0.26	96,215	122,114	0.79
1720	7.9	46,980	372,889	153	12	18	43,889	0.26	128,007	171,896	0.75
1721	9.3	41,163	384,255	134	18	27	77,428	0.25	128,085	205,513	0.62
1722	13.7	43,101	591,225	140	20	30	126,418	0.24	191,279	317,696	0.60
1723	10.2	40,366	412,187	131	25	38	117,634	0.24	129,373	247,006	0.52
1724	8.7	60,448	527,299	197	25	38	100,491	0.23	160,482	260,973	0.62
1725	11.5	56,181	648,783	183	8	12	42,571	0.23	191,368	233,939	0.82

Year											
1726	5.1	61,915	317,823	202	18	27	42,577	0.22	90,807	133,383	0.68
1727	5.4	58,748	315,304	191	13	20	32,151	0.22	87,212	119,363	0.73
1728	7.4	58,016	430,872	189	17	26	58,178	0.21	115,304	173,482	0.66
1729	6.8	66,483	450,310	216	27	41	84,271	0.21	116,514	200,785	0.58
1730	7.3	61,972	449,879	202	31	47	103,699	0.20	112,470	216,169	0.52
1731	5.8	62,068	360,137	202	30	45	80,211	0.20	87,238	167,449	0.52
1732	5.1	60,334	307,167	196	38	57	89,147	0.19	72,052	161,199	0.45
1733	3.5	52,635	185,459	171	28	42	45,462	0.19	42,098	87,560	0.48
1734	2.8	54,960	156,482	179	30	45	39,360	0.18	34,350	73,709	0.47
1735	5.8	52,597	304,757	171	23	35	61,409	0.18	64,645	126,055	0.51
1736	6.3	58,283	367,516	190	18	27	52,302	0.17	75,274	127,576	0.59
1737	5.3	67,359	356,434	219	28	42	68,274	0.17	70,433	138,707	0.51
1738	11.4	57,117	651,938	186	35	53	184,086	0.16	124,179	308,265	0.40
1739	6.0	69,094	411,612	225	26	39	71,373	0.16	75,503	146,876	0.51
1740	2.2	63,557	139,681	207	22	33	22,280	0.15	24,650	46,929	0.53
1741	4.6	62,485	284,336	203	21	32	44,034	0.15	48,221	92,255	0.52
1742	6.1	62,856	385,542	205	12	18	33,917	0.14	62,763	96,680	0.65
1743	13.3	56,703	756,895	185	7	11	43,057	0.14	118,128	161,185	0.73
1744	4.1	69,909	288,030	228	21	32	39,869	0.13	43,039	82,908	0.52
1745	8.7	35,610	310,444	116	9	14	36,155	0.13	44,349	80,504	0.55
1746	9.0	35,213	317,000	115	15	23	62,224	0.12	43,227	105,452	0.41
1747	8.5	54,007	457,589	176	9	14	35,138	0.12	59,461	94,599	0.63
1748	9.6	54,909	529,347	179	14	21	62,192	0.11	65,425	127,617	0.51
1749	7.8	66,912	520,193	218	13	20	46,571	0.11	61,028	107,599	0.57
1750	6.2	67,985	422,269	221	22	33	62,967	0.10	46,919	109,886	0.43
1751	7.9	62,073	488,352	202	20	30	72,506	0.10	53,733	126,239	0.43
1752	8.1	72,410	586,067	236	26	39	96,969	0.10	63,852	160,821	0.40
1753	9.9	70,524	697,335	230	25	38	113,909	0.10	75,223	189,131	0.40
1754	7.6	74,559	565,046	243	23	35	80,320	0.10	60,345	140,665	0.43
1755	4.3	81,930	348,935	267	21	32	41,213	0.10	36,890	78,103	0.47
1756	6.2	64,980	404,441	212	10	15	28,681	0.09	42,325	71,006	0.60

52 • *The slave trade & commercial agriculture in an African context*

	Average price of slaves in constant pounds Sterling	Total slaves carried off from sub-Saharan Africa	Total value of slaves (Col 1 x col 2)	Estimated number of slave-ships leaving Africa (Col 2/300.6)	Number of ships sent to carry only African produce	Estimated number of produce ships (Col 5 x 1.5)	Total value of produce on produce ships ((Col 6/col 4) x col 3))	Ratio of TV of produce on slave ships to TV of all cargo on slave-ships	Total value of produce on slave-ships (Col 3 x (1-col 8) x (col 8))	Total value of produce carried from Africa (Col 7 + col 9)	Produce on slave-ships as share of all produce carried from Africa (Col 9/col 10)
	1	2	3	4	5	6	7	8	9	10	11
1757	6.0	51,364	309,614	167	14	21	38,887	0.09	32,070	70,957	0.45
1758	5.2	55,410	289,421	180	26	39	62,579	0.09	29,670	92,249	0.32
1759	8.7	50,991	443,710	166	18	27	72,176	0.09	45,014	117,190	0.38
1760	9.1	58,717	532,907	191	18	27	75,279	0.09	53,496	128,775	0.42
1761	8.5	68,258	578,614	222	14	21	54,686	0.09	57,471	112,157	0.51
1762	8.6	60,222	516,049	196	21	32	82,922	0.09	50,710	133,632	0.38
1763	15.5	59,890	931,070	195	23	35	164,766	0.09	90,508	255,275	0.36
1764	10.4	86,407	899,851	281	16	24	76,781	0.09	86,524	163,305	0.53
1765	11.5	92,245	1,058,345	300	19	29	100,450	0.09	100,650	201,100	0.50
1766	10.5	98,293	1,029,787	320	23	35	111,036	0.09	96,851	207,888	0.47
1767	14.8	89,940	1,331,306	293	25	38	170,521	0.09	123,813	294,334	0.42
1768	16.1	84,531	1,360,504	275	26	39	192,828	0.08	125,104	317,932	0.40
1769	16.5	84,261	1,389,184	274	25	38	189,927	0.08	126,289	316,216	0.40
1770	11.8	94,913	1,122,857	309	24	36	130,835	0.08	100,907	231,741	0.44
1771	14.6	87,686	1,282,537	285	38	57	235,896	0.08	113,922	349,817	0.33
1772	17.7	95,578	1,694,594	311	29	44	204,249	0.08	148,762	353,011	0.42
1773	15.9	88,732	1,410,759	289	25	38	175,831	0.08	122,382	298,214	0.41
1774	16.9	98,492	1,666,287	321	26	39	202,691	0.08	142,825	345,516	0.41
1775	15.7	92,908	1,463,142	302	33	50	239,475	0.08	123,901	363,375	0.34
1776	9.7	87,146	846,012	284	25	38	111,836	0.08	70,769	182,605	0.39
1777	9.9	66,847	665,079	218	11	17	50,431	0.08	54,950	105,380	0.52
1778	7.0	60,825	427,480	198	13	20	42,101	0.08	34,880	76,981	0.45
1779	16.5	37,758	624,175	123	14	21	106,644	0.07	50,289	156,933	0.32

Year	C1	C2	C3	C4	C5	C6	C7	C8	C9	C10
1780	23.5	39,694	932,100	129	3	5	32,462	0.07	74,144	106,606	0.70
1781	21.8	48,480	1,058,222	158	3	5	30,175	0.07	83,096	113,271	0.73
1782	15.8	52,401	828,868	171	4	6	29,155	0.07	64,241	93,397	0.69
1783	32.7	62,262	2,036,257	203	3	5	45,211	0.07	155,748	200,959	0.78
1784	10.0	104,364	1,043,874	340	5	8	23,045	0.07	78,783	101,828	0.77
1785	17.0	95,152	1,619,015	310	6	9	47,043	0.07	120,549	167,592	0.72
1786	25.7	95,528	2,457,237	311	14	21	165,942	0.07	180,475	346,417	0.52
1787	21.4	101,305	2,168,192	330	12	18	118,348	0.07	157,056	275,404	0.57
1788	17.7	104,040	1,843,927	339	11	17	89,836	0.07	131,709	221,545	0.60
1789	19.7	92,413	1,815,964	301	13	20	117,714	0.07	127,885	245,599	0.52
1790	32.3	112,049	3,624,036	365	11	17	163,942	0.06	251,575	415,517	0.61
1791	19.1	107,578	2,053,271	350	16	24	140,720	0.06	140,477	281,197	0.50
1792	28.2	115,518	3,259,693	376	17	26	221,049	0.06	219,755	440,803	0.50
1793	5.7	99,562	568,824	324	4	6	10,531	0.06	37,780	48,310	0.78
1794	25.1	67,471	1,694,529	220	5	8	57,865	0.06	110,857	168,722	0.66
1795	14.6	69,237	1,010,821	225	9	14	60,547	0.06	65,123	125,670	0.52
1796	24.4	68,234	1,666,764	222	4	6	45,024	0.06	105,728	150,752	0.70
1797	22.8	75,153	1,716,986	245	2	3	21,055	0.06	107,212	128,267	0.84
1798	26.0	71,267	1,851,065	232	2	3	23,937	0.06	113,753	137,690	0.83
1799	24.6	88,084	2,166,296	287	3	5	33,998	0.06	130,985	164,983	0.79
1800	19.2	86,301	1,659,809	281	3	5	26,587	0.06	98,725	125,312	0.79
1801	21.9	79,220	1,738,701	258	5	8	50,568	0.06	101,707	152,274	0.67
1802	20.3	98,941	2,010,300	322	5	8	46,813	0.05	115,620	162,433	0.71
1803	16.8	97,721	1,640,850	318	10	15	77,374	0.05	92,764	170,137	0.55
1804	29.0	88,675	2,575,867	289	4	6	53,542	0.05	143,104	196,646	0.73
1805	27.0	89,348	2,415,364	291	6	9	74,741	0.05	131,828	206,570	0.64
1806	29.5	102,163	3,012,482	333	11	17	149,464	0.05	161,482	310,946	0.52
1807	15.7	112,576	1,768,137	366	12	18	86,849	0.05	93,060	179,909	0.52
1681–1807	10.6	61,129	762,158	199	14	21	63,577	0.2	96,327	159,904	0.60

2

São Tomé & Príncipe
The first plantation economy in the tropics

GERHARD SEIBERT

On two occasions the small archipelago of São Tomé and Príncipe, a former Portuguese colony located in the Gulf of Guinea, played an important role in the history of tropical commercial agriculture. During the Age of Discoveries, in the sixteenth century, the islands became a major sugar-producer and the first plantation economy in the tropics. After some two centuries of economic decay, in the mid-nineteenth century, the archipelago emerged as Africa's first cocoa-producer and in the early twentieth century, for a few years, even became the world's largest cocoa-producer. This paper focuses on the first period which coincided with the settlement and colonization of the hitherto uninhabited tropical islands and seeks to put the rise and fall of São Tomé's early commercial agriculture in a wider social and political context.

As far as English-language secondary sources are concerned the article draws on the theses of the historian Robert Garfield on São Tomé's early history (1972, published 1992) and of the anthropologist Pablo Eyzaguirre on the island's plantation economy (1986).[1] More recently, several Portuguese scholars have provided important additional insights into the archipelago's history in the sixteenth and seventeenth centuries. Among these authors are the historians Arlindo Caldeira, particularly on the slave trade, slavery and slave resistance (1997, 2000, 2004, 2006, 2008), Luís Pinheiro on economy and politics (2005), Pedro Cunha on the local economy (2001), and Cristina Serafim on the economic decline in the seventeenth

---

[1] Robert Garfield, *A History of São Tomé Island 1470–1655: The Key to Guinea* (San Francisco CA, 1992); Pablo B. Eyzaguirre, 'Small Farmers and Estates in São Tomé, West Africa' (PhD dissertation, New Haven CT: Yale University, 1986).

century (2000).² The most relevant findings of these historians have been considered in this paper together with those of earlier Portuguese scholars of São Tomé and Príncipe. In their analyses the different scholars, who largely draw from the same primary sources, do not differ as far as the key issues of the archipelago's early history are concerned. Differences are predominantly restricted to details, such as the quantities of sugar produced.

In the context of this volume, it is evident that sugar cultivation was not introduced into the Gulf of Guinea islands as an alternative to the slave trade, but on the contrary, it was based on a system of slave labour that served as a prototype for later plantation complexes in the Americas. From the beginning, the majority of the local population consisted of slaves from the African mainland, since few whites were willing to settle voluntarily in the distant unhealthy tropical islands. São Tomé and Príncipe was already a slave society before the trans-Atlantic slave trade began and similar slave societies appeared on the other side of the Atlantic. The first section of this chapter describes São Tomé's settlement and colonization, and the emergence of the Creole society, and analyses the social-political environment of the local economy dominated by the slave trade and the sugar industry. The second section deals with slavery, slave resistance and the emergence of a hostile 'maroon' (escaped slaves) community that became a threat to the local sugar industry. In addition this section presents the development of São Tomé's slave trade which was the mainstay of the local economy until sugar production began around the 1520s and thereafter always remained an equally important economic activity.

The last section depicts the rise of São Tomé's sugar industry and examines the various reasons for its gradual decline beginning at the end of the sixteenth century. Thanks to the tropical climate, fertile

---

[2] Arlindo Manuel Caldeira, *Mulheres, Sexualidade e Casamento no Arquipélago de S. Tomé e Príncipe (Séculos XV a XVII)* (Lisbon, 1997); id., *Viagens de um Piloto Português do Século XVI à Costa de África e a São Tomé* (Lisbon, 2000); id.,'Rebelião e Outras Formas de Resistência à Escravatura na Ilha São Tomé (séculos XVI a XVIII)', *Africana Studia* 7 (2004); id. 'A Estratégia Inicial da Colonização Portuguesa no Golfo da Guiné' (paper presented at the V Congresso de Estudos Africanos no Mundo Ibérico, Covilhã, 4–6 May 2006); id.,'Uma Ilha Quase Desconhecida: Notas Para a História de Ano Bom', *Africana Studia* 17 (2006); 'Tráfico de Escravos e Conflitualidade: O Arquipélago de São Tomé e Príncipe e o Reino do Congo Durante o Século XVI', *Ciências & Letras. História da África: Do Continente à Diáspora*. 44 (2008); Luís da Cunha Pinheiro,'O Povoamento: O Arquipélago do Golfo da Guiné: Fernando Pó, São Tomé, Príncipe e Ano Bom', 'As Estruturas Político-administrativas e os Seus Órgãos: O Arquipélago do Golfo da Guiné: Fernando Pó, São Tomé, Príncipe e Ano Bom', and 'O Século XVI, uma Economia bem Sucedida. A Economia: A Produção Açucareira, o Comércio e o Regate, a Fiscalidade e as Finanças: O Arquipélago do Golfo da Guiné: Fernando Pó, São Tomé, Príncipe e Ano Bom', *A Colonizaçāc Atlântica*, II, ed. Artur Teodoro de Matos (Lisbon, 2005); Pedro José Paiva da Cunha, 'A Organização Económica em São Tomé (de Início Do Povoamento a Meados do Século XVII)' (MA diss., University of Coimbra, 2001); Cristina Maria Seuanes Serafim, *As Ilhas de São Tomé no Século XVII* (Lisbon, 2000).

soils, and abundant rainfalls the islands offered favourable conditions for the cultivation of sugar cane. However, this was not the case with regard to sugar production, since, due to the high humidity of the tropical climate, the quality of São Tomé's sugar was inferior to that of Madeira and Brazil. The article argues that the principal factor in São Tomé's economic decline was the emergence of the sugar industry in Brazil, where both the production conditions and the quality of sugar were significantly better than in the archipelago. Attracted by the promising economic prospects, the São Tomé planters left for Brazil which had become a large-scale sugar-producer in the 1580s. Other internal and external factors including political instability, maroon assaults, slave revolts and sea-borne attacks by the Dutch contributed to and hastened São Tomé and Príncipe's economic decline. By the end of the seventeenth century the plantation economy in the archipelago had virtually ceased to exist and was replaced by cultivation of foodstuffs, mainly for subsistence, but with a surplus sold to passing slave-ships, while the slave trade continued on a smaller scale until the abolition of the slave trade in the early nineteenth century.

Settlement and society

Exactly when the Gulf of Guinea islands were discovered is not known, though most authors believe that the islands of São Tomé (859km²) and Príncipe (142km²) were first sighted by the Portuguese navigators João de Santarém and Pedro Escobar on 21 December 1471 and 17 January 1472 respectively. The first attempt to establish a settler colony in São Tomé, which was densely covered by tropical forests, was only made a few years after the island's discovery, during the reign of King João II (1481–95), following the establishment of the fort São Jorge da Mina (Elmina) on the Gold Coast in 1482 and the Portuguese arrival in Kongo in the following year. The fort at Elmina was set up to support existing regional trade networks.[3] Although the islands were small, they seemed well-suited for colonization, since they were uninhabited and out of reach of potentially hostile African settlements on the mainland. For the same reasons of security the Portuguese did not colonize the larger neighbouring island of Fernando Po (Bioko) which had an indigenous population, the Bubi, and lies close to the mainland.

The Portuguese crown expected São Tomé to become a settler colony, a supplier of sugar and food for Elmina and a safe haven for ships returning to Europe from Elmina and those sailing to and from

[3] Caldeira, 'A Estratégia Inicial', 16.

India. However, the Portuguese later realized that the latter objective was not viable, since the most favourable sea routes did not pass by São Tomé.[4] The colonization of the archipelago followed a pattern already used in the previously discovered archipelagos of Madeira, the Azores and Cabo Verde. The king appointed noblemen as captains of the islands, granting them extensive privileges in exchange for the colonization of the territory. Privileges and tax exemptions for settlers were a favourite strategy for attracting them to the new territories.

However, the settlement of the Gulf of Guinea islands proved to be difficult due a lack of food and the insalubrious tropical climate. Despite the fertility of the volcanic soil, there was a scarcity of food because Mediterranean crops like wheat, rye, barley, grapes and olive trees would not grow in the inappropriate tropical climate.[5] Initially there were almost no tropical food crops in the archipelago either. Therefore, to start with food had to be imported along with the slaves from the Niger Delta.[6] While the settlers gradually adapted to a different diet, tropical diseases, particularly malaria, made São Tomé a dangerous place for centuries. Consequently, in Portugal the island quickly gained a reputation as the 'white man's grave' and few settlers went there voluntarily. Many appointed office-holders delayed their departure from Portugal or tried to limit their stay to a minimum once they had arrived on the island.[7] Due to a lack of voluntary colonists, from the beginning deported convicts constituted a significant proportion of the Portuguese settlers. Possibly even during the height of the economic boom, around 1570, the entire European population did not exceed five hundred.[8]

In December 1485 the Portuguese king appointed São Tomé's first captain (*donatário*), João de Paiva (1485–90), who set up the essential public offices and granted the white colonists certain privileges to attract settlers to populate the island, including free trade in slaves and other goods on the coast between the Rio Real (New Calabar River) in the Bight of Biafra and the Kongo kingdom. The letter of appointment already mentioned sugar production as an integral part of São Tomé's colonization project.[9] At the time, the Portuguese already had

[4] Ibid., 22.
[5] Cunha, 'A Organização Económica', 23; Iolanda Trovoada Aguiar, 'São Tomé e Príncipe: Plantas e Povos, Origens e Consequências', *Actas do VI Congresso Luso-Afro-Brasileiro de Ciências Sociais: As Ciências Sociais nos Espaços de Língua Portuguesa, Balanços e Desafios*, ed. Rui Centeno and António Custódio Gonçalves (Porto, 2002), 361.
[6] Catarina Madeira Santos, 'A Formação das Estruturas Fundiárias e a Territorialização das Tensões Sociais: São Tomé, Primeira Metade do Século XVI', *Studia* 54/55 (1996), 56.
[7] Caldeira, 'A Estratégia Inicial', 23.
[8] Caldeira, 'Tráfico de Escravos', 75.
[9] Luís de Albuquerque, *A Ilha de São Tomé nos Séculos XV e XVI*. Biblioteca da Expansão Portuguesa (Lisbon, 1989), 47.

experience in both the African slave trade and in sugar production using slave labour.[10] However, the first settlement attempt made in the northwest of the island between 1486 and 1490 apparently failed due to tropical diseases and a shortage of food.

It was not until 1493 that the third captain, Álvaro de Caminha (1493–99), succeeded in establishing the first settlement at a bay in the northeast of the island, which later became São Tomé town. The king granted Caminha civil and criminal jurisdiction over the island and the power to appoint treasury and justice officials. He also awarded the settlers additional incentives including trade in all goods – with the exception of gold – from the island, and also the mainland from the Rio Real and Fernando Po to the territory of the Manicongo. The settlers had to pay a quarter of this trade to the crown, which in turn tithed to the Church. In 1500 the king appointed António Carneiro, a royal knight, as captain of Príncipe (1500–45), where settlers enjoyed the same privileges as those in São Tomé. The Carneiro family owned the captaincy of Príncipe until 1753 when the island returned to the crown, while São Tomé reverted to the crown in 1522, after the captain, João de Mello (1512–22), had been removed from office due to allegations of corruption. Thereafter, São Tomé was ruled by a governor appointed by the crown.[11]

In addition to many convicts, the colonists who arrived with Caminha in São Tomé in 1493 included dozens of young Jewish children who had been separated from their parents by force.[12] Every five Jewish children were allotted a couple of slaves to help look after them, while every settler received a female and male slave to work for him.[13] The crown deliberately encouraged mixed-race unions between white settlers and African slaves to safeguard the settlement of the island. Thus miscegenation was widespread in São Tomé and Príncipe and Cabo Verde, though this was not a common Portuguese practice in Africa at that time. In sixteenth-century Arguim and Elmina the crown prohibited unions between Portuguese men and African women.[14] Furthermore, in São Tomé miscegenation was more widespread among convicts, and decreased the higher the status in the social hierarchy of the settler community.[15] Convicts enjoyed the same priv-

[10] Eyzaguirre, 'Small Farmers', 34.
[11] Prior to Jan. 1584 the ruler's title was 'captain'.
[12] For an overview of the different contemporaneous Portuguese and Hebrew sources of this event see Gerhard Seibert, '500 years of the manuscript of Valentim Fernandes, a Moravian book printer in Lisbon', *Iberian and Slavonic Cultures: Contact and Comparison,* ed. Beata Elżbieta Cleszyńska (Lisbon, 2007), 85–86.
[13] Albuquerque, *A Ilha de São Tomé,* 73.
[14] Isabel de Castro Henriques, 'Ser Escravo em S. Tomé no Século XVI: Uma Outra Leitura de um Mesmo Quotidiano', *Revista Internacional de Estudos Africanos* 6 & 7 (1987), 182.
[15] Cunha, 'A Organização Económica', 49.

ileges as the other settlers and could participate freely in all economic activities. In the early period, settlers acquired slaves in the Niger Delta and in Kongo. Supposedly the early population also included a few free Africans from the continent who served as brokers in the slave trade.[16]

Slavery in São Tomé was not necessarily a permanent condition, since slaves were manumitted from the outset. The first recorded individual letters of manumission for slaves in São Tomé are mentioned in Caminha's will in 1499. As early as 1515 a royal decree granted collective manumission to the African wives of white settlers and their mixed-race offspring. Another royal decree, in 1517, freed the male slaves who had arrived with the first colonists. These royal decrees constituted the beginning of a free African population in São Tomé, called Forros.[17] Later, freed slaves assimilated into the free African sector. In 1520 a royal charter allowed free mulattoes to hold public offices in the local council provided that they had property and were married. In 1528 this decree was confirmed.[18] In 1546 another royal decree equated them with white settlers, allowing them to vote and hold office on the city council.[19] Consequently, from the outset, free mulattoes and blacks were able to ascend the social ladder, played a significant role in shaping the emerging Creole society, and participated actively in local politics and the economy.

Due to the shortage of voluntary settlers, deported convicts were often also appointed to public positions. The crown frequently pardoned these convicts in exchange for services rendered. According to contemporary chronicles, the early settlement, known as *povoação*, grew from 250 'hearths' in 1510 to between 600 and 700 hearths in the mid sixteenth century.[20] The 1510 chronicle compiled by the Lisbon-based German book printer Valentim Fernandes mentions another fifteen settled places and six plantations belonging to the captain of São Tomé. With the exception of one plantation, the entire south of São Tomé was unoccupied.[21] In April 1535 a royal charter granted the settlement city rights. São Tomé's town council was dominated by wealthy sugar-plantation owners. The town council was

[16] Ibid., 44.
[17] The term is derived from the Portuguese *carta de alforria*, meaning letter of manumission.
[18] Rui Ramos, 'Rebelião e Sociedade Colonial: "Alvoroços" e "Levantamentos" em São Tomé (1545–1555)', *Revista Internacional de Estudos Africanos* 4 & 5 (1986), 24; Izequiel Batista de Sousa, *São Tomé et Príncipe de 1485 à 1755: Une Société Coloniale Du Blanc au Noir* (Paris, 2008), 41.
[19] Cunha, 'A Organização Económica', 52.
[20] Teresa Madeira, 'Estudo Morfológico da Cidade de São Tomé no Contexto Urbanístico das Cidades Insulares Atlânticas de Origem Portuguesa' (paper presented at the Colóquio Internacional Universo Urbanístico Português 1415–1822, Coimbra, 2–6 March 1999), 8.
[21] Francisco C. Cunha Leão, 'Cartografia e Povoamento da Ilha de São Tomé (1483–1510)', *Revista do Instituto Geográfico e Cadastral* 5 (1985), 87.

frequently engaged in power struggles with the governor or the bishop who, in turn, were also in conflict with each other. The frequent disputes between the three parties resulted in considerable political instability in São Tomé. In addition, there were frequent quarrels within the institutions, often between white Portuguese from the mother country and local Creole officials. The high mortality rate among Portuguese officials also contributed to political instability, as it often created a power vacuum. In the period from 1548 to 1770 the city council ruled in the event of the governor's absence or death. In the twenty-seven year period from 1586 to 1613 São Tomé was ruled by eighteen governors, including both those appointed by the crown and interim rulers elected by the town council.[22]

One of the captain's powers was to distribute land to the settlers under the *sesmaria* system. Under this land-grant system the grantees became owners of the land after five years, provided it was cultivated successfully.[23] Otherwise, the land could be withdrawn and granted to somebody else on the same conditions. Besides these private lands, there were also crown-owned plantations, first to cultivate food crops to feed the slaves and subsequently to produce sugar.[24] In 1528 a total of 1,440 slaves produced food crops on three crown-owned plantations.[25]

During the first years of Caminha's rule, settlers and slaves alike starved, the former since they depended on food supplies from Portugal and the nearby continent, the latter because there was a scarcity of local food crops. The settlement imported flour, wine, olive oil and cheese from Portugal for the white inhabitants. Food shortages continued at least until 1499 when starving settlers were sent to Príncipe.[26] Only the oil-palm (*elaeis guineensis*) and one yam species (*dioscorea cayenensis*) already existed in São Tomé when it was discovered.[27] The Portuguese soon introduced domestic animals, such as cattle, pigs, sheep, goats, donkeys, ducks and chickens, as well as sugar-cane (*setcherum officinarum*), maize (*zea mays*), other yams (*dioscorea minutiflora* and *discorea alata*), figs (*ficus carica*), orange and lemon trees (*citrus spp.*) and plantain (*musa paradisiaca*).[28] One banana species possibly already existed in São Tomé, while other varieties (*musa sapientum*) were later introduced from Brazil. Coconut (*cocos nucifera*), manioc (*manhiot esculenta*) and sweet potatoes (*ipornea batatas*) were also success-

---

[22] Cunha, 'A Organização Económica', 33.
[23] From 1522 when São Tomé reverted to the crown the king's factor exercised this right.
[24] Cunha, 'A Organização Económica', 22.
[25] Ibid., 29.
[26] Ibid., 27.
[27] Hélder Lains e Silva, *São Tomé e Príncipe e a Cultura do Café* (Lisbon, 1958), 54, 56.
[28] Ibid., 57ff.

fully introduced from the Americas in the sixteenth century.[29] It was only in the 1510s that São Tomé was able to provide São Jorge da Mina fort with food supplies.[30] Irrespective of the availability of food in the archipelago, the mortality rate among white settlers due to tropical diseases always remained high and, demographically, whites constituted a very small minority of the population.

The colonization of São Tomé also marked the beginning of missionary activity in the region, since the expansion of Catholicism had been an integral part of the project from the outset. The Catholic Church participated actively in the local economy, both the slave trade and the sugar industry. The first Catholic priest arrived with Caminha in 1493. The island's first two churches, São Francisco, which was part of the monastery with the same name, and Santa Maria, were both constructed with stones and bricks from Portugal during Caminha's captaincy.[31] Possibly even before 1500 the mother church *Nossa Senhora da Graça* was erected near the foundations of the church Santa Maria.[32] The first Agostinian missionaries arrived as early as 1499.[33] By 1504 the Catholic Church had also established the charitable institution *Santa Casa de Misericórdia* and its hospital, in response to the high morbidity rate. In 1514 the Portuguese king obliged the settlers to baptize newly arrived slaves within six months of their purchase.[34] By 1519 there were three Catholic brotherhoods, which along with kinship were the dominant form of collective solidarity at the time.[35] At the request of the Forros, in 1526 the king allowed free blacks to establish the Catholic Brotherhood of Our Lady of the Rosary, which was given the right to engage in trade in slaves, spices and gold with Kongo and Elmina. Later, King João III (1521–57) granted the brotherhood the right to demand and obtain the freedom of their slave members.[36] These concessions demonstrate the growing importance of the Forros in local society and economy.

In 1534, Pope Paul III (1534–49) established a diocese in São Tomé, the second in Africa, following Ribeira Grande in Santiago, Cabo Verde, in 1533. The new diocese was subordinated to the diocese of Funchal (Madeira) until 1597, thereafter to Lisbon. Its jurisdiction extended from the River Santo André (Sassandra) near Cape Palmas

---

[29] The origin of the coconut is disputed, as it might also have come from South-East Asia. See Aguiar, 'Plantas e Povos', 36.
[30] Cunha, 'A Organização Económica', 18.
[31] Pinheiro, 'O Povoamento', 258.
[32] Madeira, 'Estudo Morfológico', 9.
[33] Pinheiro, 'As Estruturas Político-administrativas', 280.
[34] Cunha, 'A Organização Económica', 85.
[35] Ramos, 'Rebelião e Sociedade', 25.
[36] Eyzaguirre, 'Small Farmers', 42–43.

to the Cape of Good Hope, including Elmina and the Kongo. São Tomé's second bishop's local representative was the vicar general João Baptista (1542–52). Having been involved in continuous conflicts with the local clergy, who feared that the vicar general would succeed in controlling the See, he finally left the island for Kongo[37] Following the creation of the diocese of São Salvador in Kongo in 1596, the geographical jurisdiction of São Tomé was restricted. In 1677 the diocese of São Tomé was separated from Lisbon and became part of the archdiocese of Bahia, testifying to São Tomé's strong ties with Brazil during that period. As already pointed out, on several occasions the bishops and other members of the Catholic Church were involved in political and financial conflicts with the secular authorities.

The prosperity of the islands attracted the interest of other European powers, which had ended the Portuguese monopoly along the African coast. Following the attack on São Tomé by French corsairs in 1567, the Portuguese decided to defend the town with a fort. The construction of the São Sebastião fort, close to the city to the east, was completed in 1575. The Spanish domination of Portugal and its colonies from 1580 to 1640 increased attacks by foreign pirates and corsairs. The Dutch occupied Príncipe in 1598, but the decimation of their strength from the initial five hundred men to fewer than a hundred by disease and fighting forced them to abandon the island after only four months. In 1599 a Dutch fleet occupied and looted São Tomé. The settlers fled into the interior, from where they launched attacks on the Dutch. After the Dutch commander and some 1,200 of his men had succumbed to tropical diseases, the Dutch left the island after three weeks. However, they did not abandon their intention to seize the island. In 1641 a Dutch fleet conquered São Tomé, and until January 1649 the Dutch West India Company (WIC) occupied its fort and the harbour, from where they maintained control of the local sugar and slave trades. This occupation even resulted in a short revival of the declining sugar industry. In 1702 the French attacked Príncipe and seven years later they occupied São Tomé town for one month, demanding a huge ransom.[38] Thereafter external attacks ceased as a result of the archipelago's economic decline.

---

[37] Garfield, *History of São Tomé*, 107–9.
[38] Gerhard Seibert, *Comrades, Clients and Cousins: Colonialism, Socialism and Democratization in São Tomé and Príncipe*, 2nd edn (Leiden, 2006), 29.

Slavery in São Tomé

As already noted, from the outset the settlers owned African slaves to work for them, both in their households and on plantations.[39] In 1510, Fernandes estimated the total number of resident slaves at two thousand. Many settlers owned more than fourteen slaves for food cultivation. Besides, household and plantation slaves, slaves held temporarily in São Tomé for re-export constituted a third category representing a significant proportion of the local population, but due to their limited stay, they did not play an active role in the formation of the local Creole society. According to an account by an anonymous Portuguese pilot from the mid-sixteenth century, slaves were employed as couples, who built their own wooden houses and were allowed to work one day per week (Saturday) on their own provision plots to cultivate yams and other food crops for their personal needs.[40] Some authors have interpreted this to mean that there was only one single category of slaves in São Tomé.[41] Tenreiro, an author influenced by the lusotropicalist ideology of his time, even claimed that the labour regime in São Tomé was not real slavery but more akin to serfdom.[42] Setting the source in its historical context, however, Henriques argues that the mild slave regime only existed in the early stage of the local plantation economy.[43] Following the extension of the sugar plantations, the labour regime became more oppressive for captives both on privately and crown-owned estates, where they lived in *sanzalas* (slave quarters on plantations).

This slave labour system can be considered a forerunner of the plantation system that later developed in Brazil and the Caribbean. Garfield estimates a total of 9,000 to 12,000 slaves in São Tomé during the height of the sugar industry, based on an estimate of between sixty and eighty sugar mills with an average of 150 slaves each.[44] From the beginning, slaves ran away and tried to survive in the inaccessible mountainous interior of the island. Runaway slaves as well as those who committed suicide were documented as early as 1499 in Caminha's will. Between 1514 and 1527, out of 12,904 slaves imported and registered by the royal treasury (*fazenda real*), 670

---

[39] Eyzaguirre, 'Small Farmers', 38.
[40] For an English translation, see John William Blake, trans., ed., *Europeans in West Africa, 1450–1560* (London, 1942), 145ff.
[41] Ramos, 'Rebelião e Sociedade', 32.
[42] Francisco José Tenreiro, *A Ilha de São Tomé* (Lisbon, 1961), 70.
[43] Henriques, 'Ser Escravo'.
[44] Garfield, *History of São Tomé*, 80.

escaped, the equivalent of 5 per cent.[45] Slaves fled either into the island's interior or by dugout canoe to the open sea. Some of these canoes were carried by the sea currents to Fernando Po: former slaves from São Tomé and Príncipe and their descendants were sighted around Ureka in the south of the island by the Spanish in 1778 and by the British in 1827.[46]

Many of the runaway slaves in São Tomé died of starvation, since there were few food crops and hardly any edible animals available in the mountainous tropical forests. On the other hand, the lack of food on the sugar plantations due to insufficient food crop production, between 1531 and 1535 for instance, was one of the main reasons for slaves to escape.[47] The maroon settlements in the inaccessible interior of the island were known as *macambos*. The first forms of runaway slave organizations appeared in the 1530s when maroon gangs attacked settlers and plantations. More isolated plantations were abandoned as a result of the insecurity caused by the frequent maroon attacks.[48] In late 1531 the local Portuguese authorities complained about settlers and blacks being killed in the fight against the maroons and feared that the whole island might be lost if the problem were not solved.[49] In response, in 1533 the local authorities waged a 'bush war' (*guerra do mato*) against the maroons with militia units commanded by a bush captain (*capitão do mato*), a post maintained until the last quarter of the eighteenth century.[50] In 1685 the position of bush captain was held by a free black.[51] The financial burden of the bush war was shared equally between the royal treasurer and the town council.[52]

In 1547, one of the military expeditions returned with forty recaptured slaves.[53] As early as 1549 two men from the maroon community appeared in the town where they were taken in by the wealthy mulatto planter Ana de Chaves and claimed to have been born free men. With the support of Ana de Chaves they sent petitions to the king that the local authorities should not consider them as captives, but as free men, a request that the monarch approved.[54] The highest incidence of marronage (slaves running away) coincided with the sugar boom in the mid-sixteenth century when the number of plantation slaves had increased significantly. In 1574 runaway slaves from the *macambo*

[45] Caldeira, 'Rebelião e Outras Formas de Resistência', 109.
[46] Caldeira, 'Uma Ilha Quase Desconhecida', 106, n. 20.
[47] Ramos, 'Rebelião e Sociedade', 3.
[48] Santos, 'A Formação das Estruturas Fundiárias', 81.
[49] Caldeira, 'Rebelião e Outras Formas de Resistência', 109.
[50] Ibid., 110.
[51] Ibid., 111.
[52] Ibid., 112.
[53] Ramos, 'Rebelião e Sociedade', 35.
[54] Caldeira, 'Rebelião e Outras Formas de Resistência', 121.

attacked the town, but were expelled by the settlers.[55] As a result of the bush war, between 1587 and 1590 the runaway slaves were almost defeated. Many slaves were recaptured; others returned voluntarily due to the difficult living conditions in the *macambos*. However, they succeeded in reorganizing themselves and in 1593 again caused concern to the authorities. In that year, after having organized a military action against the maroons the governor claimed he had 'extinguished almost all the rebel slaves'.[56] Despite the military actions, the settlers failed to reoccupy and inhabit the south and western part of São Tomé, which remained insecure due to the proximity of the maroons.[57] In 1693 bush captain Mateu Pires carried out the last large military action against the maroons, who had captured slave women in the plantations. Thereafter assaults by the maroons decreased drastically and confrontations between the maroons and the settlers took place only sporadically. In the eighteenth century the maroons were known as *Angolas* or *Angolis* and they have been known as *Angolares* since the early nineteenth century. Until 1878, when their territory in the south of São Tomé was occupied by the colonial authorities, the *Angolares* enjoyed political autonomy and they have largely preserved their distinct cultural identity until today.

Armed revolt on a large scale as a form of slave resistance occurred three times in the archipelago's history. The greatest slave revolt occurred in July 1595, when the local government was weakened by a conflict between the governor and the bishop. It was headed by Amador, a slave born on the island, who succeeded in uniting five thousand slaves who raided plantations and burned sugar mills and settlers' houses. They also attacked the town three times, but without success. After three weeks, they were defeated by the settlers and the militia. At least two hundred slaves were killed in the last battle, while Amador and the other rebel leaders were executed. The majority of the slaves received a general amnesty from the governor and returned to their masters. Reportedly about sixty out of the island's eighty-five sugar mills had been destroyed during Amador's revolt, which was one of the greatest slave uprisings in Atlantic history.[58] Smaller slave revolts occurred in 1617 and in 1709, when São Tomé was occupied for almost a month by French corsairs. During the occupation the wealthy setters sought refuge on their estates, where they armed the slaves to help defend them against a possible

[55] Lains e Silva, *São Tomé e Príncipe*, 86.
[56] Caldeira, 'Rebelião e Outras Formas de Resistência', 111.
[57] Ibid., 113.
[58] 'Relatione Uenuta Dall' Isola di S.Tomé', *Monumenta Missionária Africana: África Ocidental (1570–1599)*, III, ed. António Brásio (Lisbon, 1953), 521–3. See also Gerhard Seibert, 'São Tomé's Great Slave Revolt of 1595: Background, Consequences and Misperceptions of One of the Largest Slave Uprisings in Atlantic History. *Portuguese Studies Review* 18/2 (2011), 29–50.

attack by the French. After the corsairs had left in exchange for a large ransom, the armed slaves staged a revolt and tried to invade the town, but they were easily defeated by the settlers.

The slave trade

The Portuguese began to trade slaves from the kingdom of Benin to Elmina around 1480, before the colonization of São Tomé.[59] Due to an increase in this regional slave trade, they established a trading factory in Benin in 1486. The average duration of the slaving voyage to Benin and back to Elmina was two to three months.[60] The direct coastal slave trade between Elmina and Benin continued until about 1515. In the beginning of São Tomé's colonization, the principal commercial activity of the settlers was the slave trade, which was also necessary to recruit labour for the local economy. The first slaves were bought in the Slave Coast, the Niger Delta and the island of Fernando Po. Subsequently, the São Tomé settlers bought slaves in Kongo (Soyo) and Angola. The slaves were sold in São Tomé or re-exported from there to Portugal and to Elmina. Slaves belonging to the Portuguese king were branded with a cross on their right arms.[61]

According to Fernandes' manuscript, around 1510 some five thousand slaves were kept for re-export in São Tomé. Reportedly, in that period Portugal imported between 10,000 and 12,000 slaves. In 1551, about 9,950 out of Lisbon's total population of 100,000 were slaves and from 1578 to 1583 slaves represented one-fifth of the city's population.[62] In 1516, over eleven months, the royal factory (*casa da feitoria*) in São Tomé received a total of 4,072 slaves in fifteen shiploads from the mainland for re-export. In the same period, Fernão de Melo, the captain, had purchased another 234 slaves.[63] The first slave ship from São Tomé to Elmina was reported in July 1499.[64] From 1514 to 1518 António Carneiro of Príncipe had the monopoly of the trade with Benin and the supply of slaves to Elmina, at the time a centre of the gold trade. In the last two years of his contract Carneiro shipped 300–400 slaves to Elmina.[65] In 1519 São Tomé became the centre of the

[59] Caldeira, 'A Estratégia Inicial, 17.
[60] John L. Vogt, 'The Early São Tomé-Príncipe Slave Trade with Mina, 1500–1540', *International Journal of African Historical Studies (IJAHS)* 6/3 (1973), 453.
[61] Cunha, 'A Organização Económica', 79.
[62] Jorge Fonseca, *Escravos e Senhores na Lisboa Quinhentista* (Lisbon, 2010), 88; Lains e Silva, *São Tomé e Príncipe*, 79.
[63] Cunha, 'A organização económica', 67.
[64] Vogt, 'Early São Tomé-Príncipe slave trade', 456.
[65] Ibid., 457.

slave trade from the Niger Delta to Elmina where the slaves were employed as porters in the trade into the interior.[66] At that time, São Tomé had a subordinate position in relation to Elmina and served as a support post for the fort as a food and slave supplier.[67] Elmina's factor was allowed to return slaves who did not arrive in good physical condition.

The regional slave trade between São Tomé and Elmina lasted until 1540, by which time the Spanish Caribbean (Cartagena and Vera Cruz) had become a lucrative slave market and the Portuguese gold trade in Elmina had diminished considerably. During the height of the trade, between the late 1520s and early 1530s, some five hundred slaves a year were shipped to Elmina.[68] In 1533 eighty slaves on the royal ship *Misericórdia* revolted between São Tomé and Elmina, killing almost the entire crew. Later, some of the rebellious slaves recaptured at the Forcados River (Niger Delta) were recognized due to the royal brand-mark.[69] The mortality of the slaves shipped to Elmina was considerably lower than that of slaves sent to Portugal. On twenty-two ships from São Tomé to Portugal out of a total of 2,202 slaves, 806 died during the voyage, while during the short voyage to Elmina 360 out of 383 slaves arrived alive.[70]

São Tomé's slave trade with the kingdom of Kongo began in the early sixteenth century, and subsequently that with Angola. In 1532, however, the Manicongo prohibited the direct slave trade between Angola and São Tomé because it affected the number of slavers bound for Kongo, and thereafter the kingdom became an intermediary between Angola and the São Tomé traders.[71] The São Tomé traders acquired 1,449 slaves in the three-year period from 1525 to 1528. Between 1532 and 1537 the number of slaves increased to 15,844, most of them from Kongo.[72] In 1553 King João III reconfirmed the prohibition of the trade with Angola, and Soyo, at the mouth of the Zaire River, became the port of export for Portuguese vessels. In the mid-sixteenth century, over fourteen months, between twelve and fifteen slave ships left Kongo for São Tomé, the smaller ones carrying 400 slaves and the larger ones 700 slaves each.[73] The trans-Atlantic slave trade from São Tomé to the Spanish Americas began in 1525.[74] Subsequently,

---

[66] Caldeira, *Mulheres, Sexualidade e Casamento*, 19.
[67] Cunha, 'A organização económica', 88.
[68] Vogt, 'Early São Tomé-Príncipe slave trade', 466.
[69] Cunha, 'A organização económica', 80.
[70] Ibid., 82.
[71] Ibid., 99.
[72] Pinheiro, 'O século XVI', 347.
[73] Cunha, 'A organização económica', 100.
[74] Vogt, 'Early São Tomé-Príncipe slave trade', 466.

most slaves re-exported from São Tomé went to the Caribbean and Brazil. Between 1532 and 1536 São Tomé re-exported an average of 342 slaves to the Antilles every year.[75] Caldeira estimates the number of slaves re-exported in the first half of the sixteenth century as between 5,000 and 10,000 a year.[76] Before 1580 São Tomé accounted for 75 per cent of Brazil's imports, predominantly comprising slaves.[77]

At the beginning of the seventeenth century São Tomé ceased to be an important slave trade entrepôt. In 1614 the São Tomé settlers lost access to the slave markets in Kongo because they had become a threat to other Portuguese commercial interests. Thereafter, their operations were restricted to the Gulf of Guinea, with the Cape of Lopo Gonçalves as the southern limit. With the appearance of the French, English and Dutch, and the occupation of Elmina by the Dutch in 1637, the island's traders were cut off from their previous supply markets.[78] During the Dutch occupation of Luanda and São Tomé (1641–48), the WIC directly controlled the trade in the region. Furthermore, from the mid-seventeenth century Angolan slaves were shipped directly to Brazil and the Spanish Americas and São Tomé's access to the market in Luanda was disrupted.[79] Subsequently, traders from São Tomé traded slaves predominantly at nearby markets in Gabon and Calabar. The re-export of slaves continued, but on a much smaller scale than in the sixteenth century. In the seventeenth century, slave-ships going from Bahia to the Mina coast were obliged to call at São Tomé to pay taxes and purchase food supplies. In 1710 the king reaffirmed this obligation to prevent tax evasion by the ships that sailed directly to Bahia, without stopping at São Tomé. In the eighteenth century British and Dutch slavers preferred to call at São Tomé to purchase provisions and water, while the French and Brazilians went predominantly to Príncipe to buy livestock, fruit and yams.[80] The commodities received in exchange from the ships were resold on the nearby mainland, particularly the Gabon coast, Benin, Warri and Calabar.[81] From the mid-eighteenth century, São Tomé's direct maritime connection with Lisbon largely disappeared and communication went through Angola or Bahia.[82]

[75] Pinheiro, 'O século XVI', 352.
[76] Caldeira, *Mulheres, Sexualidade e Casamento*, 20.
[77] Eyzaguirre, 'Small farmers', 58.
[78] Caldeira, *Mulheres, Sexualidade e Casamento*, 22.
[79] Caldeira, 'Rebelião e outras formas de resistência', 104.
[80] Cristina Maria Seuanes Serafim and Lúcia M.L. Tomás, 'Os séculos XVII–XVIII: O lento declinar da economia. A economia: a produção açucareira, o comércio e o regate. A fiscalidade e as finanças. O Arquipélago do Golfo da Guiné: Fernando Pó, São Tomé, Príncipe e Ano Bom', *A Colonização Atlântica*, II, ed. Artur Teodoro de Matos (Lisbon, 2005), 374.
[81] Eyzaguirre, 'Small farmers', 93.
[82] Caldeira, *Mulheres, Sexualidade e Casamento*, 28.

Sugar cultivation

Between 1520 and 1530 the export of sugar became as lucrative a business as the slave trade. Sugar-cane and people skilled in its cultivation and processing came to São Tomé from Madeira, where Genoese and Sicilians as well as Portuguese were engaged in the sugar industry. The cultivation of sugar-cane, which was concentrated on São Tomé's northern flatlands, started immediately after Caminha's arrival. When the German traveller Hieronymus Münzer visited João II in 1494, the Portuguese king told him that sugar planted in São Tomé would grow three times faster than in Madeira.[83] The production of molasses is mentioned in Caminha's will of 1499. Sugar-cane was planted and harvested year round and took five months to grow. The original forest was gradually cut back for the expanding sugar plantations in the island's northern third between Ponta Figo and Santana, while the rest of the island remained covered by primary tropical forest that was largely inaccessible to the settlers. The cultivation of sugar-cane in São Tomé proved successful due to fertile volcanic soils, the tropical climate, sufficient rainfall, and, most important, the availability of cheap slave labour from the neighbouring African continent. The sugar plantations were grouped around the sugar mills, called *engenhos*, which were built next to streams to power them, a technique already used in Madeira. The island was highly suitable for this technique, since it had a total of twenty-seven streams and seven small rivers.[84] Generally the streams also marked the boundaries between the different plantations.[85] In addition to streams to power the mills, the island also provided sufficient firewood to dry the sugar for export.

Most plantations were privately owned by royal officials and settlers and several belonged to the crown and the Catholic charity *Misericórdia*. In 1535 the crown owned six large plantations that were run by the royal factor.[86] A few estates were the property of absentee landlords resident in Portugal. Wealthy plantation owners erected wooden fortresses on their estates and maintained private armies of armed slaves. On their plantations the owners exercised great power, while the local authorities were unable to enforce their authority there.[87]

---

[83] Virginia Rau, 'O açúcar de S. Tomé no segundo quartel do século XVII', *Elementos de História da Ilha de S. Tomé*, ed. Centro de Estudos de Marinha (Lisbon, 1971), 7.
[84] Cunha, 'A organização económica', 18.
[85] Santos, 'A formação das estruturas fundiárias', 60.
[86] Ibid., 71.
[87] Ramos, 'Rebelião e Sociedade', 33.

Armed conflicts and power struggles between rival plantation-owners occurred frequently, contributing to the conflictual political climate.[88] Due to their wealth, they constituted the most important socio-economic group in the islands. Heywood and Thornton claim that: 'Some Kongolese nobles also settled on the island and owned estates where they used slave labor.'[89] However, the source cited, the report by the anonymous pilot from the mid-sixteenth century, does not mention 'Kongolese nobles' at all. The pilot reports that between 1520 and 1550 he visited São Tomé five times and he spoke with

> a negro called João Menino, a very old man, who said that he had been taken there with the first [negroes] who went from the African coast to this island when it was populated by order of our King; and this negro was very rich and had children and grandchildren and married grand-grand-children, who already had children.[90]

It seems unlikely that João Menino was a Kongolese, given that the Africans of the first settlement established between 1486 and 1490 came from the Niger Delta, while the Kongo was first mentioned as a slave supplier to São Tomé in 1502.[91] Caldeira believes that João Menino had arrived as a slave with the first settlers around 1485 and was one of the male slaves manumitted in 1517.[92] Other authors suggest that he was one of the free Africans who had settled on the island but they do not give his origin.[93]

However, at least one Kongolese may have arrived in São Tomé prior to 1499. In his will, Caminha referred to a Pêro de Manicongo who had worked as a sailor for a settler called D. Francisco.[94] In the 1550s there was undoubtedly a Kongolese nobleman, Rodrigo de Santa Maria, who lived in São Tomé, from where he travelled frequently to Lisbon. In 1550 King Diogo I of Kongo (1545–61) accused him of having been involved in a conspiracy against him.[95] In 1561 a few Kongolese noblemen who had supported King Afonso II (1561), the successor of King Diogo I, who had been killed by his

---

[88] Pinheiro, 'A conflitualidade social'.
[89] Linda M. Heywood and John Thornton, *Central Africans, Atlantic Creoles, and the Foundation of the Americas, 1585–1660* (Cambridge, 2007), 69. See also John Thornton, 'Early Kongo-Portuguese Relations: A New Interpretation', *History in Africa* (*HA*) 8 (1981), 191.
[90] Original text reproduced in Albuquerque, *A Ilha de São Tomé*, 33; See also Caldeira, *Viagens de um piloto português*, 119.
[91] Heywood and Thornton, *Central Africans*, 68.
[92] Caldeira, *Viagens de um piloto português*, 119.
[93] Henriques, 'Ser escravo', 183; Cunha, 'A organização económica', 45; Sousa, *São Tomé et Príncipe de 1485 à 1755*, 29. Isabel Castro Henriques, *São Tomé e Príncipe: A Invenção de uma Sociedade* (Lisbon, 2000), 42.
[94] Albuquerque, *A Ilha de São Tomé*, 79.
[95] Caldeira, 'Tráfico de escravos', 73.

brother, Bernardo I (1561–66) after only a few days in power, sought refuge in São Tomé.⁹⁶ While the existence of Kongolese planters is not documented, the first mulatto plantation owners born in the island are recorded as early as 1521. In addition, in the sixteenth century wealthy local planters included mulatto women like Ana de Chaves, Catarina Alves and Simoa Godinho.

Sugar production in São Tomé had started by 1517, as the first two sugar mills appear in a document of that year.⁹⁷ According to the anonymous pilot, in the mid-sixteenth century there were some sixty sugar mills in operation.⁹⁸ Eyzaguirre believes that during the height of the sugar boom the number of mills may have reached two hundred, with an average number of fifty slaves for each mill.⁹⁹ In 1529, one rich local planter operated twelve mills on his two estates. According to Garfield, each mill had an annual production capacity of up to 5,000 arrobas (each 14.7 kg) of sugar.¹⁰⁰ This average, equivalent to 73.5 tons, is considerably higher than the approximately 15–25 tons per mill given by Schwartz.¹⁰¹ In the mid-sixteenth century planters owned 150, 200 and up to 300 slaves.¹⁰² This would have been fewer than the average of 480 slaves on the earlier royal plantations producing food crops mentioned above. Due to the extension of sugar cultivation, less land was dedicated to food crops, which in turn resulted in a shortage of food and caused famine among the slaves. The mortality rate among slaves was also high, but it was easier to replace them than the white settlers.¹⁰³

Table 2.1 Sugar mills in São Tomé, 1517–1736

Year	1517	c. 1550	1595	c.1600	1610	1645	c.1672	c.1710	1736
Number	2*	c. 60**	c. 85***	c. 120	45	54	31	18–19	7****

(Sources: unless otherwise indicated, see Cristina Maria Seuanes Serafim, *As Ilhas de São Tomé no século XVII* (Lisbon, 2000), 258; * = Hélder Lains e Silva, *São Tomé e Príncipe e a Cultura do Café* (Lisbon, 1958), 83; ** = Arlindo Manuel Caldeira, *Mulheres, Sexualidade e Casamento no Arquipélago de S. Tomé e Príncipe* (Lisbon, 1997), 17; *** = 'Relatione uenuta dall' Isola di S.Tomé', *Monumenta Missionária Africana. África Ocidental*, III, ed. António Brásio (Lisbon, 1953), 523; **** = Cristina Maria Seuanes Serafim and Lúcia M.L.Tomás, 'Os séculos XVII-XVIII: O lento declinar da economia', *A Colonização Atlântica*, II, ed. Artur Teodoro de Matos (Lisbon, 2005), 358)

---

[96] Ibid., 74.
[97] Lains e Silva, *São Tomé e Príncipe*, 83.
[98] Caldeira, *Mulheres, Sexualidade e Casamento*, 17.
[99] Eyzaguirre, 'Small Farmers', 60.
[100] Garfield, *History of São Tomé*, 73.
[101] Stuart B. Schwartz, 'Introduction', *Tropical Babylon: Sugar and the Making of the Atlantic World, 1450–1680*, ed. Stuart B. Schwartz (Chapel Hill & London, 2004), 18.
[102] Cunha, 'A Organização Económica', 46.
[103] Ibid., 47.

The fully grown canes were cut into smaller pieces that were crushed in the water-driven three-roller mill to extract the juice. The sugar-cane waste was used to feed pigs. The juice was boiled three to four times, and then, still semi-moist, put into semi-conical containers to dry and harden. However, due to São Tomé's high air humidity, this process needed to be assisted by means of wood fires, but even this failed to dry the sugar completely. About 1,175 kg of sugar cane were necessary to produce one arroba of sugar.[104] The sugar was exported in the form of sugar loaves (*pães de açúcar*), weighing 15–20 *arráteis* each (1 *arrátel* = 459 gr).[105] The finished sugar loaves were packed in boxes that weighed about 86 kg each. The Portuguese king received one tenth of the sales as taxes. In the 1510s São Tomé produced an estimated 100,000 arrobas a year. In 1527 sugar producers in Madeira were concerned about possible negative consequences caused by the competition of São Tomé's sugar.[106] By that time, Madeira's sugar production had decreased from some 300,000 arrobas in around 1450 to about 40,000.[107]

In the mid-sixteenth century thirty to forty Portuguese ships arrived annually at the port of São Tomé, remaining for six or seven months to load sugar.[108] The voyage from São Tomé to Lisbon took about fifty days, with another five to ten days for the voyage to Antwerp, a significant port for the importation of sugar into Europe.[109] Between July 1535 and November 1548, 112 Portuguese ships, almost exclusively transporting sugar from São Tomé, arrived at the port of Antwerp, on average nearly nine ships annually.[110] Between 1535 and 1551 a total of 483,652 arrobas of sugar arrived in Antwerp from São Tomé. In 1552 and 1553 sugar imports from São Tomé to Antwerp totalled 85,244 arrobas, while from 1563 to 1572 they totalled 260,000.[111] In the debt books of the Augsburg-based trading company Christoph Welser and Brothers for the period 1554–60, sugar from São Tomé was the only registered commodity that was imported through the port of Lisbon.[112] The sugar boom also attracted Spanish, Italian and French merchants to São Tomé. However, the demand for the island's sugar in Europe was due to its abundance and cheapness, rather than quality, since it was fairly dark

---

[104] Garfield, *History of São Tomé*, 70.
[105] Tenreiro, *A Ilha de São Tomé*, 225.
[106] Rau, 'O Açúcar de S. Tomé', 8–9.
[107] Garfield, *History of São Tomé*, 64.
[108] Cunha, 'A Organização Económica', 117.
[109] Garfield, *History of São Tomé*, 73.
[110] Rau, 'O Açúcar de S. Tomé', 20.
[111] Pinheiro, 'O Século XVI', 337.
[112] Rau, 'O Açúcar de S. Tomé', 22.

and not very solid. Indeed, in Antwerp São Tomé's sugar was considered the 'worst in the world' because it was moist and full of tiny black ants.[113]

Estimates of sugar production by different authors shown in Table 2.2 differ and are sometimes inconsistent, since they are calculated with figures derived from information on tax revenue, ship loads or the number of sugar mills. The highest figure is 800,000 arrobas for the years before 1578, given by Lains e Silva, who based his calculation on an estimate of forty shiploads a year of 20,000 arrobas each, and seems rather unlikely.[114] However, there is no doubt that sugar production reached its height in the third quarter of the sixteenth century and a gradual decline began after that, due to various internal and external causes. As far as the position of the sugar industry in the local economy is concerned, there is no consensus on whether it really replaced the slave trade as the principal source of income. Cunha affirms that even during the sugar boom the crown's tax income from sugar did not exceed the revenue earned from the slave trade.[115] Other authors believe that sugar became the mainstay of the local economy as of the 1520s.[116]

Table 2.2 Estimates of São Tomé's sugar production, 1517–1684

Year	1517	1529	1531	1554	c.1570	1578	1579	1580	1584
Arrobas	100,000	123,170	135,860	150,000*	800,000*	120,000* / 175,000	200,000*	20–24,000	250,000**

Year	1588	1590	1591	1602	1610	c.1624	1645	1551	c.1672	1684
Arrobas	60,000	64,000**	10–12,000	40,000	60,000	89–100,000	100,000	40,000	ca. 27,000	2,000

(Sources: unless otherwise indicated, Isabel de Castro Henriques, 'Ser escravo em S. Tomé no Século XVI: Uma outra leitura de um mesmo quotidiano', *Revista Internacional de Estudos Africanos* 6 & 7 (1987), 92 (for the period of 1517 to 1591), and Cristina Maria Seuanes Serafim & Lúcia M.L. Tomás, 'Os séculos XVII-XVIII: O lento declinar da economia', *A Colonização Atlântica*, II, ed. Artur Teodoro de Matos (Lisbon, 2005), 355 (1602-84); * = Hélder Lains e Silva, *São Tomé e Príncipe e a Cultura do Café* (Lisbon, 1958), 84-5; ** = Pablo B. Eyzaguirre, 'Small Farmers and Estates in São Tomé, West Africa' (PhD dissertation, New Haven CT: Yale University, 1986), 60)

---

[113] Garfield, *History of São Tomé*, 72.
[114] Lains e Silva, *São Tomé e Príncipe*, 84.
[115] Cunha, 'A Organização Económica', 120.
[116] Blake, *Europeans in West Africa*, 62.

Decline of São Tomé's sugar industry

As already indicated, because of the high humidity of the tropical climate the crystallization and refinement of São Tomé's sugar was poor and, consequently, its quality was inferior to that of Madeira and Brazil. As a result, already in 1533 the medical use of São Tomé's sugar was prohibited.[117] The local planters were unable to improve the quality of their product due to the lack of appropriate technology. Therefore, the prices paid for São Tomé's sugar were always lower than for that of its competitors. From 1578 to 1582, when São Tomé's annual sugar production reached its peak, prices ranged between 630 and 950 reis per arroba in Lisbon, while Madeira's sugar was traded for 2,500–3,000 reis per arroba.[118] Brazilian sugar, large-scale production of which had begun around 1533, had also achieved higher prices, with 1,400–1,850 reis per arroba.

The gradual decline of São Tomé's economy started in the last quarter of the sixteenth century when Brazil emerged as a large, fast-growing sugar-producer. In 1548 six mills existed in Brazil.[119] In contrast, in 1579 São Tomé's sugar production dropped by 35 per cent due to drought, exacerbated by impoverished soils and an infestation of worms which attacked the plants' roots. This decline in production was not offset by higher prices, as Brazil was already producing considerable quantities of sugar for the European markets at the time. By 1583 a total of 102 mills in Brazil produced 200,000 arrobas.[120] Furthermore, the Brazilian mills produced three to four times more per unit a year.[121] Equally important was the higher quality of Brazilian sugar, which was white and dry and, consequently, achieved better prices on the world market. Apart from better production conditions and higher prices, Brazil also provided planters a more stable political environment and, last but not least, a healthier climate. While competition from Brazil was the most important reason for the gradual collapse of São Tomé's sugar industry, there were also other internal and external causes.

The constant political instability caused by frequent conflicts between the political and religious authorities negatively affected the local economy. This instability was exacerbated by the assaults of the

[117] Cunha, 'A Organização Económica', 110.
[118] João Lúcio de Azevedo, *Épocas de Portugal Económico*, 2nd edn (Lisbon, 1947) quoted in Francisco José Tenreiro, 'Descrição da Ilha de S. Tomé no Século XVI', *Garcia de Orta* 1/2 (1953), 227.
[119] Garfield, *History of São Tomé*, 74.
[120] Ibid., 75.
[121] Schwartz, 'Introduction', 18.

Angolares maroons, who constituted a permanent threat to the sugar estates. A serious blow for the local sugar industry was Amador's great slave uprising of July 1595 when the rebels destroyed at least sixty sugar mills, i.e. more than 70 per cent of production capacity. The slave revolt, occurring at a time when the economic decline had already begun, accelerated the exodus of São Tomé's planters to Brazil, where they expected better conditions for sugar production.

Externally, from the late sixteenth century the archipelago was increasingly threatened by the appearance of other European powers in the Atlantic. As already mentioned, the Dutch occupied Príncipe in 1598 and looted São Tomé in the following year. Based on Dutch sources, Ratelband claims that in 1599 the occupiers destroyed sixty-four out of 118 mills existing at the time.[122] This total tallies with the 120 mills given by Serafim (see Table 2.1) for this period, but not with the number of approximately 85 existing mills mentioned in the document relating to the slave revolt in 1595.[123] Whatever the exact numbers, it is certain that within a short period the local economy suffered severe damages. Yet the settlers still tried to seek support for economic reconstruction. In response to a request from the settlers dated 24 December 1605, the king granted tax exemptions for ten years for sugar produced in rebuilt mills. For these purposes, all mills that had lain dormant for longer than two harvests were considered to have been rebuilt.[124]

Despite these efforts, the sugar industry never recovered completely from the destruction suffered during the slave revolt and the Dutch raid in 1599. However, there was a partial recovery, as the statistics for mills and sugar production in the seventeenth century show. As already mentioned, during the Dutch occupation in 1641–8 the sugar industry even enjoyed a short revival. In this period, however, the island's governor, Lourenço Pires de Tavora (1642–45), complained that the planters had been unable to replace deceased slaves due to a shortage of boats and commodities for the barter trade on the coast.[125] The scarcity of shipping also affected sugar exports to Portugal. The increasing activity of pirates and corsairs had resulted in the loss of Portuguese ships, while others were used to protect merchant ships in other territories like Brazil, which had priority.[126] The insufficiency of maritime transport to São Tomé also led to a shortage of commodities for the slave trade and of spare parts for the sugar mills. When

[122] Klaas Ratelband, 'A Ilha de São Tomé Segundo as Fontes Holandesas da Primeira Metade do Século XVII' (paper presented at Fundação Calouste Gulbenkian, Lisbon, 9 June 1945).
[123] See note 55.
[124] Serafim, *As Ilhas de São Tomé no século XVII* (Lisbon, 2000), 197.
[125] Ibid., 200.
[126] Ibid., 210.

Governor Pedro da Silva left São Tomé in 1672 only 32 mills remained, producing about 27,000 arrobas a year.[127] Thereafter, the decline gradually continued until sugar production ceased almost completely in the early eighteenth century. In the course of the seventeenth century most planters had left for Brazil in search of new opportunities. In the same century, sugar production expanded to the Caribbean, but never returned to São Tomé because the island could not compete with the new sugar producers in the Americas. As a consequence of the settler exodus from São Tomé, the local white population almost disappeared and the mulatto population gradually became more African in character. At the same time, the island's total population also decreased as a result of the economic decline.

The plantation economy virtually disappeared and the tropical forest covered many of the former sugar estates. Some of the former plantations passed to the crown, which however did not try to exploit them. Others were appropriated by landless Forros.[128] In 1736, there were seven sugar mills in São Tomé, producing mainly liquor which was consumed locally and used in the coastal barter trade for slaves. Around 1770 the Praia Melão estate, with 150 slaves at the time, was one of the few with a sugar mill. However, in addition to sugar and liquor, it mainly produced cassava flour, palm-oil, coconuts and other fruits.[129] The sugar monoculture was gradually replaced by diversified foodstuff production, whose surplus was sold to passing slave-ships. This smallholder agriculture was complemented by the breeding of pigs and chickens. Slavery continued without the plantations, predominantly as household slavery. Meanwhile most slave-owners were also African, which enabled slaves to become incorporated into the free Forro category, either by intermarriage or by assimilation over successive generations.[130] Only in the mid-nineteenth century, following the introduction of coffee (1787) and cocoa (c. 1820) from Brazil, did the Portuguese re-colonize São Tomé and Príncipe. They established large plantations for the production of the new cash crops and the plantation economy re-emerged. One of the prominent cocoa pioneers was João Maria de Sousa e Almeida (1816–69), a former slave-trader born of Brazilian parents in Príncipe who, after the abolition of the slave trade, reinvested his capital in commercial agriculture in São Tomé. Initially, the production of coffee and cocoa was based on slave labour from the mainland. When the Portuguese abolished slavery in São Tomé and Príncipe in 1875, it was immediately replaced

---

[127] Ibid., 199.
[128] Eyzaguirre, 'Small Farmers', 85.
[129] Caldeira, *Mulheres, Sexualidade e Casamento*, 27.
[130] Eyzaguirre, 'Small Farmers', 98.

by contract labour, predominantly from Angola, to satisfy the increasing labour demands from the newly established plantations. The living and working conditions of the African contract workers on the estates were often similar to those of the slaves earlier. The local Creole population, Forros and Angolares alike, refused contract labour which they considered beneath their status as free Africans.

Conclusion

From the beginning the Portuguese expected São Tomé and Príncipe to become a sugar-producer comparable with Madeira. In fact, unlike the drought-stricken Cabo Verde Islands, the Gulf of Guinea islands offered favourable conditions for sugar-cane cultivation, since there were fertile soils, level lands, abundant rainfall, numerous streams to drive the sugar mills, and firewood to dry the sugar loaves. Besides, cheap slave labour was easily available on the nearby continent. Equally important was that the archipelago was uninhabited and out of reach to potential enemies from the mainland. A great danger, however, existed on the archipelago itself in the form of life-threatening tropical diseases for whites. Consequently, the distant islands quickly proved unsuited to become a settler colony like Madeira and the Azores and the bulk of the local population were African slaves from the neighbouring mainland.

Initially, the São Tomé settlers engaged in the slave trade, which remained an important economic activity after the beginning of sugar production in the early sixteenth century. Due to the favourable natural conditions, the cultivation of sugar was successful for almost a century. São Tomé became the first plantation economy in the tropics based on a monoculture crop and African slave labour. Commercial sugar production did not emerge as an alternative to the slave trade, but the two commercial activities were complementary. São Tomé in the sixteenth century was a prototype for later plantation economies in the Americas. However, while the island offered excellent conditions for the cultivation of sugar-cane, this was not the case with regard to sugar production. Due to high humidity the quality of São Tomé sugar was always inferior to that from Madeira and Brazil. The demand for São Tomé sugar was based on quantity, but not on quality. Consequently, when Brazil emerged as a large-scale producer of sugar, the decline of São Tomé's sugar industry began inevitably. In addition to poor quality and competition from Brazil, the decline of São Tomé's sugar industry was reinforced by political instability and assaults by runaway slaves. The great slave revolt of 1595 accelerated the emigration of São Tomé planters to Brazil that had begun earlier. The end of the Portuguese trade monopoly in the region due to the appearance of the French, Dutch and English in the seventeenth century also contributed to the island's economic downturn. In addition, the war between the Dutch and Spanish resulted in a shortage of Portuguese ships both for sugar exports and the regional trade, because they were sent to Brazil where their support was considered more urgent. As a result of these external and internal factors, the plantation economy disappeared and the island lost its role as an important slave trade entrepôt. Largely abandoned by the Portuguese and virtually controlled by the local Creole elite, São Tomé and Príncipe became predominantly a food supply station for foreign slave-ships in the Gulf of Guinea on their way to the Americas.

3

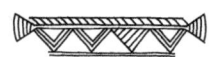

The export of rice & millet from Upper Guinea into the sixteenth-century Atlantic trade

TOBY GREEN

The connection of Atlantic slavery to production is a topic of importance to the early modern history of Atlantic Africa, a subject which after a couple of decades of neglect is now being studied with renewed concentration by historians. The legacy of the nineteenth-century transition to 'legitimate' trade and the imposition of a cash-crop economy not only involved a rupture from preceding mixed agricultural economies but also the obscuring of how these economies operated and how they interacted with and were related to the expansion of Atlantic trade from the fifteenth century onwards.[1] New studies, located primarily in Upper Guinea, have illustrated not only how some African societies changed their methods and crops of production, but also how existing techniques were then transferred to the Americas.[2]

What has emerged is that the relationship between production and Atlantic slavery is highly complex. Where commercial agriculture is concerned, there could be two perspectives that would ask us to see how or whether commercial agriculture was seen as a viable alternative to the trans-Atlantic slave trade in the early period of Atlantic trade, and if not why not. From the African perspective we could ask whether, given that we know how African societies did reshape their

---

[1]   See Robin Law, ed., *From Slave Trade to 'Legitimate' Commerce: The Commercial Transition in Nineteenth-Century West Africa* (Cambridge, 1995).
[2]   This especially relates to rice production. See Judith A. Carney, *Black Rice: The African Origins of Rice Cultivation in the Americas* (Cambridge MA, 2001); Walter Hawthorne, 'Nourishing a Stateless Society During the Slave Trade: The Rise of Balanta Paddy-Rice Production in Guinea-Bissau', *Journal of African History (JAH)* 42/1 (2001), 1–24; id., *Planting Rice and Harvesting Slaves: Transformations Along the Guinea-Bissau Coast, 1400–1900* (Portsmouth NH, 2003); Edda L. Fields-Black, *Deep Roots: Rice Farmers in West Africa and the African Diaspora* (Bloomington IN, 2009). See also the AHR Exchange on 'The Question of "Black Rice"', *American Historical Review* 115/1 (2010), 123–71.

productive systems in the Atlantic era, societies in Atlantic Africa could substitute agricultural exports for a trade in slaves, and if so under what conditions this happened. From the European perspective we could ask why the possibility of commercial agricultural production in Africa to go with that undertaken in the Atlantic islands and in the New World was not pursued.[3]

The latter question is the easier to deal with and in a sense also particularly relevant to the early period. After all, it was in this early period that the initiation of the trans-Atlantic trade followed the development of plantations on the Atlantic islands of the Canaries, Madeira and São Tomé.[4] It would then have been logical, on one level, to seek to extend this pattern into Africa, just as it was subsequently extended into the New World. Indeed, clearly in São Tomé commercial agriculture *was* seen as an alternative to the trans-Atlantic slave trade in the sixteenth century since, as many studies have shown, the early trans-Atlantic slave trade was centred on the Upper Guinea region, while slaves from Kongo and Angola in the first half of the sixteenth century were destined primarily for the sugar plantations of São Tomé and the gold trade at Elmina.[5]

This shows however that in this early period, even where commercial agriculture was seen as an alternative to the trans-Atlantic slave trade, it relied on slavery. There were historical, practical, and sociological reasons for this relationship between the commercial Atlantic plantation and slavery. From the practical perspective there was the simple fact that, as Sidney Mintz and others have shown, the work of the sugar plantation was extremely hard physically, much harder than that of alternative crops.[6] Slavery had been associated with sugar plantations in the eastern Mediterranean in the thirteenth and fourteenth centuries, expanding after the plague-induced mortality of the latter

---

[3] An analysis of why this did not take place is a key component of David Eltis, *The Rise of African Slavery in the Americas* (Cambridge, 2000).

[4] A good general study on the parameters of Atlantic slavery and the plantation system in the 16th century is António de Almeida Mendes, 'Esclavages et Traites Ibériques Entre Méditerranée et Atlantique (XVe – XVIIe Siècles): Une Histoire Globale' (unpublished PhD thesis, École des Hautes Études en Sciences Sociales, Paris, 2007). See also Vitorino Magalhães Godinho, *Os Descobrimentos e a Economia Mundial*. (4 vols, Lisbon, 1981).

[5] J. Bato'ora Ballong-wen-Menuda, *São Jorge da Mina, 1482–1637: La Vie d'Un Comptoir Portugais en Afrique Occidentale*, (2 vols, Paris/Lisbon, 1993), shows that Elmina procured slaves from Benin and São Tomé, with 800 slaves taken annually from São Tomé to Elmina in the early 16th century (I, 160, 344–57). Ivana Elbl, 'The Volume of the Early Atlantic Slave Trade, 1450–1521', *JAH* 38/1 (1997), 31–75, shows that by 1520 most slaves from the Gulf of Guinea were bought by the settlers of São Tomé. John Vogt, *Portuguese Rule on the Gold Coast (1469–1682)* (Athens GA), 1979, 57–8, 70, describes the inter-regional trade between Benin, Elmina and São Tomé. See also Toby Green, *The Rise of the Trans-Atlantic Slave Trade in Western Africa, 1300–1589* (Cambridge and New York, 2012).

[6] Sidney Mintz, *Sweetness and Power: The Place of Sugar in Modern History* (New York, 1985), 26.

fourteenth century, and thus the Portuguese in the Atlantic merely borrowed from this tradition.[7]

Thus commercial plantations, had they been instigated in West Africa in the fifteenth and sixteenth centuries, could only have operated for the Atlantic market within a paradigm of slavery controlled by Europeans. Such a system would have been very difficult to implement in a political landscape controlled by Africans, not Europeans. This was in fact a key reason why commercial plantations located in Africa and directed towards the Atlantic economy never developed in this period. Such production would have required a colonial presence and, unlike in India and the Americas, the Portuguese were unable to achieve such a presence until their incursions in Angola in the late sixteenth century.

In Upper Guinea, this political factor was vital in preventing the development of commercial plantations for the Atlantic trade. In particular, there is the reality that European traders in Upper Guinea functioned as guests of host communities and had no political control in the region. As John Thornton has shown, after some initial violent incursions in the 1440s, the Portuguese realized very swiftly that they did not have the military wherewithal to defeat, let alone conquer, Upper Guinean societies.[8] Where trading outposts with a heavy European presence were established, as at Bugendo in the early sixteenth century and Cacheu in the late sixteenth century, this always depended on alliances with Upper Guinean lineages and subordinated resident European and Caboverdean traders to African kings. At Bugendo, attacks on resident Europeans were frequent.[9] In Cacheu in the 1670s, meanwhile, whenever the king of the Pepel people wanted to impose his will he took control of the water source and the Europeans resident there had no option but to comply.[10]

In the final analysis, the place of slave labour in the construction of the Euro-Atlantic economic system was inseparable from the place of usable land. Slave labour in this system was valuable only where fertile land existed that could be easily controlled by Europeans. Such land did not exist in Upper Guinea, where the land was controlled by African lineage heads. Where it did exist in the African Atlantic, in São Tomé, commercial plantations were indeed developed in the early period.[11]

[7] Ibid., 27–9.
[8] John K. Thornton, 'Early Portuguese Expansion in West Africa: Its Nature and Consequences', *Portugal the Pathfinder: Journeys from the Medieval Toward the Modern World 1300 – ca. 1600*, ed. George D. Winius (Madison WI, 1995), 121–32 (see 122).
[9] See André Alvares d'Almada's account, in *Monumenta Misonária Africana: África Ocidental*, ed. António Brásio, 15 vols (Lisbon, 1952–1988) (MMAI), III, 305.
[10] Arquivo Histórico Ultramarino (AHU): Guiné, Caixa 2, doc. 48.
[11] See Gerhard Seibert, this volume, chapter 2.

Instead of looking at European plantations, therefore, this chapter concentrates on how and why Atlantic trade affected the ways in which Upper Guinean societies reshaped their productive systems, and on whether this reshaping saw agricultural exports substituted for the slave trade as a means to access valued trade goods. This was not a question of Africans producing sugar for export, which does not appear, as far as the evidence goes, to have been contemplated in this early period, but of the production of millet and rice for the provisions trade. What emerges from this is that there was a relationship between commercial agriculture and Atlantic slavery in this early period. While in subsequent periods some peoples of Upper Guinea could and did develop a trade in alternative products, especially in hides and beeswax, to secure access to trade goods from the Atlantic, the evidence for this process in the sixteenth century is somewhat different.[12] For some societies, commercialization of agricultural production was not directly connected to the slave trade and was solely a means to secure access to trade goods. However, such commercialization was always at least indirectly connected to the Atlantic slave trade. My suggestion here is that this pattern was analogous to that which had already been developed for the long-distance trans-Saharan trade, and thus that continuities were central in influencing how Upper Guinean societies reshaped their agricultural production following the onset of Atlantic trade.

Patterns of production in Upper Guinea and the connection to long-distance trade

On 4 August 1519 Alessandro Geraldini, a renowned humanist and theologian, set out from Seville for the New World having been appointed as the new bishop of Santo Domingo on the island of Hispaniola. Having called in at the Cape Verde peninsula (present-day Dakar), he continued southwards and spent some time among the Sereer people who inhabited the Saluum delta. There one of the Sereer's ritual leaders, called Naasamón by him, described how the kingdom was rich because of the large quantities of rice, palm-wine, fruits, cattle and the plentiful fish stocks.[13]

[12] On wax, see Michael W. Tuck, 'Everyday Commodities, the Rivers of Guinea, and the Atlantic World: The Beeswax Export Trade, c.1450–c.1800', *Brokers of Change: Atlantic Commerce and Cultures in Pre-Colonial Western Africa*, ed. Toby Green (Oxford, 2012), 285–304. On hides, see Toby Green, 'Further Considerations on the Sephardim of the Petite Côte', *History in Africa* 32 (2005), 165–183; Peter Mark and José da Silva Horta, *The Forgotten Diaspora: Jewish Communities in West Africa and the Making of the Atlantic World* (Cambridge and New York, 2011).

[13] Alessandro Geraldini, *Itinerario por las Regiones Subequinocciales* (Santo Domingo, 1977), 100.

Geraldini quotes Naasamón as citing the cultivation of rice as being the most important factor, and rice was certainly a well-established crop in Upper Guinea long before the start of Atlantic trade. Duarte Pacheco Pereira's *Esmeraldo de Situ Orbis*, written c. 1505, described how the Biafada people of the region around the Grande river, where Bissau is located today, cultivated much rice and millet.[14] Millet was also cultivated and sold at the Jolof markets near the Senegal river, according to Valentim Fernandes, another contemporary source.[15] The impression of long-standing cultivation which these three sources provide confirms the important picture which a new generation of historians such as Judith Carney and Walter Hawthorne have provided of the longevity of rice production in this part of West Africa, and of its long-term significance to the region's productive economy.[16]

This is also the picture which emerges from linguistic work on this area. In her book *Deep Roots*, Fields-Black uses glottochronology to argue that rice had been grown south of the Gambia river for centuries prior to the fifteenth century.[17] Her research suggests that an underlying knowledge-base related to production techniques in the tidal swamplands of this part of the African coast was harnessed to improved access to iron offered by Mande incomers, beginning the process of enhancing rice production, which accelerated with the still-greater access to iron which began with the onset of Atlantic trade. As this process of Mande migration towards the coastal rice-producing areas became most pronounced from the second half of the thirteenth century onwards, we could pinpoint this era as one when this enhanced production might have begun.[18] This is a similar conclusion to that of Judith Carney, who argued in *Black Rice* that Mandinka incomers expanded rice cultivation from here into the highlands of present-day Guinea-Conakry between the twelfth and the sixteenth centuries, and also that there was some connection between the emergence of Sahelian empires and the domestication of rice.[19]

Why should we consider this pre-Atlantic history of rice cultivation in order to understand the place of agricultural production in Upper Guinea in relation to early Atlantic trade? This pre-Atlantic history may offer an example of how the place of agricultural production as

---

[14] MMAI, I, 648.
[15] Ibid., I, 673. Marvin Miracle gives a strong argument to suggest that this '*milho zaburro*' may in fact have been maize imported from the Americas, although the first decade of the 16th century – when this account was compiled – would have been very early for this to have been so: Marvin P. Miracle, 'The Introduction and Spread of Maize in Africa', *JAH* 6/1 (1965), 39–55 (see 39–40).
[16] Carney, *Black Rice*; Fields-Black, *Deep Roots*; Hawthorne, 'Nourishing a Stateless Society'.
[17] Fields-Black, *Deep Roots*, 30.
[18] On the Mande migrations, see Green, *Rise of the Trans-Atlantic Slave Trade*, chapter 1.
[19] Carney, *Black Rice*, 39–40.

a commercial export in the early Atlantic era was not revolutionary, but rather was analogous to pre-existing trends relating to long-distance trades. For prior to the fifteenth century agricultural production was probably also connected to the long-distance caravan trades across the Sahara, as well as offering the cultivation of a subsistence crop.

Some evidence for this emerges in the writings of some of the observers mentioned above. We saw there that millet was for sale at Jolof markets near the Senegal river, for instance. Such crops formed part of the commercial exchanges of the region, therefore, and it is interesting to ask who would have bought such supplies. It is likely that at least some of this marketized production would have found its way into the caravans plying the extensive trans-Saharan trade in the pre-Atlantic era. The western route from Oualata to Sijilmassa in southern Morocco brought significant numbers of North African traders to this part of West Africa, as is evidenced through the accounts of Cadamosto of the court of Damel of Cajor in the 1450s and the fact that Diogo Gomes met a trader from the Algerian town of Tlemcen on the Upper Gambia in the late 1440s.[20] Clearly, these long and arduous journeys required food for the return journey, and moreover this food was needed not only for the traders but also for the slaves who also travelled in these caravans. According to Arabic sources from the tenth century cooks from sub-Saharan Africa were highly prized by North African traders at Awdaghust, and travelled with the caravans of traders.[21] Ibrahima Seck believes that the word 'couscous' may itself have a Jolof origin, which may suggest at least some small trade of foodstuffs north across the Sahara.[22] It is of course uncertain whence this millet came for sale into the Saharan trade, but given that we know that trading networks crisscrossed the wider region long before the inception of Atlantic trade, such export of surplus may have influenced societies further south from the Senegal river; certainly, early seventeenth-century sources indicate that there was a trade in kola nuts from Sierra Leone into the trans-Saharan networks, and therefore we know that agricultural exports from this more southerly region were purveyed along the routes of the western caravan trade, even if kola nuts themselves were more valued and highly prized and easier to transport in bulk than cereal crops.[23]

Thus it is important to remember that long before the Atlantic era

---

[20] On Gomes, see MMAI, I, 195; on Cadamosto, ibid., I, 326–7.

[21] Ibrahima Seck, 'The French Discovery of Senegal: Premises for a Policy of Selective Assimilation', *Brokers of Change*, Green, 149–70.

[22] Ibid.

[23] Biblioteca da Ajuda (BA): Códice 51-IX-25, fol. 122v: kola nuts are sent '*por muito proveitosa mercadoria pa Berberia* [as a very profitable commodity to Barbary]'.

there was a long-distance trade connected to Upper Guinea which, as with the trans-Atlantic trade, required food to sustain both traders and slaves, food which must have been produced as a surplus and, on the evidence of Fernandes, was commercially available in the markets of the region. Such considerations support Carney's view that the development of rice production in Upper Guinea was central to the expansion of Mande power under Mali from the twelfth century onwards. Clearly, without the ability to create a surplus with which to feed the caravan traders and the slaves they took with them to North Africa, such a trade would not have reached the proportions that it did. This pre-history, linking agricultural production in Upper Guinea to a long-distance trade connected to slavery, was very important. For it meant that when an analogous trade took shape in the Atlantic world in the late fifteenth and early sixteenth centuries, the requirements of that trade for surplus agricultural produce were nothing new.

Early transfers of agricultural surpluses into the Atlantic trading networks

The first signs of this Atlantic transfer of a productive surplus occurred with the colonists of the Cabo Verde Islands. These were settled by the Portuguese in 1462, and quickly became central to the Portuguese project in West Africa.[24] The intensification of the trade in Upper Guinea in the early sixteenth century was partly facilitated by the settlement of the islands and the more frequent exchanges which this settlement facilitated. For it took much less time for Caboverdeans to travel to and from the African coast than it did traders from Portugal and thus, as Trevor Hall showed, Caboverdeans very soon came to dominate in the Atlantic trade of Upper Guinea.[25]

But the Caboverdean islands were arid. Unlike the Canaries, Madeira and São Tomé, they could not support commercial sugar plantations because they did not have the climate for them. They were what António Correia e Silva has called 'an insular Sahel'.[26] There was a transitory attempt to grow sugar there but by the early sixteenth century this had failed and the crop had collapsed.[27] Moreover, the islands' economy could only be developed through the importation of

[24] See esp. Green, *Rise of the Trans-Atlantic Slave Trade*, chapter 3.
[25] Trevor P. Hall, 'The Role of Cape Verde Islanders in Organizing and Operating Maritime Trade Between West Africa and Iberian Territories, 1441–1616' (2 vols, unpublished PhD thesis, Johns Hopkins University, Baltimore, 1992).
[26] António Leão Correia e Silva, *Histórias de um Sahel Insular*, (Praia, 1996).
[27] Stuart B. Schwartz, *Sugar Plantations in the Formation of Brazilian Society: Bahía, 1550–1835* (Cambridge/New York, 1985), 12–13.

agricultural produce. Islanders were only permitted to trade produce which had been raised on the islands, but this required a growing population for which the islands' ecology and rainfall were too precarious to offer a reliable source of food. Moreover, had all this food been grown in the archipelago, the amount of land that would have needed to be set aside for this purpose would have made the economic potential of the islands for raising other produce to trade to and from Africa severely limited.

In this situation, Caboverdean islanders turned to West Africa to supply staple foods. This position was established by the early sixteenth century at the latest, that is before the trans-Atlantic era had taken hold. By the 1490s, taxes to the *almoxarife* who managed the state finances on Cabo Verde were paid in millet and rice, showing that these were being imported to the islands by this date.[28] As Maria Manuel Torrão has shown, of the fourteen ships officially logged as returning from Upper Guinea to Cabo Verde in 1514, ten brought millet or rice; in the following year 1515 eleven out of sixteen did so, and by 1528 all the fourteen ships returned with millet and seven of them also brought rice.[29] Clearly these are very important data for they show not only that, as the textual evidence we have looked at from Fernandes, Geraldini and Pacheco Pereira showed, millet- and rice-growing were well established in Upper Guinea by the early sixteenth century, and nothing new, but also that surplus agricultural production was already being hived off from the West African market into the Atlantic trade.

This pattern of food supply for Cabo Verde continued throughout the sixteenth century. When Martín de Centinera visited Cabo Verde en route to the Americas in 1572, he bought maize there which had probably been grown in West Africa before being sold on the islands.[30] Later, during the drought of 1609–11, which devastated Cabo Verde and contributed to the terminal decline in the archipelago's place in the Atlantic trading system, the islands' troubles were partially relieved through the transport of some supplies from the Mandinkas of the Gambia.[31] Moreover, it is of central importance that this agricultural surplus from Upper Guinea supplied both the regional and the global trades. In 1514, ship captains sailing between São Tomé and Lisbon

---

[28] *Archivo Histórico Portuguez*, ed. Anselmo Braancamp Freire et al., 11 vols (Lisbon, 1903–1918), I, 95.
[29] Maria Manuel Ferraz Torrão, 'Actividade Comercial Externa de Cabo Verde: Organização, Funcionamento, Evolução', in, *Historia Geral de Cabo Verde*, Vol. I, ed. Luís de Albuquerque, and Maria Emília Madeira Santos (Lisbon, 1991), 237–345 (see 265–7).
[30] Archivo General de Indias (AGI): Patronato 29, ramo 26.
[31] Sociedade da Geografia (SG), 'Etiópia Menor: Descrpção Geographica da Provincia da Serra Leoa' by Manuel Alvarez, folio 8v.

were instructed to call at Senegambia to provision their ships with millet, and it is quite probable that this custom was widespread until the latter sixteenth century when Portuguese settlement began in Angola and provisions began to be imported from Brazil, while the first plantations known as *arimos* were developed along the rivers north of Luanda.[32]

Thus when we look at the interconnection of agricultural production and the Atlantic economy in the sixteenth century, it is clear that the two were connected in Upper Guinea from the beginning. Although there were no commercial plantations on the Cabo Verde islands to connect them to the later plantation-mode of production of the Caribbean, the Caboverdean economy was a slave-oriented one, not only through the export trade to the Americas, as will be discussed later in this chapter, but also through the way in which slave labour was fundamental to all the productive activities which facilitated the economy in the first place, producing and tending the products which were used by Caboverdean traders in their exchanges on the African mainland; work was principally concentrated around cloth manufacture, the cotton for which was cultivated largely on the island of Fogo.

Throughout the sixteenth century, therefore, the Caboverdean economy was extracting produce from Upper Guinean agricultural systems and using it to feed its slave-oriented export system. While this may have offered some continuity from the preceding trans-Saharan pattern, however, it was a continuity which represented something new. This Atlantic dimension brought about additional pressures on local production systems, and the extra demand for food supplies may have accelerated changes in production techniques in Upper Guinea, since we know from Walter Hawthorne's work that the arrival of the Portuguese was connected with new tools used in cultivation.[33] Moreover, this system forged a direct link between procuring both 'surplus' produce and 'surplus' labour from Africa and the development in Atlantic islands of a slave-based productive system; this was a fundamental difference to what had occurred in the trans-Saharan trade, where the slaves had been predominantly female and their worth not tied to the productive systems of the Maghreb.[34]

---

[32] MMAI, IV, 76. On *arimos*, see e.g. Beatriz Heintze, ed., *Fontes Para a História de Angola do Século XVII, Vol. 1: Memórias, Relações e Outros Manuscritos da Colectânea Documental de Fernão de Sousa (1622–1635)* (Stuttgart, 1985), 114.
[33] Hawthorne, *Planting Rice*, 43–8.
[34] António de Almeida Mendes, 'Slavery, Society and the First Steps Towards an Atlantic Revolution in Western Africa (15th–16th Centuries)', *Brokers of Change,* ed. Green, 239–58.

The onset of the trans-Atlantic slave trade and the effects on production

The initial intersection of Upper Guinean agricultural exports and the Atlantic economy had an essentially localized context. Caboverdean traders inserted themselves within the pre-existing economic circuits of West Africa and represented, if anything, a continuity from the trans-Saharan trade rather than something abruptly new. Caboverdean traders were swiftly integrated into the cultural and economic space of the West African coast, and thus from the Upper Guinean perspective this initial trade of crops to the islands was probably seen as part of a regional trade.[35]

However, this picture was soon to change. The sixteenth century of course saw the onset of the trans-Atlantic slave trade, and as scholars have long known the initial focal point for this trade was the Caboverdean region.[36] In this part of the chapter, we see how the demand for agricultural exports from Upper Guinea expanded along with this trans-Atlantic dimension, and that the essentially local demand which the Caboverdean economy had initiated from an Atlantic perspective was thus translated into a more global one. In the succeeding and final part of this chapter we will then see how this may have related to changing agricultural practices in the Upper Guinean region, and the demand for trade goods from the Atlantic.

Examples illustrate the importance of agricultural exports from Upper Guinea in feeding the slaves who were sold into the sixteenth-century trans-Atlantic trade. Importantly for the argument here, all these cases emerge from documents from Latin America, which emphasizes the global nature of these influences on Upper Guinean agricultural exports by this time. The first case relates to Blas Ferreira, who was the agent of Duarte de Leão, contractor for Guiné in the 1550s and 1560s. In 1565, Ferreira was tried by the Spanish authorities for trading slaves to Cartagena de las Indias in the Nuevo Reino de Granada (present-day Colombia) without a licence. Ferreira described how he had been sent by Leão and his fellow contractor Antonio Gonçalves de Gusmão to the River São Domingos in 1563 to procure slaves. There they spent the rainy

---

[35] Torrão, 'Actividade Comercial Externa de Cabo Verde', 237, shows that neither Cabo Verde nor the Upper Guinea coast can be studied in isolation in this period from an economic perspective.
[36] Philip D. Curtin, *Economic Change in Precolonial Africa: Senegambia in the Era of the Slave Trade* (Madison WI, 1975), 13; Hugh Thomas, *The Slave Trade: The History of the Atlantic Slave Trade* (London, 1997), 117.

season, and procured provisions, before proceeding to the Sierra Leone region where they obtained 180 slaves. This was all according to the 'mode, use and custom of the contractors [i.e. Leão and Gonçalves]'.[37]

Ferreira's evidence demonstrates that by the 1560s at the latest – and probably considerably before – slaves were supplied with food for the Atlantic crossing from the Bugendo region. A later source, the account of Pierre du Jarric of the work of Jesuit missionaries in Upper Guinea in the first decade of the seventeenth century, gives some idea of how this trade operated. Jarric recounts how Portuguese traders went to Sierra Leone to buy kola and then traded this throughout Upper Guinea for gold, slaves and provisions, according to their particular requirements.[38] Most likely, therefore, Ferreira picked up provisions which had already been secured by agents of the contractors in Bugendo through such a regional system of exchange.[39] Moreover, the use of agricultural surpluses from Upper Guinea to feed the slave trade was a widespread phenomenon, as is confirmed by other evidence from the period. In 1566, for instance, Baltasar Barbosa was instructed to buy food for the slaves on his ship *Santiago* in Upper Guinea 'so that they are well treated'.[40] Twenty-two years later, in 1588, Juan de Narria loaded a ship in Upper Guinea with provisions including eight barrels of rice.[41]

The extent of the impact of this trade on Upper Guinean productive systems is implied also by the extent of the trade of the contractor, Duarte de Leão. We know that Leão was connected to a large contraband trade in slaves. When his finances collapsed in the 1570s, the Portuguese Crown complained that not only did Leão and his fellow-contractor Gusmão owe the Crown large sums after the end of their contract in 1570, but that 'they were not left with enough property in this Kingdom to pay what was owing'.[42] The money owed proceeded from 'slaves which were taken without registers or licences'.[43] Numerous cases were taken out against Leão and Gusmão in the years that followed for taking slaves without registers, and a letter of c. 1580

---

[37] AGI: Justicia 878, no. 2.
[38] Pierre du Jarric, *Histoire des Choses Plus Memorables Advenues Tant ez Indes Orientales que Autres Païs de la Descouverte des Portugais*. (3 vols, Bordeaux, 1610–1614), III, 369.
[39] It is known that Duarte de Leão had an extensive network of such agents in the region at the time: see Green, *Rise of the Trans-Atlantic Slave Trade*, chapters 5 & 7.
[40] AGI: Justicia 996, no. 2, ramo 3, fols. 12v–13r.
[41] AGI: Escribanía 2B, no. 3 (this case has no folio numbers). The general importance of African production for the provisions trade has recently been highlighted by Judith A. Carney and Richard Nicholas Rosomoff, *In the Shadow of Slavery: Africa's Botanical Legacy in the New World* (Berkeley CA, 2009).
[42] BA: Códice 49-X-4, 223r: '*lhes não fiquou fazenda neste Reyno q baste para paguamento do q devem…*'
[43] Ibid.: '*escravos q se levarão sem registros, nem licenças*'.

emphasized the large debts which Leão still owed to the Crown from the Caboverdean contract.[44]

Such evidence suggests that the contraband slave trade from Upper Guinea in the sixteenth century was high. In general, historians of the early Atlantic slave trade have underestimated the volume of this early trade. In 1969, the late Philip Curtin proposed an average of 1,098 slaves a year leaving Senegambia and Upper Guinea for America and Europe between 1526 and 1550.[45] There have been some moves towards a reconfiguration of the estimates in the past decade, with David Eltis suggesting a doubling of Curtin's estimate of 75,000 slave exports for the sixteenth century to 150,000.[46] The new Trans-Atlantic Slave Trade Database proposes a revised estimate of 196,940 slaves shipped to the Americas between 1501 and 1590 at an annual average of 2,188, of which 143,316 came from the part of Africa examined by this book, producing an annual average of 1,592.[47] However these estimates are still far too low, and there is indeed extensive evidence on the true extent of the contraband trade in the second half of the sixteenth century which suggests that Walter Rodney's 1970 estimate of an annual average of 5,000 slaves exported from Upper Guinea after 1550 is much nearer the mark.[48]

Clearly, this export trade had impacts on agricultural productivity and the export of surplus produce in Upper Guinea. If all these slave-ships required food to sustain their cargoes on the Atlantic crossing, as they did, then the new Atlantic slave trade represented a shift in the demand structure for agricultural produce. Some fairly detailed estimates of the volume of this trade are possible. Historians are generally agreed that one kilogram of grain per person per day is a reasonable estimate of the rations allotted to slaves during their periods of

---

[44] In addition to the aforementioned case from Cartagena, see also AGI: Patronato 291, fol. 145r, a case about slaves taken to Cartagena without being registered; BA: Códice 49-X-4, fol. 223r. For more details on this, see Green, *Rise of the Trans-Atlantic Slave Trade*, chapter 7.

[45] Philip D. Curtin, *The Atlantic Slave Trade: A Census* (Madison WI, 1969), 101.

[46] David Eltis, 'The Volume and Structure of the Transatlantic Slave Trade: A Reassessment', *William and Mary Quarterly*, 58/1 (2001), 17–46 (see 23–4); see also Rolando Mellafe, *Negro Slavery in Latin America*, trans. J.W.S. Judge (Berkeley CA, 1975), 72–3, who suggests doubling the ratio of slaves/tonnage of shipping allowed for by Curtin to account for the contraband trade. Curtin, *Atlantic Slave Trade*, 24 n.13, criticizes this because it is double the ratio permitted by the contracts. However, Mellafe's strategy brings us much nearer the reality than Curtin's approach.

[47] www.slavevoyages.org, accessed 13 March 2013. Much of this new research was carried out by António de Almeida Mendes: 'Esclavages et Traites Ibériques entre Méditerranée et Atlantique', and 'The Foundations of the System: A Reassessment on the Slave Trade to the Spanish Americas in the Sixteenth and Seventeenth Centuries', *Extending the Frontiers: Essays on the New Transatlantic Slave Trade Database*, ed. David Eltis and David Richardson (New Haven CT and London, 2008), 63–94.

[48] Walter Rodney, *History of the Upper Guinea Coast, 1545–1800* (Oxford, 1970), 98; for detailed supporting evidence, see Green, *Rise of the Trans-Atlantic Slave Trade*, chapters 6 and 7.

captivity and transport to the Americas.⁴⁹ Research from the account books of Manuel Bautista Pérez, a slave-trader in Upper Guinea in the 1610s, shows that he purchased just over 25,000 kilograms of millet locally to feed the between 300 and 400 slaves that he purchased, an average of roughly 70 kilograms per head.⁵⁰ Thus, with an annual export of around 5,000 slaves to the Americas, one could estimate the purchase of approximately 350,000 kilograms of millet from Upper Guinea. In addition to this, one would need to take account of the trade in grain to the Cabo Verde islands, which was certainly no less than 300,000 kilograms per year (and may have been considerably more), and also the export of other grains such as couscous which were also purchased as provisions for the slave-ships.⁵¹ Writing in the 1660s, Francisco Lemos Coelho made it clear that rice was exported in large volumes from the Gambia river in the Badibu region (although precise figures are not available), which means that these would have to be added to the millet exports to get an overall picture of the agricultural exports of the region; he added that the most productive villages in Upper Guinea produced approximately 330,000 kilograms of millet.⁵²

Thus the new data on the contraband slave trade in this era combined with what we know of the provisions required for the slave trade, make it clear that the rapid growth of the Atlantic trade meant that not a small quantity of surplus agricultural produce had to be hived off for the provisioning of slave-ships. This required increased agricultural productivity, something which was a characteristic of the birth of the Atlantic era in Upper Guinea, as the work of Carney, Fields-Black and Hawthorne on rice production has shown.⁵³ What we can now do in the final section of this chapter is begin to chart changes in the production cycles of Upper Guinean communities in tune with the new pattern of Atlantic slaving and the demand for agri-

---

⁴⁹ Miracle, 'Introduction and Spread of Maize', 43; Linda A. Newson and Susie Minchin, *From Capture to Sale: The Portuguese Slave Trade to Spanish South America in the Early Seventeenth Century* (Leiden and Boston MA, 2007), 82.
⁵⁰ Newson and Minchin, *From Capture to Sale*, 82.
⁵¹ On the Cabo Verde trade, Maria Manuel Ferraz Torrão, 'Rotas Comerciais, Agentes Económicos, Meios de Pagamento', *Historia Geral de Cabo Verde*, II, ed. Luís de Albuquerque and Maria Emilia Madeira Santos (Lisbon, 1995), 19–123 (see 36), notes that in the seven months between March and Sept. 1610 alone four traders from Santiago purchased 16,000 *alquieres* of millet from the Bugendo region (approximately 180,000 kilograms – hence my estimate of at least 300,000 kilograms for the year); while Newson and Minchin, *From Capture to Sale*, 82, cite the existence in Bautista Pérez's accounts of evidence on his purchase of couscous and rice.
⁵² Damião Peres, ed., *Duas Descrições Seiscentistas da Guiné* (Lisbon, 1953), 153.
⁵³ Hawthorne, *Planting Rice*; Fields-Black, *Deep Roots*. This documentary evidence confirms the link made by Carney, *Black Rice*, 69, between Atlantic slavery and surplus agricultural production in Upper Guinea.

cultural produce from the sixteenth-century Atlantic economy proposed thus far. While clearly these two processes were deeply connected, what remains to be discussed is whether, at this early stage, such increased agricultural productivity was yet an alternative to the Atlantic slave trade, or rather so deeply connected to it that the two processes could not yet be separated.

Cycles of production in Upper Guinea and the Atlantic trade

So far, we have seen two important developments relating to the volume of agricultural exports from Upper Guinea in the sixteenth century. The first was the development of a new society in Cabo Verde which required the import of crops from the African mainland to guarantee food supply. The second was the development of long-distance transportation of slaves to Iberia and, increasingly, to the Americas, which required the extensive export of crops from Upper Guinea to sustain the slaves on the Middle Passage. The evidence of the account books of Manuel Bautista Pérez, as cited above, has shown that slaves did receive reasonable rations of food on the Middle Passage, and thus we must conclude that both these developments considerably increased the volume of agricultural exports from Upper Guinea in the sixteenth century. The interesting question, therefore, becomes how Upper Guinean societies adapted to these new conditions to meet the rising demand for agricultural produce.

A first noteworthy observation is that it was a question of choice as to how much of this food was supplied by Upper Guineans. In the late sixteenth and early seventeenth centuries a series of droughts struck Cabo Verde and famine afflicted the islands. The three major episodes were in 1580–3, 1590–4, and 1609–11[54]: the one that began in the 1580s was so bad that the news reached Europe and alms were sent in the form of flour and grain.[55] However, as has been clear in this chapter, the droughts were not the sole cause of the famine which afflicted Cabo Verde for, in any case, the islands were not self-sufficient in food supplies. The droughts decimated the population of Caboverdean livestock from which hides could be procured for export and also the cotton plantations on Fogo from which the *panos di terra* were woven. This crucial economic impact severely prejudiced Cabo

---

[54] António Carreira, *Cabo Verde: Formaçao e Extinçao de Uma Sociedade Escravocrata (1460–1878)* (Porto, 1972), 191.
[55] Archivio Segretto Vaticano, Secretaria di Stato di Portogallo, vol. 1, fol. 408v. See also AHU: Cabo Verde, caixa 1, doc. 23, where the governor of Cabo Verde, Francisco Ruiz de Sequeira, mentions that millet was sent to the islands during the 1583 famine.

Verde's ability to exchange the islands' produce for goods on the African mainland and it was this which led to famine conditions on the islands. As the dry spell continued in the seventeenth century and the islands remained unable to secure a trading advantage on the African coast, the ability to procure sufficient food supplies remained weak and this contributed to the islands' decline. Thus it is clear that food supplies were procured through a process of commercial exchange and therefore that Upper Guinean societies took active decisions to seek to meet the new demand for agricultural exports which arose with the Atlantic trade. In this sense, it is clear that from the perspective of those Africans who participated in the trade, the commercial agricultural trade was seen as an internal productive stimulus, and not simply as an external stimulus.

Evidence for an increase in agricultural production can be derived not only from the evidence for the role of agricultural exports in feeding Caboverdean society and the slaves on the Middle Passage, but also from evidence of commercial and social changes in West Africa which appear to have begun at around this time. In this context, it is very important that the evidence of Blas Ferreira, mentioned above, claimed that Duarte de Leão usually obtained provisions from Bugendo. Such provisions resulted from two developments of the sixteenth century: the increase in commercial exchanges along the Upper Guinean coast, and associated socio-political changes which followed in the Bugendo region.

To take the case of commercial exchange in the first instance, it is clear that the increase in Atlantic-oriented trade, not only in slaves but also in wax, hides and ivory, increased the flow of commercial networks up and down the Upper Guinean coast. The importance of navigational craft in this is emphasized by the accounts of Baltasar Barreira from the first decade of the seventeenth century, who found his progress brought to a halt when forced to travel overland, as it was almost impossible for him to follow the overland paths in the region of Sierra Leone because of the thickness of the jungle.[56] Thus although there had long been trade along the coast, some of it carried out by waterborne craft, the sudden increase in shipping heralded by Atlantic traders accelerated the processes of exchange. As we have seen, kola from Sierra Leone was used by Portuguese traders to exchange for provisions further north, and the Caboverdean André Donelha also described how in the late sixteenth century rice for Bugendo was procured from the Nuñes river area.[57]

[56] Jarric, *Histoire des Choses Plus Memorables*, III, 424.
[57] *Monumenta Missionária Africana: África Ocidental: Segunda Série*, ed. António Brásio, 7 vols (Lisbon, 1958–2004), V, 153.

Thus some of the kola bought by the Portuguese in Sierra Leone was exchanged for provisions in the Nuñes river area, and likely elsewhere along the coast, and this was one source of the provisions used to feed the slaves who departed Bugendo on the Middle Passage.[58] Clearly, therefore, for those societies involved in this intra-regional trade which supplied agricultural provisions eventually used in the export slave trade, the relationship of their productive work to slavery was tenuous. Few of the slaves for the Atlantic slave trade of the sixteenth century came from the Nuñes area, and thus it is likely that the Baga and Cocolí peoples of this region did not see their productive work as connected to the demands of slaving but as a means of securing access to Atlantic trade goods. In this sense, agricultural exports were a strategic choice made by these peoples to secure access to new trade goods, a choice which for them was not connected to Atlantic slavery.

But another corollary of Ferreira's 1565 account of securing provisions in the Bugendo region is that this might suggest that it was also here, in the Casamance region, that increased productivity was also marked. Significantly, such a view is supported by the fact that there were important political and social changes in this era which may have facilitated an increase in production. As mentioned already, some of this increased production may have been facilitated through increased availability of iron and of iron-edged tools which helped communities to clear more land for agriculture. Certainly, iron was a key import to Upper Guinea as early as the fifteenth century, with European traders greatly accelerating the access to iron which had been opened up by Mandinka smiths, and the work of Hawthorne and others suggests that this iron was prized in part because of its usefulness to the manufacture of agricultural implements; but some of this increased production may only have been possible through developing new means of social production, through solidifying the use of age-grades in agricultural labour for instance, which Walter Hawthorne has shown was a key strategy to secure rice production among the Balanta of present-day Upper Guinea.[59]

As Eric Wolf showed, shifts in the production and distribution of surpluses have generally reflected political shifts in human societies.[60] In Upper Guinea, the development of age-grades presented a new system of production which emerged at the same time as this expansion of commercial agricultural exports, and was therefore connected to it. As Hawthorne showed, it offered a more efficient and produc-

---

[58] See Newson and Minchin, *From Capture to Sale*, 83.
[59] Hawthorne, *Planting Rice*, 161–3.
[60] Eric R. Wolf, *Europe and the People Without History* (Berkeley CA, 1982), 82.

tive means of cultivation. According to Wolf's thesis, the development of new social structures for production is accompanied by new forms of the manipulation of political authority, and thus it is of great significance that, in the region around Bugendo, the king of the Kassanké in this era was known to have greatly increased his power and to have created a new and more authoritarian political structure.

The Kassanké monarch of this era – 1560s–80s – was called Masatamba and his kingdom, Cassamansa was at that time the main purveyor of slaves from Upper Guinea into the Atlantic trade. The Caboverdean trader André Donelha wrote that Masatamba had sold ten to fifteen slaves for one good horse in the 1570s.[61] Donelha's near-contemporary, André Alvares d'Almada, wrote that many whites lived in Bugendo 'because of the high volume of trade', and that 'in this river of São Domingos there is a higher slave trade than in all the other [rivers] of Guinea'.[62] Something of the extent of Masatamba's slave trade was shown by the English captain Edward Fenton, who wrote in 1582 that Masatamba possessed 5,000 horses.[63] Even if these horses were not all procured in exchange for slaves, they indicate a heavy trade.[64]

Masatamba ruled the Kassanké at a time when the political structure of the kingdom was becoming ever more authoritarian. Almada wrote that whole families were sent into the Atlantic slave trade by Masatamba if found guilty of certain crimes while, according to Donelha, Masatamba had condemned a woman and her husband into slavery for a false accusation of rape.[65] The people who lived in Masatamba's kingdom were so scared of the consequences of mistreating Europeans that they would bring something dropped on a road to the royal palace the following day.[66] Such capricious enslavement for the Atlantic trade points to a new and more coercive social environment. Most importantly, this new political environment accompanied the development of new social means of production connected to the expansion of agricultural exports. In a political system where people could be enslaved for the slightest crime, and enslavement was growing as a response to the opening of the trans-Atlantic trade, resistance to

[61] MMAI,V, 141.
[62] Ibid., III. 3, 303–4, 307: '*muitos dos nossos, por causa do muito trato*'; '*Neste Rio de São Domingos ha mais escravos que em todos os outros de Guiné*'.
[63] E.G.R. Taylor, ed., *The Troublesome Voyage of Captain Edward Fenton, 1582–1583*. (Cambridge, 1959), 106. See also José Lingna Nafafé, *Colonial Encounters: Issues of Culture, Hybridity and Creolisation: Portuguese Mercantile Settlers in West Africa* (Frankfurt-am-Main, 2007), 80.
[64] The reliability of this figure is questioned by George E. Brooks, *Landlords and Strangers: Ecology, Society and Trade in Western Africa, 1000–1630* (Boulder CO, 1993), 233.
[65] MMAI, III, 293–4, & V, 140–1.
[66] Ibid.,V, 140.

new patterns of social production which made greater physical demands was dangerous. Thus the increase in agricultural exports which the Atlantic slave economy required was facilitated by the increasing social insecurity which Atlantic slavery itself had promoted, and by the political changes which accompanied both the new trade and the new modes of production.

Such an interpretation is borne out by the fact that we know that much of the millet bought for consumption both in Cabo Verde and on the Atlantic ships was grown in the Bugendo region. Four traders coming to buy millet for Cabo Verde in 1610 went to the São Domingos river region where Bugendo is located.[67] Cacheu was importing around 260,000 kilograms of millet and rice annually directly from Bugendo by 1615.[68] Moreover, as Newson and Minchin show, the major supplier of grain to the slave-trader Manuel Bautista Pérez in the 1610s, Nicolao Rodrigues, lived in Bugendo.[69] Clearly, therefore, not only was surplus grain and rice being traded from the Nuñes river area and elsewhere to the Casamance region in the sixteenth century for export, but the region itself was a prime region where this grain for export had begun to be grown.

When considering whether commercial agriculture was an alternative to the Atlantic slave trade in sixteenth-century Upper Guinea, therefore, what is central is the differentiation of the relationship of different peoples to the Atlantic world. In areas like the Nuñes river, where enslavement for the Atlantic trade was not commonplace, peoples could and did expand their production for the Atlantic trade to secure access to European trade goods. However, the transfer of this agricultural surplus was ultimately organized through Bugendo, where increasing production for export should not be seen as an alternative to the Atlantic slave trade but rather as connected to it. Here the export of surpluses was linked to Atlantic slavery through the activities of Masatamba, king of the Kassanké and the main exporter of slaves for the trans-Atlantic trade in this era. Moreover, the development of new social means of production was connected to the Atlantic slave trade through the political changes accompanying the development of a more authoritarian kingship under Masatamba.

It turns out, then, that in some contexts the opening to Atlantic trade had not only pressed forward technological innovations and the enhanced use of iron tools, but also social changes related to agricultural production. What we have seen is that changes to both labour

---

[67] Torrão, 'Rotas Comerciais', 36.
[68] SG, Alvarez 'Etiópia Menor', chapter 5. Alvarez describes the import of 400 *moios*, where 1 *moio* is equivalent to around 10 *alqueires*, and 1 *alqueire* to a little over 11 kilograms.
[69] Newson and Minchin, *From Capture to Sale*, 80.

organization and the question of labour itself were connected, as the trans-Atlantic slave trade developed in the early sixteenth century and affected Upper Guinea. While in some societies, the new use of surpluses for export was not directly connected to the Atlantic slave trade, there was always an indirect connection, since these surplus exports were subsequently used for feeding slaves on the Middle Passage or for export to Cabo Verde. As these exports expanded, it would be hard to separate them from Atlantic commerce, as Africa's agricultural exports were tied to the exports of labour and other commodities which came to characterize the Atlantic system and what Jean-François Bayart calls the extraversion which went with it.[70]

[70] Jean-François Bayart, 'Africa in the World: A History of Extraversion', *African Affairs* 99 (2000), 217–67.

4

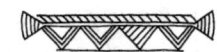

'Our indico designe'
Planting & processing indigo for export, Upper Guinea Coast, 1684–1702

COLLEEN E. KRIGER[1]

Indigo was a major trade commodity in antiquity and in the Indian Ocean trade, as one of the dyestuffs falling under the generic categories of 'spices' or 'drugs'. It continued to be important in the era of Atlantic trade as planters in the Americas vied with the Mughal Empire and other Asian indigo producers for a competitive edge in world markets. Dyeing with natural substances, which were often extremely variable in quality, was highly skilled work, often looked upon as an 'art' or a 'mystery' that few could master.[2] But with the chemical revolutions of the late nineteenth and early twentieth centuries, the bright lights of the laboratory and the instruments of scientific analysis displaced that aura of mystery. Invention of the synthetic dyes to which we are so accustomed today has almost entirely obscured the history of dyestuffs and their past technological and economic significance, especially for indigo. What was once so highly valued and in such great demand is now rarely recognized as such. This paper seeks to retrieve, examine, and better understand one small but informative part of this history – the attempt in the late seventeenth century by the Royal African Company of England (RAC) to establish plantations and processing of indigo on Africa's Upper Guinea coast.[3]

---

[1] Thanks to the editors, and also to Adam Jones for their useful comments on earlier drafts of this chapter.
[2] See, for example, Susan Fairlie, 'Dyestuffs in the Eighteenth Century', *Economic History Review (EHR)* new series 17/3 (1965), 488–510.
[3] The RAC indigo experiment has been noted only very briefly in the literature. Davies mentioned it in a single paragraph; Kup, noting that there were constant RAC experiments with indigo, devotes barely a page to it; and Rodney's one-page chronology is neither accurate nor complete. K.G. Davies, *The Royal African Company* (New York, 1975; reprint of first edition, 1957), 221; A.P. Kup, *A History of Sierra Leone, 1400–1787* (Cambridge, 1961), 93–4; Walter Rodney, *A History of the Upper Guinea Coast 1545-1800* (Oxford, 1970), 169–70.

Before the advent of Britain's cotton revolution in the mid-eighteenth century, indigo dyestuff was considered a desirable import primarily for her woollen industry. The seventeenth century saw woollens manufacture become a vigorously expanding export sector thanks in large part to new product lines – called 'new draperies' – presumably introduced by immigrant artisans from the Low Countries. Having the advantage of being lighter in weight than the standard broadcloths, and therefore more widely acceptable in warmer climates, these new woollens were very well suited to world trade. Clothiers produced them in a range of different widths, lengths, weights, and prices. So, as the reach of English trade grew, so did more workshops arise in England's countryside, producing plain, undyed 'new draperies' according to specifications. Production was concentrated in three main regions: the west – primarily Devonshire and Wiltshire, East Anglia and Yorkshire. Specialist dyers and finishers, especially in London, thus profited in turn from the bulk orders that came their way from the RAC and other merchant groups.[4]

English exporters nevertheless faced fierce competition for market share in seventeenth-century world trade. Quality mattered in addition to cost, and it was the Dutch above all who enjoyed a reputation for offering goods of a reliably high level of quality. Their textiles were especially admired. England's merchants would therefore have to raise the standard of quality of their export woollens while also keeping costs competitive. As indigo grew more available in the sixteenth and seventeenth centuries, it came to be recognized as far preferable to the European woad plant as a source of blue dye, mainly because of its much greater potency. Dyers found that they could use only four pounds of indigo for what would have required up to two hundred pounds of woad. The use of indigo thus made it possible to produce attractively coloured woollens in a more cost-effective way, and this was so not only for the clearest and deepest blues but for other colours as well. Dyers worked their special magic, creating their richest and most permanent hues by overdyeing, which involved steeping yardage in two or three separate vats of different colours in sequence. Hence there were shades of greens, greys, browns, violets and blacks that also called for indigo dye.[5] Charges for shipping processed indigo over long

---

[4] Florence Montgomery, *Textiles in America, 1650–1870* (New York, 2007); Davies, *Royal African*, 174–8.

[5] Fairlie, 'Dyestuffs', 491, 496. Specific dye recipes are notoriously hard to come by, as most dyers kept them as trade secrets. For a useful resource, especially on the combinations and sequences of overdyeing, see Elijah Bemiss, *The Dyer's Companion* (New York, 1973; republication 2nd edn 1815; 1st edn, 1806). A packet of manufacturers' samples of British export woollens, which was part of the cargo of a ship seized on a voyage to Madeira in January 1746, shows a palette of colours including black, three shades of brown, four shades of blue, and four shades of green, most if not all of which would have required an indigo dye bath: The National Archives, London (TNA), HCA 32/125.

distances from the tropical zones where it grew were therefore offset by its superior performance in the dye vat, and woad subsequently lost much of its market. Access to indigo thus developed into an important factor in textile production.[6]

With demand for indigo on the increase in England, seventeenth-century supplies of it came first from far-off East India and then also from the Caribbean. Already by the 1620s, indigo dye was a chief trade item for the newly established English and Dutch East India Companies. Crops of Asian indigo, primarily *indigofera tinctoria*, were brought to harvest and processed for export in Gujarat, northern India, and the hinterlands of the Coromandel Coast, with much of the best indigo destined for the Middle East. India's remaining supplies were therefore not enough to satisfy the steady demand for indigo in Europe, a condition which was partly ameliorated by the development of other sources, particularly in the Americas. Spain had led the way in the sixteenth century by establishing large-scale indigo plantations and processing works in New Spain and Guatemala, based on an American species of indigo, *indigofera suffruticosa*. Over the course of the seventeenth century, indigo became a cash crop in and around the Caribbean, but the more lucrative sugar industry began to eclipse it from mid-century onward. Jamaican planters had already been producing modest amounts of indigo when England captured the island in 1655, but even as production of the dyestuff rose over the next two decades the total exports from India and Jamaica combined with smaller amounts coming from the Leeward Islands were together still not enough to meet the home demand. Additional supplies had to be sought from either Spanish or French colonies. It was not until the 1740s, when South Carolina began to produce indigo for export, that Britain had herself another supplier.[7]

A plan to have RAC factors develop indigo plantations and processing works on the Upper Guinea coast as a more convenient (and hopefully regular) source of supply clearly made a good deal of sense, at least in principle. In the 1680s and 1690s, the Company had three main stations there: James Island at the mouth of the Gambia River; Bence Island in the estuary just to the north of the Sierra Leone peninsula; and York Island, situated between Sherbro Island and the mainland. All three were engaged in supplying slaves to Jamaica, Barbados, and the Leeward Islands, while at the same time they were

---

[6] For tables showing the drop-off in pastel (woad) exports from the Azores and imports of it into Exeter, England over the 17th century, see T. Bentley Duncan, *Atlantic Islands: Madeira, the Azores, and the Cape Verdes in Seventeenth-century Commerce and Navigation* (Chicago, 1972), 89–90.

[7] Dauril Alden, 'The Growth and Decline of Indigo Production in Colonial Brazil: A Study in Comparative Economic History', *Journal of Economic History* 25/1 (1965), 38–45; K.N. Chaudhuri, *Trade and Civilisation in the Indian Ocean* (Cambridge, 1985), 200.

also trading directly with London, sending ships back home laden with camwood (a red dyestuff), bees-wax, hides and ivory. All three stations were also located in areas where it was known that 'wild' and cultivated varieties of indigo did very well. Moreover, the timing was right: prices for indigo were rising during this period – doubling, reportedly, between 1687 and 1692.[8] Producing indigo on a large scale to add to the cargoes destined for London must have seemed a very promising prospect indeed.

In practice it was beset with many difficulties. I will be focusing on three of the most important ones: establishing and managing an indigo crop with limited supplies and expertise, the particularly challenging apparatus and technology of intensive indigo-processing, and problems in maintaining a willing and able labour force. Overcoming these difficulties was by no means easy, but nevertheless, by the end of the seventeenth century, some samples of processed indigo were sent to London where they were judged to be of very good quality. Who, exactly, was responsible for this apparent success – 'experts' hired from the West Indies or the RAC's West African slaves (gromettos) who did most of the work – cannot be determined. But several times after 1702, and just as the 'indico designe' was beginning to bear fruit, the island stations were attacked and plundered by various groups of private traders. In the end, out of all the serious obstacles encountered by the RAC, it was, in my view, the chronic insecurities caused by war with France and armed conflicts with 'interlopers' in this region that proved most insurmountable.

Planting indigo

In the mid-1680s, RAC officers in London were in the early stages of formulating their 'indico designe'. They began by gathering intelligence – samples of local indigo were sent to London from James Island, and Captain James Jobson described to them how the 'indico weed' should be made up for London dyers. What he described, however, was the West African method of processing the indigo leaf and turning it into a dyestuff, not the way it was done on a large scale in south Asia and the West Indies.[9] Moreover, it is not clear what kind

---

[8] *Calendar of State Papers, Colonial Series, America and West Indies, 1689–1692* (London, 1901), 718–19. In May of 1691 the RAC Committee of Goods remarked on the 'dearness' of indigo in the dyeing of sayes and perpetuanas, which were important export woollens: TNA, T70/127, 19 May 1691.

[9] TNA, T70/546, Homebound cargo laden on *The Delight*, 15 Dec. 1684; T70/50, fol. 10, RAC (London) to Cleeve (Gambia), 4 June 1686.

of indigo samples were sent. Two major species of indigo grew in West Africa, each classified in a different genus and each one suited to particular ecological conditions. The Asian indigo that had originated in India, *indigofera tinctoria*, was the most widespread, having been introduced in Africa during early Muslim trade if not before. It was planted from seed and cultivated, sometimes on a large scale, in the drier savannah regions of West Africa. In forested and well-watered locales another indigo plant, *lonchocarpus cyanescens*, grew in the wild and may have been indigenous to Africa. It had larger leaves than the domesticated *indigofera* and grew like a vine, climbing up trunks and along branches of trees for support. Young plants were sometimes tended and nurtured in the wild or transplanted for convenience. When planted from seed, it grew into tall shrubs seven or eight feet high.[10]

Establishing indigo production for export was far from easy, as the RAC soon learned.[11] Planting alone was more of a problem than they had imagined and it took at least ten years or more for their factors on the Guinea coast to successfully bring an indigo crop to harvest. One of the first problems they encountered was getting a supply of good quality seed. The Company wrote to their agent in Gambia that even though the plant grew 'naturally' (i.e. 'wild' indigo) in the vicinity, he should plant from seed in order to strengthen and improve it. They launched their indigo project in earnest in 1691 by hiring two brothers, Richard and William Bridgman, and sending them from the Leeward Islands to the Guinea coast. Richard in particular had been recommended as someone who had experience in raising and 'making' (processing) indigo in Jamaica, and he had with him a box of indigo seed. Initially, the plan was to sow indigo on James Island and also on York Island, for better security of materials, equipment, and management. But not long after his arrival, Bridgman had surveyed his limited options and chose the much larger Tasso Island in the Sierra Leone estuary as the most promising locale for the project. The factors were instructed to send a ship to Jamaica with 150 slaves, and request the factors there to send back to them supplies of indigo seed, tools, and anything else they needed.[12] But this part of the plan never got off the

---

[10] André Álvares de Almada et al., *Brief Treatise on the Rivers of Guinea*, trans. and ed. P.E.H. Hair, typescript, Department of History, University of Liverpool, July 1984 [c. 1594], Chapter 13, 3–4; Jean-Baptiste Labat, *Nouvelle Relation de l'Afrique Occidentale*, 5 vols (Paris, 1728), I, 74–7; J. M. Dalziel, *The Useful Plants of West Tropical Africa* (London, 1937), 243–9.

[11] Dunn erroneously describes indigo production for export as 'relatively easy' and 'foolproof'. Richard S. Dunn, *Sugar and Slaves: The Rise of the Planter Class in the English West Indies, 1624–1713* (Chapel Hill NC, 1973), 168.

[12] TNA, T70/50 fols 126–30, RAC (London) to Gibson (Sherbro), 22 Sept., 24 & 28 Oct. 1691; T70/11 fol. 165, Gibson (Sherbro) to RAC, 24 Jan. 1692. It is not clear whether the indigo seed was the American species (*indigofera suffruticosa*) or the Asian species (*indigofera tinctoria*).

ground, as Jamaica merchants petitioned in 1692 against the RAC plan to start up indigo works on the Guinea coast. Henceforth, the Company turned instead to the Leeward Islands for assistance.[13] The first planting in 1692 failed, Bridgman claiming the seed was bad. Another supply of seed arrived (it is not clear from where) but it, too, failed in the 1693 planting season. Seed was sent from Montserrat in 1694 in time for planting, accompanied by at least one more (unnamed) expert, and a crop was finally brought to harvest in the fourth planting season of 1695.[14]

These kinds of trials and errors are echoed in other accounts of indigo-growing. A description from Jamaica, recorded in 1687, noted the importance of inter-cropping indigo with maize – five rows of indigo, then one of maize – to offer protective shade and shelter. Indigo was deemed a 'tendor plant' that could easily be damaged by too much direct sun or exposure to the wind.[15] Jean-Baptiste Labat observed indigo works in the French Antilles, noting that the planting was particularly hard and tedious work that could easily be destroyed by infestations of caterpillars.[16] Another description, from South Carolina, spells out some of the problems Eliza Lucas experienced in learning to grow indigo in the early 1740s. She admitted that at the outset she had been ignorant of when to plant it and what kind of soil was preferable, but in her case she was fortunate to have her father in Antigua, ready to send her supplies of seed as needed. Her first planting, in 1740, died from frost, and the next year's crop failed again. At last in the third year her indigo plants survived to bear seed, which she then distributed to other planters.[17] Sensitivity of the crop to environmental conditions and their fluctuations clearly presented ongoing obstacles to planters interested in getting into the business of producing indigo. Having access to a range of climatic conditions was shown to be advantageous in the West Indies. An archaeological survey of indigo works that operated in Guadeloupe in the seventeenth and eighteenth centuries suggests that the preferred locations for growing

---

[13] TNA, T70/83 fols 57–8, RAC Court of Assistants, London, 13 Sept. 1692; T70/50 fol. 139, RAC to Corker (Sherbro), 13 Oct. 1692; *Calendar of State Papers*, 718–9.
[14] TNA, T70/11 fol. 167, Bridgman (Sherbro) to RAC (London), 10 April 1692; T70/50 fols 136–7, RAC to Gibson (Sherbro), 27 Sept. 1692; T70/588 fol. 77, arrival of *The Experiment* at York Island, Sherbro, 17 Feb. 1694; T70/50 fols 156–7, RAC to Corker (Sherbro), 16 Oct. 1694; T70/11 fol. 170, Corker (Bence Island) to RAC, 7 Oct. 1695.
[15] David Buisseret, ed., *Jamaica in 1687. The Taylor Manuscript at the National Library of Jamaica* (Kingston, 2008), 258.
[16] Jean-Baptiste Labat, *Nouveau Voyage aux Isles de l'Amerique*, 6 vols. (Paris, 1722), I, 281–3.
[17] Elise Pinckney, ed. *The Letterbook of Eliza Lucas Pinckney, 1739–1762* (Columbia SC, 1997; reprint of 1972 edition), 8, 16, 17, 22; H. Roy Merrens, ed. *The Colonial South Carolina Scene, Contemporary Views, 1697–1774* (Columbia SC, 1977), 145–6; Harriott Horry Ravenel, *Eliza Pinckney* (New York, 1909), 102.

and processing indigo were near the coast, where yearly rainfall averaged 1250mm or less.[18] Much heavier levels of rainfall than that, along with very low elevation and swampy terrain, help to explain why attempts to grow indigo on York Island were such a chronic failure. The roots kept rotting, causing the leaves to fall before the plant reached maturity. RAC officers back in London, however, insistently proffered remedies such as drainage systems 'for the better increase of that profitable commodity'.[19]

Processing indigo: from plant to dyestuff

Unlike tropical dyewoods, which were cut into billets and shipped to Europe for processing there, indigo had to be processed on site, near to where it was grown and harvested. In that respect indigo production was more like tobacco or sugar, and indeed, one finds that indigo was often a start-up crop that planters in the Caribbean used to build up the necessary capital for investing in sugar production.[20] It presented its own special challenges. Harvesting and processing indigo intensively for global trade required investments in equipment and the construction of large sturdy cisterns where the extraction of the dyestuff actually took place. Most importantly, turning indigo leaves into an easily transportable and workable dye substance is a complex procedure, the chemistry of which was not fully understood until the early twentieth century.[21]

Throughout the RAC project on the Upper Guinea coast, there was an underlying tension between two contrasting indigo technologies: one, a West African indigo technology that had been widely practiced there for centuries; and the other, an intensive Asian technology relying on large-scale processing cisterns, that had been developed in India and transferred to the West Indies. It appears that the latter had also been established in the Cabo Verde Islands sometime in the sixteenth century. André Álvares de Almada, an Afro-Portuguese merchant from the islands who plied the Guinea coast during the 1560s, describes several ways of making indigo that he had observed

---

[18] Tristan Yvon, 'La Production d'indigo en Guadeloupe au XVIIème et XVIIIème siècle ou l'archéologie d'une des premières industries du Nouveau Monde' (http://medieval-europe-paris-2007.univ-paris1.fr/T.Yvon.pdf Accessed 14 March 2013), 4.
[19] Christopher Fyfe, *A History of Sierra Leone* (Cambridge and London, 1962), 4; TNA, T70/51 fols 65–6, RAC (London) to Coats (Sherbro), 9 July 1700.
[20] Dunn, *Sugar and Slaves*, 126, 128, 168–71.
[21] See K.K. Trivedi, 'Innovation and Change in Indigo Production in Bayana, Eastern Rajasthan', *Studies in History* 10/1 n.s. (1994), 55–6.

at home and in his travels.[22] His account indicates that planters on the Cabo Verde Islands had adopted the intensive large-scale indigo technology, but whether it was transferred there directly from India or from the West Indies or other Atlantic islands is unclear. What is most noteworthy is that Álvares de Almada associates it with the islands only, and not with the mainland.

A careful analytical reading of Álvares de Almada's brief and compressed account makes it possible to separate out the several different practices he observed and, in particular, to appreciate the distinctive features of the West African indigo technology. He states that domesticated indigo, *indigofera tinctoria*, was cultivated on the Jolof coast, between Cape Verde and the Senegal River. At the proper time, before the plant went to seed, workers gathered and crushed the leaves, formed the paste into small balls, and dried them. When a dyer wished to make an indigo vat dye, he or she would take these balls of semi-processed indigo and place them in a pot with water where after some days it would be fermented. The account omits the critical phase of stirring and beating, which precipitates the dyestuff itself, but continues on with the addition of salt which helps to create an effective vat dye. All of this was done in a large pot. Álvares de Almada goes on to describe indigo dyestuff circulating as a commodity currency farther south on the coast, between the Casamance and Nuñes rivers, though in this case the source of indigo was the 'wild' variety, *lonchocarpus cyanescens*. Here too, workers gathered and crushed the leaves, forming the paste into medium-sized loaves that were then wrapped in leaves for trade. The description of how the vat dye was prepared follows along the same lines as for the Jolof coast. But buried in the account, and easy to overlook, is also a short digression about the indigo dye made on Santiago Island. Large quantities of domesticated indigo ('true indigo') were grown there and processed into indigo 'tablets' – the characteristic result of the intensive large-scale indigo technology. Islanders exported some of this dyestuff to Seville and Cadiz, at least in the late sixteenth century.[23]

Unfortunately, the account does not include a full description of intensive indigo processing as it was carried out in the Cabo Verde Islands. Other sources indicate that it was neither well-established nor widespread there and that it coexisted alongside the West African

[22] Álvares de Almada *Brief Treatise*.
[23] Ibid., chapter 2, 18–19 & chapter 13, 3–4. He describes the Nuñes indigo as growing like a vine, and having large leaves, which are characteristic features of *lonchocarpus cyanescens*. For photographic images of *indigofera tinctoria* and *isatis tinctoria* (woad), see Jenny Balfour-Paul, *Indigo in the Arab World* (Richmond UK, 1997), Plates 1(b) and 1(c). For photographic images of *lonchocarpus cyanescens*, see Anne-Chantal Gravellini and Annie Ringuedé, *Bleus et Ocres de Guinée* (Saint-Maur-des-Fossés, 2005), 24.

indigo technology. Under instructions from Portugal, the Governor of the islands tried to expand the production of indigo for export in 1704 and 1708 by encouraging the construction of large-scale processing cisterns but the outcome of this effort is not clear. Whatever the case, the volume of indigo exports seems not to have reached the hoped-for levels. Meanwhile, female slaves apparently worked indigo there in the familiar West African way, pounding the indigo in wooden mortars and drying the small loaves of indigo paste in the sun as was done on the mainland.[24] This method continued to be practised and was not displaced by the intensive large-scale indigo technology.

The major difference between these two indigo technologies is in the way the processing of the plant is carried out. In mainland West Africa, only the indigo leaves were processed. They were pounded into a paste in a wooden mortar and pestle and then formed into balls or loaves and dried for storing or trading. Thus the plant material was composted, or only partly fermented. In contrast, the Indian and West Indian technology, the one the RAC had in mind, took the entire indigo plant, stems and all, fully through fermentation in a large cistern. Workers then took the liquid solution through oxidation in another large cistern, a stage that would normally have been carried out in the dye bath proper. In other words, the plant material was completely transformed into the gummy dyeing agent itself. That substance was drained and then dried in moulds, becoming hard cakes or 'tablets'. Dyers using such dyestuff had to reverse these changes in order to create a workable dye bath.[25] This technical difference in processing the indigo plant, which might seem to be a minor detail, was actually quite significant and very likely one of the main factors that contributed to serious labour problems for the RAC.

The 'mystery' of indigo dye truly was mysterious, although that did not discourage self-styled experts from promoting themselves as capable overseers of indigo-processing. The RAC hired an unknown number of indigo specialists in the 1690s to forward their 'indico designe' on the Guinea coast, some at very good salaries. When Richard Bridgman signed on with the company in 1691, he was given 'great wages' as encouragement, starting at £50 for the first year and then doubling to £100 per year if he met with success. Another professed indigo expert, George Bainham, was hired in 1692 on a proposed four-year contract at £35, with conditional increases of £5 per year. The company sent other indigo specialists to Sherbro in 1694, one of them noted not by name but simply described admiringly as an 'understanding man in that mystery' from Montserrat.[26] However,

[24] Duncan, *Atlantic Islands*, 219–21.
[25] Balfour-Paul, *Indigo in the Arab World*, 47, 84–5.

such contracts did not necessarily pan out. During his first year on the Guinea coast, Richard Bridgman asked the RAC to send him a still so he could supply the factory with rum and sugar. He later left the company to become an independent trader and in 1697 married the successful Luso-African wholesaler Señora Isabella Gunn.[27]

The RAC made other significant investments in their 'indico designe'. In October of 1691 they wrote to their factor at Sherbro that they were purchasing an 80-ton vessel to ship indigo processing equipment to Guinea, and to be on hand to send slaves to the plantations in exchange for indigo supplies and expertise. The ship, *The Experiment*, left London on 9 February 1692, laden with parts and instructions for assembling the framed cisterns in which the indigo dye would be extracted. Their detailed instructions have apparently been lost, but from the surviving summary it is clear that RAC officers were not very well informed about the technical aspects and difficulties of the work. Among their high hopes for the plan was a presumption that the climate in Guinea would produce better indigo than what was grown in the West Indies.[28]

Records of voyages of *The Experiment* are disappointingly patchy. When it arrived in Sherbro is not clear, but in October of 1692 the RAC wrote to the factor there suggesting that the ship be sent to the Leewards laden with slaves and with a formal request that supplies of indigo seed be sent on the return voyage. Whether they did so is unlikely, given that *The Experiment* was listed in the York Island inventory, recorded 2 May 1693. Sometime during its extremely long layover, probably spent engaging in the coasting trade, Captain May, the ship's commander, died and was replaced by another RAC ship captain named Chilper. Exactly when just over a hundred slaves were embarked on *The Experiment* is not known either, but the ship arrived in Antigua in August of 1693 with 86 surviving slaves.[29] It then spent

[26] TNA, T70/50 fols 132–3, RAC (London) to Gibson (Sherbro), 9 Feb. 1692; fol. 140, RAC to Corker (Sherbro), 25 Oct. 1692; fols 156–7, RAC to Corker (Sherbro), 16 Oct. 1694. Bainham is listed as an indigo planter at Sherbro in Nov. 1692, and he was still on salary in Feb. 1693, but it is not clear from the records how long he worked on the project: T70/322, fols 132, 206, RAC *Home Journal*, 4 Nov. 1692 & 11 Feb. 1693.
[27] TNA, T70/11 fol. 167, Bridgman (Sherbro) to RAC (London), 10 April 1692; fol. 178, RAC to Bridgman (Sherbro), 14 Jan. 1698.
[28] TNA, T70/50 fols 132–3, RAC (London) to Gibson (Sherbro), 9 Feb. 1692.
[29] TNA, T70/50 fol. 139, RAC (London) to Corker (Sherbro), 13 Oct. 1692; fol. 146, RAC to Corker (Sherbro), 12 Sept. 1693; T70/588, inventory, Sherbro, 2 May 1693; T70/946, fol. 35, sale of 86 Negroes, Antigua, 20 Aug. 1693. The voyage of *The Experiment* from London to Sherbro and then on to the West Indies is listed twice in the Trans-Atlantic Slave Trade Database (TASTD), as voyages #9705 and #20780. Presumably, the reason for the confusion is the death of one captain and then his replacement by another, giving the impression that these were two separate voyages. David Eltis, Stephen Behrendt, David Richardson, and Herbert Klein, eds *Voyages: The Transatlantic Slave Trade Database* (Atlanta GA, 2008 and 2009) (www.slavevoyages.org accessed 14 March 2013).

the next four or five months in the Lesser Antilles, arriving back in Sherbro the following February carrying rum from Barbados, sugar and tobacco from Antigua, and four hogsheads of indigo seed from Montserrat. It is not known how many other Atlantic crossings *The Experiment* made, if any, before the April 1696 York Island inventory listed it as 'laid up rotten'.[30]

The Experiment did indeed serve its purpose of transferring an intensified indigo-processing technology to West Africa by delivering indigo seed, equipment, advice, expert managers, artisans, and labourers to the Guinea coast at various times between 1692 and 1696. But when, exactly, the first attempts to process indigo occurred, and where, is not clear. The earliest reference in the sources to the construction of great cisterns for intensive indigo processing comes in 1687, with complaints from the agent at Sierra Leone that he was having trouble cutting planks for making them. RAC records list unassembled equipment for constructing cisterns in the May 1693 inventory for Sherbro, but not for the Sierra Leone factory. By June of 1697 a sample of processed indigo dye was ready to be sent to London for their inspection, probably the product of the indigo works at Tasso Island. London responded with compliments on its very good quality, followed by a flurry of requests. The Sherbro factor was to offer the (un-named) workman in charge a lifetime service with the Company if he so wished. London also urged that the workman teach others the skills, and suggested that any ships coming to Sherbro from the Leewards carry over at least one more 'Whiteman skilled in making indico'. Obviously spurred on by the indigo sample, the RAC set about trying to calculate its potential profits. They requested information about the costs and inputs of making indigo and how many vats of the dye might be expected from one crop. 1698 was probably the peak year of the 'indico designe'. Inventories for both Sherbro and Sierra Leone listed cisterns and other indigo processing equipment, along with 246 lbs of indigo cakes. Factors shipped this batch of indigo to London in April of 1699, along with 88 tons of camwood and 400 small sugar-canes.[31] So in this relatively short period of fifteen years, from 1684 to 1699,

[30] TNA, T70/588 fol. 77, Arrival of *The Experiment*, York Island, 17 Feb. 1694; T70/589 fol. 1, inventory, York Island, 1 April 1696. The voyage of *The Experiment* from the Lesser Antilles to Sherbro is not included in the TASTD.
[31] TNA, T70/11 fol. 139, Case (Sierra Leone) to RAC (London), 1 Aug. 1687; T70/588, inventory, York Island, 2 May 1693; T70/11 fol. 172, Corker (Sherbro) to RAC, 21 June 1697; T70/589, Journal E, fol. 22, homeward goods laden on *The Companion*, York Island, 21 June 1697; T70/50 fol. 176, RAC to Corker (Sherbro), 21 Dec. 1697; T70/590 fols 3, 9, 12–13, inventories, York Island and Sierra Leone, 22 Oct. 1698; fol. 170, revision of 12 April 1699 accounts, York Island, 1 June 1700.

the 'indico designe' did produce some promising results in the form of processed indigo dye, at least twice.

Major gaps in the company records have imposed severe limits on our knowledge of what happened next. Over the following several years, the correspondence from London to the forts becomes increasingly desperate in tone, as they urged factors to send supplies of a variety of tropical products. Acknowledging problems with 'root rot' in the indigo fields, cotton was suggested as a possible commodity to grow or purchase, along with locally spun cotton yarn. By late 1702, the RAC suggested that the factor at Sierra Leone buy or hire workers skilled in spinning and weaving cotton and that their products be sent to London. In addition to indigo, they expressed interest in supplies of ginger, sugar-cane, pepper, spice, gum, trees and drugs, insisting that in the garden on Bence Island or at Tasso Island they 'set, sow, and plant all things that might be improvable' for trade.[32]

Technology, labour and economies of scale

Although the 'indico designe' achieved some success, it is not possible to determine how and by whom that success actually came about. Indigo was apparently grown from seed, but what variety, how, and how much is unclear. It would not be surprising if some of the indigo that was processed into dye was in fact not grown by the company at all but was the local 'wild' indigo – the RAC in London had themselves suggested on at least two occasions that their factors try and transplant 'wild' indigo or simply gather up as much of it as they could.[33] Similar questions surround the extraction process. It appears that workers assembled equipment for processing indigo on Tasso Island and perhaps also on York Island, and that at least one skilled manager was able to produce some limited quantities of indigo dyestuff, but just what the apparatus looked like or what sequence of procedures workers followed is not at all clear. Several reports on the project, sent to London in 1694, 1695 and 1696, are

[32] TNA, T70/51 fols 65–9, RAC (London) to Coats (Sherbro) and Lewis (Sierra Leone), 9 July 1700; fols 136–41, RAC to Freeman (Sierra Leone), 4 Aug. 1702; fols 147–8, RAC to Freeman (Sierra Leone), 9 Sept. 1702. In late 1703 the Sherbro factory was in disarray, the recently appointed factors Barker and Greenaway having been sent home after it was discovered they were embezzling Company goods by falsifying the account books: T70/14, Sherbro to RAC (London), 16 Feb. 1704.

[33] TNA, T70/50 fol. 139, RAC (London) to Corker (Sherbro), 13 Oct. 1692; T70/51 fols 42–4, RAC to Loadman (Sherbro), 2 Jan. 1700. Labat described how André Brüe of the Compagnie du Sénégal planted some wild indigo in the garden at Fort Saint-Louis but had to pull it up because it multiplied and spread so vigorously that it choked out the other plants. Labat, *Nouvelle Relation*, I, 77.

apparently lost.³⁴ In short, it is important not to overstate the company's success or to imagine that their production of indigo dye could have been sustained over the long term.

To get an idea of what the process probably entailed, an unusually detailed contemporary description of indigo dye extraction in Jamaica provides a vivid picture of both the apparatus and the sequence of procedures:

> Now whilst we have bin building our indigo work, the plant is grown to maturity and fully ripe for cutting, which as sone as they perceive they cutt it up cloce to the ground with sharp hooks which they have for that purpose, and carry it forthwith to the work, and fill the stupe cistern or vatee therewith; then with rammers of guacum the Negroas beat it, poun it and drive it close together, lain after laine until the cistern is ¾ full therewith. Then they fill the residue of the cistern with water and soe lett it stand in stupe for that day and night and untill the next day noon; then they tread it with their feet and take out the hearb, wringing it drie betwixt their hands and afterwards throw it to the dunghill, haveing take ye hearb clean out of the watter both stalks, leaves, etc.; they stir all their water together with poles, and soe opens the cock and lets it run into the drain cistern, it lokeing now of a blewish green colloure, and soe they fills up their stupe cistern with fresh hearb, in preparation to worke another vate. Now it being in the drain cistern they beat the water with flat poles (made for that purpose) untill it changeth its colloure and lookes cleare like brandy, then to know when 'tis of a due hight, they tast it, which when it tast strong and fiery on the tunge 'tis beat enough, and then they lett it stand for the indigo to settle all night. Now on the morrow when the indigo is setled to the botom like blew mudd, and the watter of the cistern looks cleare like water, then they open the cock of the drain cistern and lett it runn clean out gently into the water cistern, where 'tis keep for some time to save the second setling of the indigo, if there remains any behind.
>
> This being done they take all the mud out of ye drain cistern, and spred it in broad thin cakes in boxes which they have for that pourpose, and soe drie itt in the sun, and soe it becomes a pure shining blew hard gummy substance, called indigo, and thus they work it vatte after vate until they have finished their hearb...³⁵

One particularly valuable feature of this description is that it includes some of the specific signals that managers of the process had learned to look out for – indicators such as colour and taste – that told them when to move on to the next step. Such indicators were crucial pieces of information but they were not commonly known and are often

---

³⁴ TNA, T70/11 fol. 168, Corker (Sherbro) to RAC (London), 21 May 1694; fols 169–70, Corker (Sherbro) to RAC, 15 Jan. 1695; fol. 171, Corker (Sherbro) to RAC, 26 April 1696.
³⁵ Buisseret, *Jamaica in 1687*, 258–60.

missing from these kinds of accounts. Nevertheless, the above description is limited to what was thought to be correct apparatus and procedure at this particular time and place, and it is typically very schematic. Contemporary descriptions and especially diagrams of indigo works consistently present a relatively uniform, standardized, and ideal plan for extracting indigo dye. Studies by historical archaeologists are therefore essential for showing how adaptable and variable the technology could be in actual practice. Indigo works in seventeenth- and eighteenth-century Guadeloupe, for example, exhibited cisterns that were half the size noted in the Jamaica description, and often included two pairs of cisterns per site.[36] Such variations suggest that at least in some locales there was some experimentation in constructing the apparatus to further increase the intensity of production.

The main feature that made this indigo dye extraction technology an intensive one is the two-cistern design. Twentieth-century studies of this design as it had been employed in India explain the significance of dividing the process into two physically separate stages – the 'steeping' cistern and the 'draining' or 'beating' cistern, or what I call the 'split' process. This indigo-processing technology was well established in seventeenth-century India (and probably earlier), and it became the model for intensive indigo extraction in the West Indies. From what we now understand of the chemistry, a specific chemical change took place in each cistern. Steeping was actually a process of fermentation that transformed the colouring agent in the indigo leaves, indican, into a sugar (indoxyl glucose) that then dissolved into the water. How long this stage lasted would vary, depending on the ambient temperature.[37] After fermentation, the water solution was then drained off into the second cistern. There, 'beating' the liquid subjected the indoxyl to oxidation, which converted it into the insoluble indigo dyestuff. Particles of indigo settled into a sludge at the bottom of the cistern. Draining the clear water off left the dyestuff to be gathered up, drained, and dried into cakes. Having two separate cisterns meant that while indoxyl was being beaten in the second cistern, the next batch of indigo plants could be fermenting in the first one. Hence the process was continuous, and cut production time in half.[38]

[36] Yvon, 'La Production d'indigo en Guadeloupe', 8–12.
[37] This critical technical point about temperature was not fully appreciated until scientific studies of the technology were carried out. In practice, if the indigo were left to ferment too long, the quality of the dye would be compromised and impurities would be created that affected the strength of the final dye product. Imperfect fermentation probably accounts for at least some of the variations in quality of indigo dye that are so frequently noted in the sources. Trivedi, 'Innovation and Change', 56–7.
[38] Trivedi, 'Innovation and Change', 55–64. See also Balfour-Paul, *Indigo in the Arab World*, especially 84–9.

As practised in the West Indies, the two-cistern indigo extraction process depended upon supervised slave labour. Slaves cut the indigo plants, carried them to the indigo works, loaded them into the steeping cistern, and either pressed them down under the water with logs or battens or, as in the Jamaican description, crushed them down with their feet. Tending the fermenting indigo, which let off disgusting fumes and attracted swarms of flies, was easily the most unpleasant of the indigo processing tasks.[39] Beating the liquid in the second cistern was also unpleasant and taxing, requiring continuous labour adjacent to the stinking fermentation cistern. Other slaves collected the gummy dyestuff out of the receptacle and poured it into sacks for draining, while still others cut the drained and dried dye into cakes. Standard diagrams of indigo processing are extremely misleading if taken at face value, especially in the way they represent labour. Working conditions are deceptively sanitized, slaves appear healthy and obedient, and only sometimes are slaves shown being closely watched by an overseer.

Labour problems the RAC experienced with regard to their 'indico designe' may well be related to some of the more repellent aspects of the work. The Company gromettos, slaves owned by the RAC who were on salary, were the designated workforce for planting, harvesting, and processing the indigo, and some of them registered their attitude toward the project by running away. At the beginning of the 1692 planting season, when Richard Bridgman initiated the indigo works at Tasso Island, four skilled gromettos – all carpenters – deserted the company. Then in December of 1693 the company lost eighteen gromettos – fifteen men and three women ran away, making off with four muskets, four fowling shotguns, and two pairs of pistols. There is no record of them having been returned for bounty. Just when the 'indico designe' was making headway in extracting the dye in 1696, the RAC factor at Sherbro wrote to London that the relatively large size of Tasso Island gave opportunities for the gromettos to run away. He did not provide any numbers. The Company wrote back suggesting only that the workers be given encouragement to continue the project. The next year the same factor wrote that the local population in general was afraid of French attacks and were not bringing camwood down to the shore. He added that when French ships put in at Sierra Leone, the RAC gromettos ran away from the indigo works at Tasso, complaining also that their own methods of working with indigo were preferable. Henceforth the company tried to resolve the labour problem by recommending the use of gromettos from else-

---

[39] An 18th-century observer recommended that, because of the stench and flies, an indigo work should be located at least a quarter of a mile from any domicile. Bernard Romans, *A Concise Natural History of East and West Florida* (New York, 1775), 139.

where. They were of the opinion that slaves from the Gold Coast would be best, but the workers they ended up sending were not described. What was becoming increasingly evident was the importance of having a workforce that was completely dependent on the company, with limited options for escape.[40]

The gromettos' complaints about the indigo work were, from their perspective, quite understandable. Compared to the way indigo was customarily grown or gathered and processed in seventeenth-century West Africa, certain aspects of the labour regime the RAC was probably imposing with their intensive 'indico designe' could have seemed both unappealing and unnecessary.[41] Gromettos who were familiar with tending fields of indigo grown from seed would not have been surprised by the considerable labour involved in bringing indigo plants to harvest. They might, however, have preferred inter-cropping, if that was not RAC practice. Those gromettos who were used to the gathering of 'wild' indigo, an excellent source of dye, might have questioned why they had to grow it from seed in the first place. The West African technology of indigo-processing was done on a smaller scale, in smaller batches, and in several discrete stages. Workers collected fresh leaves, then placed them in a mortar and pounded them with a pestle into a gummy pulp. The final stage was to form the paste into balls or loaves which would either be dried and stored or dried and wrapped for marketing. In this case it was not necessary to constantly tend a large cistern of fermenting indigo. Completion of fermentation was done later at the time of the actual dyeing, and the fermenting indigo dye bath was covered, reducing the smell and the numbers of flies and other insects. What is especially striking about the intensive two-cistern process is the fact that the crushing of the whole indigo plants, stems and all, during steeping – either with logs, battens, or the slaves' own feet – was not only unnecessary but created impurities in the dyestuff that lowered its strength and quality.[42] In short, some of the labour inputs of the intensive extraction process could justifiably be viewed as rather extravagant and even wasteful compared to the West African process. A worker familiar with

[40] Some of them were from Allada, according to a report from Sherbro in 1698. TNA, T70/11 fol. 166, Gibson (Sherbro) to RAC (London), April 1692; T70/588 fol. 70, grometto runaways, York Island, 20 Dec. 1693; fol. 171, Corker (Sherbro) to RAC, 26 April 1696; T70/50 fol. 172, RAC to Corker (Sherbro), 28 Jan. 1697; T70/11 fol. 172, Corker (Sherbro) to RAC, 21 June 1697; T70/50 fol. 176, RAC to Corker (Sherbro), 21 Dec. 1697; T70/51 fols 9–10, RAC to Corker (Sherbro), 13 Sept. 1698; T70/11 fol. 174, Corker (Sherbro) to RAC, 26 Nov. 1698.

[41] Several references to the trade in semi-processed indigo along the Upper Guinea coast in the mid-17th century can be found in Francisco de Lemos Coelho, *Description of the Coast of Guinea*, trans. and ed. P.E.H. Hair, Typescript, Department of History, University of Liverpool, October 1985 [1684], chapter 4 no. 1, chapter 8 no. 26, & especially chapter 9 no. 15.

[42] Trivedi, 'Innovation and Change', 64–5.

both processes, comparing them and having a choice, would have good reason to prefer the latter.

Conclusions

The RAC experiment in indigo planting and processing can be viewed as either a partial success or a failure. What is more important, though, is what it tells us about how commerce along the Upper Guinea coast was viewed in the late seventeenth century, at least by the Company officials in London. For them, it was a trade in slaves and also a trade in other important commodities. The Company was willing to make considerable investments in support of these other commercial sectors. Slave-trading at this time was embedded in a broader conceptual field of commodities-production and -processing for export that was more variable than the familiar 'triangular trade' diagrams would suggest.

As the era of RAC monopoly waned, the company struggled to redefine itself. The 'indico designe' presents one example of their competing visions of possibility during this 'frontier' period in Atlantic trade. One part of the plan was to engage in a back-and-forth trade between the Upper Guinea coast and the Caribbean – slaves for technology transfer and logistical support. Slaves from Upper Guinea would be shipped in the Company ship *The Experiment* to the Leeward Islands in exchange for supplies, equipment, expertise, and labour that would be sent to build up and operate an intensive indigo-growing and -processing works on the Guinea coast. In the other part of the plan, processed indigo dyestuff would be shipped along with other goods such as camwood, bees-wax, and ivory to London, with returning shipments bringing the overseas commodities that were essential for trading operations. The plan is a product of its particular time, on the threshold of the enormous economic and technological changes of the eighteenth century.

The most probable immediate causes of its failure were the chronic insecurities of the time and the Company's severely limited options for locating their indigo works. James Island on the Gambia would have been a better site for indigo plantations, based on its yearly rainfall average, but it was a relatively small island and was particularly vulnerable to French attacks.[43] Bence Island was also small, and was seized temporarily by French warships in 1704; York Island, though better

---

[43] French ships captured and plundered James Island in 1695, and it was not resettled by the British until 1699. It was then captured and ransomed twice by the French, in 1702 and 1704, and the RAC abandoned Gambia in 1709. Davies, *Royal African Company*, 271–4.

protected, was swampy. But even if the Company had been able to acquire and maintain more secure locations for their 'indico designe' they still would have had to overcome two other major difficulties. First, as was the case with many merchants eager to trade in indigo, RAC officials were unfamiliar with the technology and the technological challenges they would be facing. Hence they were forced to rely on 'experts' who had their own reasons for claiming expertise. Some of them were knowledgeable, others not. Second, Company factors had limited control over labour – in addition to high mortality rates among the Company men, there were also instances of them deserting the Company or quitting and going home. Gromettos, too, fled when they could, especially for their own safety and also for preferable working conditions. In short, for all these reasons, the 'indico designe' was not sustainable.

What did continue on the Upper Guinea coast and elsewhere in West Africa was the other indigo technology – the processing of indigo leaves in a mortar and dyeing in relatively small-scale pots. It remained a skill well into the twentieth century, though synthetic substitutes eventually replaced it. In those very rare testimonies we have from master dyers, they insisted that their own dyestuff, produced from fermented indigo leaves, was far superior to synthetic indigo. Similarly, before synthetics, dyers often chose to work with freshly processed indigo rather than what was available through long-distance trade.[44] Quality of colour – as perceived by skilled artisans and their more discerning customers – was undoubtedly a factor contributing to the persistence of what might otherwise be considered an uneconomic technology. To that advantage we can now add what was a comparatively less onerous and less unpleasant labour process. One salutary finding of this study is the voice, though muffled, of the gromettos that comes through in the RAC sources. 'The natives insist on their own methods of working with indico.'[45] We also now have a better understanding of why there was such variability in the quality of processed indigo cakes and, by extension, how much variation there could be in pricing. Built in to the technology as it was understood and practised in the West Indies and in nineteenth-century India was an inappropriate and unnecessary use of labour that generated impurities in the final product. Despite all the well-meaning instructions, diagrams and advice, there was and is only one reliable proof of indigo's quality – the colour.

[44] Nancy Stanfield et al., 'Dyeing Methods in Western Nigeria', *Adire Cloth in Nigeria*, ed. Jane Barbour and Doig Simmonds (Ibadan, 1971), 23; Dalziel, *Useful Plants*, 245; Colleen E. Kriger, *Cloth in West African History* (Lanham MD, 2006), chapter 4.
[45] TNA, T70/11 fol. 172, Corker (Sherbro) to RAC (London), 21 June 1697.

5

'There's nothing grows in the West Indies but will grow here'
Dutch and English projects of plantation agriculture on the Gold Coast, 1650s–1780s

ROBIN LAW

The idea of the promotion of commercial agriculture in West Africa as a substitute for the export of slaves is familiar in the context of the Abolitionist movement, from the late eighteenth century onwards. But the alternative of employing slaves in cultivation in Africa, rather than transporting them to the Americas, existed from the beginning of maritime contacts between sub-Saharan Africa and Europe. Most if not all of the crops enslaved Africans were employed to cultivate in the Americas could also be grown in West Africa. Of the crops grown on American plantations, some were not introduced into West Africa until relatively late – notably coffee.[1] But others were either already established in West Africa when European traders first arrived in the fifteenth century, or were introduced there soon afterwards.[2]

Consider, first, rice, which became a major export from South Carolina – a variety of rice (*oryza glaberrima*) was indigenous to West Africa, and indeed was purchased by European traders for provisions for the Middle Passage.[3] Admittedly it was Asian rice (*oryza sativa*) which was cultivated for export in Carolina; although 'Guinea rice' was also introduced there, it was grown by slaves only for their own subsistence.[4] It might be supposed that African rice, although accept-

---

[1] Varieties of coffee, notably *coffea robusta*, were indigenous to West and Central Africa, but not commercially exploited before the end of the 18th century. The variety cultivated in the Americas was *coffea arabica* (originally, despite its name, from Ethiopia), which was introduced from Brazil to islands off the West African coast in the 1780s.
[2] Stanley Alpern, 'The European Introduction of Crops into West Africa in Pre-colonial times', *History in Africa (HA)* 19 (1992), 13–43; 'Exotic Plants of Western Africa: Where They Came from and When', *HA* 35 (2008), 63–102.
[3] See Toby Green, this volume, chapter 3.
[4] Judith A. Carney, 'African Rice in the Columbian Exchange', *Journal of African History (JAH)* 42/3 (2001), 377–96.

able for slaves' provisions, was not saleable in Europe. However, Asian rice seems to have been introduced, presumably by Europeans, into West Africa by the 1570s, when a white form of rice was reported being cultivated in Sierra Leone.[5]

Second, indigo, cultivated in the West Indies and South Carolina – the principal commercially cultivated species, *indigofera tinctoria*, originated from India, and was introduced into the Americas by Europeans, although there was also a form indigenous to America, *indigofera suffruticosa*. *Tinctoria* indigo had also been introduced into West Africa, overland from the Muslim world, and became widely cultivated in the savannah area of the interior, being already noted in the Senegal area in the late sixteenth century.[6] It apparently did not spread into the forest area of coastal West Africa, but there was another variety, *lonchocarpus cyanescens*, which was indigenous to that area, the leaves of which were harvested from wild plants.[7] Whether the Indian and American varieties were in any sense superior to the indigenous African form is unclear; but in any case, seeds from the West Indies were brought for propagation in West Africa by the 1690s, if not earlier.[8]

Third, cotton (originally from Asia, but long established in the Mediterranean) had been introduced from North Africa across the Sahara by the tenth century, and was clearly established through much of West Africa before the Europeans arrived.[9] The varieties of cotton cultivated in Africa (*glossypium arboreum* and *herbaceum*) were different from those grown in America, which were indigenous to that continent (*glossypium hirsutum* and *barbadense*), and of superior quality to the African cottons: but, as will be seen below, American varieties of cotton had also probably been introduced into West Africa by the early eighteenth century.

Fourth, the principal American plantation crop, sugar (also originally from Asia, but long established in the Mediterranean), had crossed the Sahara from North to West Africa by the twelfth century, but had apparently not yet arrived in the coastal area by the fifteenth century. The Portuguese, however, then introduced its cultivation on the Cabo

[5] Alpern, 'European Introduction', 21.
[6] André Alvares de Almada, *Brief Treatise on the Rivers of Guinea (c. 1594)*, trans. P.E.H. Hair (University of Liverpool, cyclostyled, 1984), chapter 2, paragraph 3, referring to a plant which was 'the same as that from which true indigo is made in our East Indies'.
[7] Another variety, *indigofera arrecta*, was indigenous to eastern and southern Africa, and introduced into West Africa, but perhaps only in modern times: see Colleen E. Kriger, *Cloth in West African History* (Lanham MD, 2006), 120–2 (and also personal communications).
[8] See Colleen Kriger, this volume, chapter 4, for the introduction of indigo seeds from the West Indies to Sierra Leone.
[9] Colleen E. Kriger, 'Mapping the History of Cotton Textile Production in Precolonial West Africa', *African Economic History* 33 (2005), 87–116.

Verde Islands and the island of São Tomé, the latter becoming (for a century or so) the leading sugar producer in the world, until overshadowed by the development of sugar cultivation in Brazil.[10] From the offshore islands, sugar soon spread to the West African mainland, being noted on the Gold Coast, for example, by the 1570s.[11]

Fifth, tobacco, indigenous to America and introduced from there through the trans-Atlantic trade, was being grown in West Africa by the early seventeenth century.[12]

Why, then, were these crops not cultivated in Africa for export to European markets? This is not a merely hypothetical question, since Europeans did in fact experiment with the establishment of plantations on mainland West Africa, employing African slaves locally to cultivate the same tropical crops which America supplied. Europeans were in general keenly interested in agricultural experimentation in West Africa. The European forts on the Gold Coast, and also at Ouidah on the Slave Coast (Bight of Benin) to the east, regularly maintained 'gardens', in which attempts were made to cultivate European and other introduced crops.[13] These were primarily intended to provide familiar provisions (vegetables and fruits) to locally resident Europeans, and to visiting ships, but there was also recurrent interest in the development of commercial crops, including sugar, cotton, indigo and tobacco. The focus of this chapter is on experiments in the commercial cultivation of crops in West Africa undertaken by the two most important European companies concerned in the African trade – the West India Company (WIC) of the Netherlands and the Royal African Company of England (RAC), in the seventeenth and eighteenth centuries.

These experiments have attracted only limited academic attention hitherto. The classic study of the RAC by K.G. Davies (published in 1957) mentions its experiments with West African plantations only very briefly (half a paragraph), and refers only to indigo cultivation on the Sierra Leone coast in the late seventeenth century, but not at all to the longer-lasting interest in plantations on the Gold Coast, in the first half of the eighteenth century.[14] For the Dutch case, Albert Van Dantzig's study of Dutch enterprise on the Gold Coast (1980) includes a more substantial but still brief (three pages) treatment of the WIC's experiments with plantations, but focuses narrowly on

[10] See Gerhard Seibert, this volume, chapter 2.
[11] Alpern, 'European Introduction', 15. In West Africa, the canes were chewed, rather than processed to make granulated sugar.
[12] Ibid., 30.
[13] Dominique Juhé-Beaulaton, 'Les Jardins des Forts Européens de Ouidah (Bénin): premiers jardins d'essai', *Cahiers du Centre de Recherches Africaines* (Paris) 8 (1994), 84–105.
[14] K.G. Davies, *The Royal African Company* (London, 1957), 220–1.

the years 1700–1702, with no reference to earlier or later initiatives.[15]

European projects of cultivation are also mentioned in some regional studies, albeit again very cursorily. Walter Rodney's study of the trade of Upper Guinea (1970) refers briefly (two pages) to the RAC's experiments with indigo cultivation in Sierra Leone.[16] K.Y. Daaku's study of the Gold Coast (1970) makes similarly brief reference (again, two pages) to Dutch and English projects of plantations between the 1650s and 1710s (but not later); and these earlier attempts are also mentioned even more briefly (a single paragraph) as an antecedent to post-Abolition projects of commercial agriculture in this region by Edward Reynolds (1974).[17] For the Slave Coast, my own study (1991) also mentions projects of plantations by the RAC at Ouidah between 1704 and 1724, but again a mere two pages.[18]

There have been very few studies dealing specifically with the European plantation projects. An article by myself (1991) was focused on a proposal made for the establishment of European plantations, employing slaves locally, by King Agaja of Dahomey, on the Slave Coast (probably inspired by the RAC's earlier experiments with plantations at Ouidah), which was delivered to London in 1731.[19] There seem to be only two previous studies which have treated early European projects of West African plantations more generally at any length: by Andrew Whalley, in a conference paper presented in 1996, which regrettably remained unpublished;[20] and by Joe Inikori, who devotes nine pages of his recent book on the economic impact of the Atlantic slave trade (2002) to the Dutch and English experiments after 1700, though he omits their earlier ventures in this field.[21]

This chapter first surveys the successive Dutch and English attempts to establish commercial plantations in West Africa, and then considers

---

[15] Albert Van Dantzig, *Les Hollandais sur la Côte de Guinée à l'Époque de l'Essor de l'Ashanti et du Dahomey, 1680–1740* (Paris, 1980), 141–4.

[16] Walter Rodney, *A History of the Upper Guinea Coast 1545–1800* (Oxford, 1970), 169–70.

[17] K.Y. Daaku, *Trade and Politics on the Gold Coast 1600–1720: A Study of the African Reaction to European Trade* (Oxford, 1970), 44–6; Edward Reynolds, *Trade and Economic Change on the Gold Coast 1807–1874* (Harlow and New York, 1974), 63. Some reference to early Dutch (though oddly, not English) plantation projects can also be found in Kwamina B. Dickson, *A Historical Geography of Ghana* (Cambridge, 1969), 75–6, 126–7.

[18] Robin Law, *The Slave Coast of West Africa 1550–1750: The Impact of the Atlantic Slave Trade on an African Society* (Oxford, 1991), 197–8.

[19] Robin Law, 'King Agaja of Dahomey, the Slave Trade, and the Question of West African Plantations: the Mission of Bulfinch Lambe and Adomo Tomo to England, 1726–32', *Journal of Imperial and Commonwealth History* 19 (1991), 137–63.

[20] Andrew J. Whalley, 'Dutch and English Attempts to Establish Plantations on the Gold and Slave Coasts', African Studies Association of the UK, Biennial Conference, University of Bristol, 1996. My thanks to the author for permission to draw upon this paper.

[21] Joseph E. Inikori, *Africans and the Industrial Revolution in England* (Cambridge, 2002), 383–92.

some of the general issues which they raise, including the reasons both for their initiation and for their failure.

The Dutch projects

There had been attempts by the Dutch at cultivation of crops for export in the seventeenth century, although these are not very well documented. The first Dutch West India Company had experimented with the cultivation of cotton on the Gold Coast as early as the 1650s: a report of 1658 refers to a 'cotton plantation' established at Axim and Butri, on the western Gold Coast.[22] The second West India Company in 1689 also formulated (but perhaps never implemented) plans for the cultivation of sugar, tobacco, indigo and cotton at Shama, again on the western Gold Coast.[23]

Further and more substantial efforts at cotton cultivation were initiated by Jan van Sevenhuysen, Director-General of the WIC in West Africa from 1696 to 1702, apparently from 1700 onwards.[24] In 1701 van Sevenhuysen reported that he had planted 8,000–10,000 cotton cuttings, which were growing well, and anticipated being able to supply cotton to Europe after about two years.[25] His successor as Director-General, Willem de la Palma, in 1702 reported that the planting of cotton had been 'reasonably successful', although an expanded labour force was needed 'in order to continue this work with vigour'.[26] In 1703 de la Palma actually sent a sample of cotton to the Netherlands; and the Company expressed satisfaction with its quality.[27] These plantations were established, again, initially on the western Gold Coast, at Axim, Butri, Sekondi and Shama; but in 1703 de la Palma reported that he had extended the cultivation of cotton to the hinterland of the WIC's headquarters at Elmina.[28]

---

[22] Report of Director-General J. Valckenburg, 10 July 1658, quoted in Daaku, *Trade and Politics*, 45, n.1.
[23] Letter from Elmina, 15 Nov. 1689, cited in Dickson, *Historical Geography*, 75.
[24] Daaku says from 1697: *Trade and Politics*, 44. But in 1702, the plantations were said to have been initiated 'more than a year ago': Albert Van Dantzig (ed.), *The Dutch and the Guinea Coast 1674–1742: A collection of documents from the General State Archive at The Hague* (Accra, 1978), no. 97, de la Palma, Elmina, 26 June 1702. The effort had begun by 1700, when a proposal to establish a school for the local children of Dutch officers stipulated that the curriculum should include instruction in 'the making of plantations such as those of cotton': ibid., no. 86, Resolutions of Council, Elmina, 10 March 1700.
[25] Van Sevenhuysen, Elmina, 30 May 1701, quoted in Van Dantzig, *Les Hollandais*, 142.
[26] Van Dantzig, *The Dutch*, no. 97, de la Palma, 26 June 1702.
[27] Ibid., no. 101, de la Palma, 10 Oct. 1703; WIC to de la Palma, 10 March 1704, cited in Van Dantzig, *Les Hollandais*, 143.
[28] Van Dantzig, *The Dutch*, no. 101, de la Palma, 10 Oct. 1703.

Having begun with cotton, interest soon shifted into the potential for the cultivation of other crops also, especially sugar. This is reflected in the published account of Willem Bosman, who had served as Chief Merchant under van Sevenhuysen at Elmina, and who refers at various points to the potential of establishing plantations: in particular, he says of the Ahanta (Butri-Sekondi) area: 'The sugar-canes grow here more and larger than elsewhere; so that I am not without hopes that a successful plantation may in time be here set on foot.'[29] De la Palma in 1702 also suggested the promotion of sugar cultivation, as well as cotton.[30] The intention was apparently not only to produce sugar for export to Europe, but also to use it to manufacture rum on the coast, to substitute for that imported from the Americas.[31] In 1705 de la Palma expressed his belief that, in addition to cotton, the production of sugar 'could be done on this Coast better and more successfully than in any colony of America', and also suggested that the cultivation of indigo should be taken up.[32]

In 1707 there was still confidence that cultivation of cotton, sugar and indigo was potentially viable, if sufficient labour and the necessary equipment was available.[33] The local chief agent of the RAC in 1709 reported that, although the Dutch were still trying to cultivate sugar, their attempted cotton plantation had 'come to nothing'.[34] But efforts at cultivation of cotton and indigo in fact continued in the following years.[35] In 1718 work on the indigo plantation was halted, but cultivation of cotton at Axim continued.[36] By then, however, disillusionment had evidently set in among the WIC's employees in Africa, if not in the Company in the Netherlands. When the Company in 1718 suggested the cultivation of pepper on the Gold Coast, the Director-General held that 'long experience has shown' that this was unlikely to succeed.[37] The Company, however, maintained its pressure, in 1720 urging its factors on the Gold Coast to promote the cultivation of cotton in the hinterland of Elmina.[38]

---

[29] William Bosman, *A New and Accurate Description of the Coast of Guinea* (London, 1705), 16. Bosman also suggested that sugar and indigo plantations could be established at Ouidah, on the Slave Coast: ibid., 394.
[30] Van Danztig, *The Dutch*, no. 97, de la Palma, 26 June 1702.
[31] At least according to a contemporary British report: The National Archives, London (TNA), T70/5, Sir Dalby Thomas, Cape Coast Castle (hereafter, CCC), 29 Nov. 1709: 'The Dutch are about planting sugar for the sake of making rum.'
[32] Van Danztig, *The Dutch*, no. 118, de la Palma, 5 Sept. 1705.
[33] Ibid., no. 138, Resolutions of Council, Elmina, 14 March 1707.
[34] TNA, T70/5, Thomas, CCC, 29 Nov. 1709.
[35] Daaku, *Trade and Politics*, 45.
[36] Van Danztig, *The Dutch*, no. 232, Resolutions of Council, Elmina, 25 April 1718.
[37] Ibid., no. 229, Resolutions of Council, Elmina, 4 March 1718.
[38] Ibid., no. 243, Secret Papers of Assembly of Ten of the WIC, 20 Nov. 1720.

The interest in cotton evidently persisted into (or perhaps was revived in) the second half of the eighteenth century. In 1752 it was reported that the Dutch 'for many years' had been employing 120 slaves to cultivate cotton at Axim, although the Company was critical of its local agents for 'neglecting' the project.[39] The WIC's records show that it was exporting small amounts of cotton from Axim and Shama between 1765 and 1783, though the Shama project was discontinued in 1772, and that at Axim had also been abandoned by the end of the century.[40]

A further attempt was also made to cultivate indigo, this time on the coast to the east, at Keta (in what is today south-eastern Ghana), in the 1730s. This was judged 'successful', at least to the extent that a sample of indigo was sent from there to Elmina in 1737, but the experiment came to an end when the Keta factory was destroyed in a local war in the same year.[41]

The English projects

Of the RAC's project to cultivate indigo in Sierra Leone, I will say little, since this is the subject of chapter four by Colleen Kriger. I will also leave aside the Company's interest in indigo and cotton cultivation at Ouidah on the Slave Coast in the early eighteenth century (which in any case produced no concrete results), since I have dealt with this in earlier publications.

Attempts at establishing plantations were also made on the Gold Coast. The earliest suggestion to this effect so far traced was made in 1703, by the RAC's factor at Winneba, on the eastern Gold Coast, who wrote 'about settling a plantation in those parts, which he says is fit for [sugar-]canes, ginger or cotton'.[42] This was presumably inspired by the efforts of the Dutch WIC to cultivate cotton and sugar on the Gold Coast, which had been initiated a couple of years earlier. The idea was taken up by Sir Dalby Thomas, who was sent out as the RAC's Agent-General at the end of 1703. He, however, concentrated initially on the cultivation of indigo – probably to replace the indigo project at Sierra Leone, which was suspended around this time. The cultivation of indigo was apparently done mainly at the Company's headquarters at Cape Coast Castle, but also elsewhere, including at

---

[39] TNA, T70/29, Thomas Melvil, CCC, 11 June 1752. My thanks to Silke Strickrodt for this reference.
[40] Dickson, *Historical Geography*, 126–7.
[41] Van Dantzig, *The Dutch*, no. 396, Minutes of Council, Elmina, 6 March 1738.
[42] TNA, T70/13, William Coles, Winneba, 27 July 1703.

Komenda to the west.[43] In 1706, Thomas reported that he had two acres of indigo growing, which were 'as good as any in the West Indies', and hoped to increase this to twenty in the following year, and declared that he was 'sure there's nothing grows in the West Indies but will grow here'.[44] The Company greeted his efforts with enthusiasm, even suggesting that West African indigo 'by due cultivation may suffice the whole world and may be afforded so cheap that nobody can make it but themselves'.[45] It also urged Thomas to supply samples of cotton, as well as indigo.[46] He responded with his characteristic enthusiasm in 1708 that, given an adequate labour force, 'you might have what plantations for provisions, sugar, indigo, cotton & ginger you think fit.'[47] The Company also suggested that tobacco might be cultivated on the Gold Coast (and also at Ouidah), 'to supply our ships for the use of the Negroes'; and Thomas, ever the optimist, opined that tobacco also 'might grow there', though there is no indication that this was ever in fact attempted.[48]

These projects appear, however, to have depended greatly on the personal initiative and commitment of Thomas. Some of his subordinates expressed scepticism about the venture: one questioned the degree of success which Thomas claimed, insisting that the supposed 'improvements in cotton, indico &c.' amounted to no more than 'trifles', apart from 'a little indico' cultivated at Komenda, and another complained that the Company's slaves would be better employed in the repair of its forts, which had been allowed to fall into 'a ruinous condition'.[49] When Thomas died in 1711, his surviving colleagues initially undertook to 'continue' the plantation project,[50] but nothing further seems to have been reported on the matter in the immediately following years.[51] An enquiry about prospects for cotton cultivation was apparently addressed by the Company to its factors on the Gold Coast in 1718, but the latter were unenthusiastic, implying that they would not be able to produce it any cheaper than private traders outside the Company could acquire

---

[43] Most of the references to Thomas's 'planting' activities in TNA, T70/5, are geographically unspecific, but one of his letters (15 Jan. 1708) noted that 'indigo thrives well at Commenda'. Suggestions for indigo cultivation were also made by Thomas to the RAC's factors at Ouidah on the Slave Coast, though to no practical effect: Law, Slave Coast, 198.
[44] TNA, T70/5, Sir Dalby Thomas, CCC, 6 June/1 Aug. 1706 The text (summarizing Thomas's letters in the third person) actually has 'there' (i.e. in Africa).
[45] TNA, T70/52, Court of Assistants of RAC, 12 Feb. 1707, quoted in Whalley, 'Dutch and English Attempts', 5.
[46] TNA, T70/52, RAC to Thomas, 26 June 1707.
[47] TNA, T70/5, Thomas, CCC, 24 Sept. 1708.
[48] TNA, T70/52, RAC to Thomas, 19 Oct. 1704; T70/5, Thomas, CCC, 21 May/9 June 1709.
[49] TNA, T70/5, William Hicks, CCC, 23 Dec. 1707; Charles Hayes, CCC, 13 Jan. 1707.
[50] Ibid., Seth Grosvenor & James Phipps, CCC, 24 Jan. 1711.
[51] But it should be noted that this correspondence is not recorded in full, but only in the form of brief extracts and summaries, so that any argument from silence is hazardous.

it, presumably from African suppliers: 'The cotton plantation would signify nothing without an exclusive trade.'[52]

The idea was resuscitated as part of an attempt to revive the RAC's fortunes, after several years of decline, from 1721 onwards. In that year the Company instructed its factors on the Gold Coast to use their 'utmost endeavours to improve the planting of cotton, indigo and pepper'.[53] Its local agents, however, were again unenthusiastic, expressing scepticism that either Europeans or local Africans would be interested in such cultivation, and stressing the limited potential for local production of cotton, in particular, by implication criticizing the Company's hopes as unrealistic: 'the whole year's produce at all your settlements on the Gold Coast was it to have been collected would not amount to two thousand weight [i.e. 2,000 lbs].'[54] The project was nevertheless still being actively pursued in 1730, when the Company's factors shipped 195 lbs of cotton to England, and hoped to increase output in the future.[55] The Company expressed its satisfaction and support, but by 1734 was complaining of the 'indifference' of its factors in West Africa to cotton cultivation, and thereafter the effort seems again to have petered out.[56]

A further attempt was made by the Company of Merchants Trading to Africa, which was established in 1750 to replace the RAC. Thomas Melvil, appointed as the new Company's Governor on the Gold Coast in 1751, began the cultivation of cotton and indigo soon after his arrival.[57] But when this came to the attention of the Commissioners for Trade and the Plantations in London, in 1752, they queried the propriety of these experiments, as being 'contrary to the known established policy of this trade' – the meaning of this will be returned to later – and ruled that such an innovation would require authorization from Parliament. They therefore directed the Company to instruct Melvil to 'suspend any further proceedings in this scheme until the sense of Parliament be known'.[58] Inikori assumes that Parliament did in fact rule against cultivation for export in Africa, but there is no evidence that the question was ever put to it.[59] However, the idea of commercial plantations on the Gold Coast was seemingly now defin-

---

[52] TNA, T70/6, Chief Merchants, CCC, 15 Aug. 1718.
[53] TNA, T70/53, RAC to Chief Merchants, CCC, 27 Feb. 1721. Inquiries about the feasibility of plantations were also addressed at this time to the Company's factors at Sierra Leone and Ouidah: Rodney, *History*, 170; Law, *Slave Coast*, 198.
[54] TNA, C113/274, Officers at CCC to RAC, 2 July 1722, quoted in Inikori, *Africans*, 387–8.
[55] TNA, T70/7, Chief Merchants, CCC, 25 May 1730.
[56] TNA, T70/54, RAC to Chief Merchants, 9 July 1730, 3 April 1733, 1 Aug. 1734.
[57] This was not, as assumed by Inikori, an initiative of 'private British traders': *Africans*, 390.
[58] *Journal of the Commissioners for Trade and Plantations*, vol. 9, 270–71 [14 Feb. 1752].
[59] Inikori, *Africans*, 391.

itively abandoned, until its revival in the context of the Abolitionist movement.

The character of the European projects

The available sources provide little concrete detail on the nature of these European efforts at commercial agriculture, and it should be stressed that the word 'plantation', which was regularly applied to them, in contemporary usage referred to farming of any sort, and did not necessarily imply large-scale enterprise. Nevertheless, it seems that the West African 'plantations' were in general conceived as projects of large-scale production, employing slave labour, on the model of the existing plantations in the Americas. The RAC on two occasions, in 1706 and again in 1721, did suggest that its own plantations, rather than being cultivated directly by its agents, might be divided up and rented out (to its European employees and African slaves, or to 'free natives') for small-scale production, but there is no indication that this plan was ever attempted.[60]

The land for these projects was presumably leased from (or at least occupied with the consent of) the local African authorities, but it does not appear that this presented a problem, at least on the Gold Coast. The RAC's local agents assured it in 1722, for example, that there was 'extent of ground enough that we can secure your Honours the property of'.[61]

The projects seem to have been mainly based on the exploitation of the varieties of crops available locally, though there were also attempts to introduce new species. The WIC, for example, shipped two bags of cotton seeds to Africa in 1710 – presumably of varieties cultivated in the West Indies.[62] The RAC's indigo project in the 1700s included both the improvement of local indigo and the importation of seeds from the West Indies;[63] though in the case of cotton, it seems that it was one of the locally available varieties that was used, rather than seeds imported from the Americas.[64] Both Dutch and English

---

[60] TNA, T70/52, RAC to Thomas, 14 Nov. 1706; C113/272, RAC to Chief Merchants at CCC, 7 Sept. 1721; T70/53, RAC to Chief Merchants, 23 Aug. & 18 Sept. 1723.
[61] TNA, C113/274, Chief Merchants, CCC, 2 July 1722. But the situation was more difficult at Ouidah, where it was reported that there was 'no room to plant or improve [cotton] there, the country being very populous': TNA, T70/7, Baldwyn & Barlow, Ouidah, 28 March 1723.
[62] Daaku, *Trade and Politics*, 45.
[63] See TNA, T70/52, RAC to Thomas, 19 Oct. 1704, 'the weed grows there in great plenty which may be improved when the workmen [from the West Indies] transplant the same & sow such seeds as they carry with them.'
[64] The Company, responding to complaints from Cape Coast about difficulties in cleaning the cotton, acknowledged that 'the cotton is not the West India cotton' and 'more difficult to clean': TNA, T70/53, RAC to Chief Merchants, 4 July 1723.

projects also involved the introduction of new technology, in the form of cotton-gins (to separate the seeds from the fibre) and sugar-mills (to crush the canes to produce the juice). The Dutch Director-General in 1701 requested the supply of cotton 'mills' (i.e. gins), and the WIC in 1704 arranged for the sending of 'the requisite tools' for cotton-processing from Curaçao;[65] in 1705 a request was made for a sugar-mill, though it is not clear if this was ever supplied.[66] The RAC likewise in 1721 undertook to send 'engines and utensils', including specifically cotton-gins, to its agents on the Gold Coast, although its local agents declared that those received were 'unfit for use' and 'will not work'.[67] The RAC also supplied instructions for the processing of indigo, according to the Indian technique involving the use of large cisterns for fermentation and oxidation.[68]

The labour for the plantations was drawn mainly from the existing stock of 'Castle slaves' employed locally, but additional slaves were also acquired. The Dutch Director-General in 1702 proposed to import 250 slaves from Ouidah for use in the cultivation of cotton, and further slaves were imported for this purpose in the following year.[69] For the English projects, Thomas in 1708 likewise suggested the importation of slaves from the Gambia and Sherbro, though it is not clear whether this was ever done.[70] This importation of slaves from other areas of Africa, it may be noted, was standard practice, since they were thought less likely to try to escape than ones recruited locally. The RAC in 1721 authorized its local factors to purchase 'such an additional number of slaves … as you shall judge necessary for that purpose', although it does not seem that they acted on this.[71]

Skilled labour was brought from the West Indies, also often in the form of slaves. The Dutch Director-General in 1701 requested the sending of '4 or 6 competent slaves', as well as cotton-gins, from the West Indies, since 'nobody here knows how to do this [ginning] properly'; and in 1704 a man called 'Black Peter', presumably a slave, was

---

[65] Van Sevenhuysen, Elmina, 30 May 1701, and WIC to de la Palma, 10 March 1704, quoted in Van Dantzig, *Les Hollandais*, 142–3.

[66] Van Dantzig, *The Dutch*, no. 118, de la Palma, 5 Sept. 1705.

[67] TNA, T70/53, RAC to Chief Merchants, 27 Feb., 12 Sept. 1721; T70/54, RAC to Chief Merchants, 9 July 1730; T70/7, Chief Merchants, CCC, 26 Dec. 1729, 6 Oct. 1731.

[68] A 'receipt [i.e. recipe] for making indigo' was sent to Cape Coast in 1721, and a later letter refers explicitly to the construction of cisterns: TNA, T70/53, RAC to Chief Merchants, 27 Feb. 1721, 4 July 1723; cf. also T70/27, Chief Merchants, CCC, 5 March 1724. The technique is described by Kriger, in this volume.

[69] Van Dantzig, *The Dutch*, nos 97, 101 de la Palma, 26 June 1702, 10 Oct. 1703.

[70] TNA, T70/5, Thomas to RAC, 24 Sept. 1708.

[71] TNA, T70/53, RAC to Chief Merchants, 27 Feb. 1721; the authorization was repeated on 5 Dec. 1723.

sent, together with the necessary 'tools', from Curaçao.[72] In 1712 the WIC likewise arranged for the sending of an 'indigo planter', apparently a European, from Curaçao to the Gold Coast.[73] In the English case, the RAC in 1704 directed some 'indigo makers' to be sent from the West Indies to the Gold Coast; and from 1721 repeatedly undertook to supply not only cotton-gins but also persons from the West Indies skilled in their use, although it is not clear that they ever did.[74] It is unclear whether any attempt was made to exploit existing African expertise, although it may be significant that Thomas in 1708, as noted earlier, proposed to import slaves specifically from the Gambia and Sherbro, which were centres of indigenous indigo production.[75]

It should be stressed that, although the foregoing account has concentrated on projects of European-managed plantations, both the Dutch and English companies were also interested in promoting commercial cultivation by independent African farmers living beyond their jurisdiction. The Dutch cotton plantations established from 1700 onwards were explicitly intended to serve, in part, as model farms, to encourage emulation by Africans: 'We may convince the natives, by our example, that the planting of cotton is profitable, and induce them to grow those plants for their own profit.'[76] Likewise the RAC in 1721 urged, not only the planting of export crops by its agents in West Africa, but also that they should 'encourage the Natives in doing the same'.[77] But it does not appear that this idea met with any greater success. In practice, in fact, the attitude of the RAC to African involvement was ambivalent. It instructed its factors to prevent knowledge of the process of cotton-ginning from falling into the hands of 'the natives', as this would enable them to sell their cotton to private traders, rather than to itself, and thought that the best way to ensure this might in fact be to restrict cotton production to the Company's own 'garden'.[78]

---

[72] Van Sevenhuysen, Elmina, 30 May 1701, and WIC to de la Palma, 10 March 1704, cited in Van Dantzig, *Les Hollandais*, 142–3.
[73] Daaku, *Trade and Politics*, 45: named as Jacob Van Munnikhuysen in Van Dantzig, *The Dutch*, no. 232, Minutes of Council, Elmina, 25 April 1718.
[74] TNA, T70/52, RAC to Thomas, 19 Oct. 1704; T70/53, RAC to Chief Merchants, 27 Feb., 12 Sept. 1721, 6 June 1723.
[75] On one occasion the RAC suggested, in relation to cultivation of corn, that if none of its factors had the requisite expertise, 'it's easy to buy negroes who do from those parts where it grows': TNA, T70/53, RAC to Chief Merchants, 18 Sept. 1723. But no similar suggestion has been traced in relation to cotton or indigo.
[76] Van Dantzig, *The Dutch*, no. 97, de la Palma, 26 June 1702: cf. Daaku, *Trade and Politics*, 44.
[77] TNA, T70/53, RAC to Chief Merchants, 27 Feb. 1721.
[78] Ibid., RAC to Chief Merchants, 12 Sept. 1721.

Motives

It should be stressed that these projects had no connection with any Abolitionist sentiment. Indeed, most of them were initiated at a time when there was still no significant public debate in Europe over the morality of the slave trade. By the 1730s, such a debate had begun to emerge, at least in Great Britain; but there is no evidence that projects of cultivation in Africa were in any sense a response to this developing controversy. King Agaja of Dahomey's proposal for the establishment of European plantations in his country, delivered to London in 1731, was interpreted as an Abolitionist measure by at least one contemporary British observer, but this is not supported by anything in the original records relating to it.[79] These early projects of plantations were clearly considered as a potential supplement to rather than a replacement for the slave trade. In some cases, indeed, as with the Dutch project of using locally cultivated sugar to manufacture rum on the Gold Coast and the British interest in tobacco cultivation, they were intended to produce commodities which could be used in the slave trade – rum to purchase slaves, tobacco for consumption by slaves in the Middle Passage. Nevertheless, in another sense, these projects can be seen as attempts to find alternatives to the slave trade, though the motivation for this was pragmatic, rather than humanitarian.

The earliest venture, by the Dutch in the 1650s, is poorly documented. But it is surely significant that it followed the failure of Dutch attempts to seize established sugar-producing areas in northern Brazil (occupied in 1630, but largely lost to a local rebellion in 1645, and completely evacuated in 1658) and on the island of São Tomé (captured in 1641, but retaken by the Portuguese in 1648). The idea of using slaves on plantations on the West African mainland might have seemed a rational response to these reverses; although in the longer run, the WIC opted rather to supply slaves to the American plantations of other nations.

With regard to the later experiments with African cultivation of export crops which both the WIC and the RAC undertook from the 1680s onwards, Inikori noted that these occurred at a time when both these Companies were finding the slave trade 'increasingly less profitable', because of the competition of 'private traders' operating in breach of their legal monopolies of trade.[80] In the case of the WIC,

---

[79] Viz. John Atkins, *A Voyage to Guinea, Brasil and the West Indies* (London, 1735), 121–2; see Law, 'King Agaja', 151–4.
[80] Inikori, *Africans*, 388.

Director-General de la Palma in 1702 explicitly stated that the plantation project had been taken up 'in view of the general poor condition of the trade'.[81] Inikori assumes that this refers to difficulties in the slave trade;[82] but study of the context makes clear that it was rather the decline of the gold trade which motivated a search for alternatives, including the establishment of plantations.[83] The problems in the gold trade were caused not only by competition on the European side, but also by the disruption of the supply of gold, by the escalation of warfare among African states in the hinterland of the Gold Coast from the 1690s onwards. In the longer run, the principal response of the WIC to the decline of the gold trade was in fact to concentrate energies on the slave trade, but in the initial stages the profitability of the slave trade was also in doubt, given substantial increases in the price of slaves, and this probably also reinforced interest in the idea of West African plantations, as Inikori suggests.

Likewise, with regard to the RAC's efforts at indigo cultivation in Sierra Leone, Davies suggested that these were taken up because the Company's trade in this region had proved 'disappointing', and represented 'an attempt to make the trade of this part of Africa more profitable'.[84] Here too, the existing trade which was judged unsatisfactory was not primarily in slaves, of which few were then exported from Sierra Leone, but rather in dye-wood, ivory, bees-wax and gum. However, the RAC's experiments with indigo cultivation in this region, as Rodney suggested, can be thought of as part of a more general policy of neglecting the slave trade in favour of alternative commodities, and thus as a quest for 'substitutes for slaves' which anticipated the 'legitimate trade' of the nineteenth century, albeit undertaken 'not for reasons of morality, but because the directors of the Royal African Company saw the commercial wisdom of transforming the region into one where the slave trade was absent'.[85]

The RAC's projects of cultivation of export crops on the Gold Coast from 1703 onwards, equally clearly, reflected difficulties in its established trade, in both gold and slaves. The Company had for some years been failing in competition with independent traders, and this became critical with the loss of the company's legal monopoly of the African trade in 1698. The Company in 1722 explicitly advocated plantations as part of a policy of promoting 'articles of the home return' (i.e. the shipping of African commodities to Great Britain), on

---

[81] Van Dantzig, *The Dutch*, no. 97, De La Palma, Elmina, 26 June 1702.
[82] Inikori, *Africans*, 383.
[83] Cf. Van Dantzig, *Les Hollandais*, 141.
[84] Davies, *Royal African Company*, 221.
[85] Rodney, *History*, 170.

the grounds that 'the negroe [i.e. slave] branch of [the trade] grows every day less and less profitable'.[86]

The revival of interest in West African plantations in the 1750s, on the other hand, related to the particular circumstances of the establishment of the Company of Merchants Trading to Africa in 1750. This new body was a 'regulated company', i.e. one open to membership of all interested merchants, on payment of a subscription, rather than a monopoly trading company; in fact, its charter prohibited it from trading in a corporate capacity, its role being restricted to the management of the British factories on the African coast. In this context an attempt to generate revenue by more effective utilization of the Company's assets in Africa, including the land and labour force attached to its factories, made obvious sense. The projected plantations were in this case indeed intended as a 'substitute' for the slave trade, from which the Company was legally excluded. However, the Commissioners for Trade evidently took the view that the prohibition of trading by the Company covered dealing in local produce, as well as in slaves – this seems to be the meaning of their ruling that its cultivation of crops for export was 'contrary to the known established policy of this trade'.

Reasons for failure

It has often been suggested that these projects of West African plantations were abandoned owing to the opposition of established West Indian planters who feared their competition. For example, Daaku suggested that the RAC's experiments on the Gold Coast were abandoned 'as a result of pressures from the West Indian planters'; Reynolds likewise said that experiments in plantation agriculture were dropped 'because of the opposition of planters with vested interests in the West Indies'; and Van Dantzig that the RAC's plantation projects failed because 'British West Indian planters were opposed to [them]'.[87]

None of these authors, however, cited any explicit evidence in support of their assertions,[88] and the argument is in fact questionable. The vested interests of established West Indian planters can have played

[86] TNA, C.113/272, RAC to Chief Merchants, 13 March 1722.
[87] Daaku, *Trade and Politics*, 46; Reynolds, *Trade and Economic Change*, 63; Van Dantzig, *Les hollandais*, 143–4.
[88] Reynolds & Van Dantzig cite H.A. Wyndham, *The Atlantic and Slavery* (Oxford, 1935), 23, and Eveline C. Martin, *The British West African Settlements 1750–1821* (London, 1927), 48, who both refer to the suspension of British plantation projects on the Gold Coast in 1752, but do not in fact explicitly connect this with West Indian opposition.

little role in the case of the Dutch WIC, since the Dutch in fact possessed no significant number of sugar-producing colonies in the Americas (only Dutch Guiana, now Suriname). In the British case there is on the face of it more substance to this argument. Certainly British West Indian interests did seek to oppose projects of cultivation in West Africa. Three instances of this are recorded, but all are more or less ambiguous in their implications.

First, the RAC's attempts to cultivate indigo in Sierra Leone provoked merchants and planters in Jamaica to petition the Commissioners for Trade in London against it, in 1692, claiming that the development of cultivation in Africa would threaten 'the utter undoing' of their business and 'great loss' to England nationally, and requesting an order restraining the Company from their project of planting. The RAC in response, however, maintained that in fact only 'small quantities' of indigo were currently being supplied by Jamaica, the principal supply coming from the East Indies, and that it was in the interest of English textile manufacturers to find a cheaper source of supply. The Commissioners of Trade in the event decided to take no action, declaring that the petitioners should be 'left to their legal remedy', i.e. they could pursue the matter by private action in the courts, if they so wished (and there is no evidence that they ever did).[89] The RAC's experiment in fact continued for several years longer; although it was eventually abandoned, this was due to local problems of insecurity, with attacks on Company's factories by English private traders and the French, rather than to West Indian pressure.[90]

Sir Dalby Thomas's projects of plantations on the Gold Coast from 1703 onwards also attracted hostile attention. In 1709 a Bill presented to the British Parliament for the setting up of a regulated company to replace the RAC included a provision prohibiting the cultivation in Africa, or purchase from Africans, of 'any sugar-canes, tobacco, or other commodities produced on the British Plantations in America'.[91] It is, indeed, probable that this clause was inserted in the interests of established West India planters, but on this occasion again, it does not appear that their opposition was politically effective. The Bill was never enacted: after passing its second reading, it was lost when Parliament was dissolved, and not subsequently revived,[92] and the formation of a regulated company to replace the RAC did not in fact occur until

---

[89] *Acts of the Privy Council of England, Colonial Series*, Vol. II 1680–1720 (London, 1910), no. 462, 8 Sept. 1692; *Calendar of State Papers, Colonial Series, Africa and the West Indies, 1689–92* (London, 1901), nos 2530, 2546, 29 Sept. & 11 Oct. 1692; RAC's response (undated) in TNA, T70/169, fol. 95.
[90] See Kriger, this volume, chapter 4.
[91] Text of Bill in TNA, CO267/5.
[92] Davies, *Royal African Company*, 150.

1750. Moreover, as has been seen, the RAC's experiments with West African plantations continued into the 1730s.

The decision in 1752 to suspend projects of cultivation has seemed more clearcut, at least in the perception of some modern scholars. Inikori, for example, cited it as evidence that 'the British Parliament [*sic*] thought it necessary to prohibit efforts ... to develop in Western Africa the production of commodities that could compete with those produced in the British Caribbean', which was in turn seen as an indication of 'the growing political power of the West Indian interest in England'.[93]

Looked at in detail, however, the matter is less clear. The basic objection of the Commissioners for Trade to the Company's project of cultivating commercial crops in Africa, as indicated earlier, was the legal one that the Company's charter prohibited it from trading: strictly, their ruling applied to cultivation by the Company in particular, rather than the idea of West African plantations in general. The Commissioners did, however, also raise more general issues, citing two distinct grounds for objection. First, was indeed the safeguarding of existing colonial interests: 'there was no saying where this might stop ... it might extend to tobacco, sugar and every other commodity which we now take from our colonies; and thereby the Africans ... would become planters and their slaves be employed in the culture of these articles in Africa, which they are now employed in, in America.' But second, it was a matter of security: 'our possessions in America were firmly secured to us, whereas those in Africa were more open to the invasions of our enemies, and besides that in Africa we were only the tenants of the soil which we held at the good will of the natives.'

The records of the discussions before the Commissioners suggest that it was the second, rather than the former argument which carried greater weight. Members of the Committee of the Company of Merchants interviewed by the Commissioners dismissed the argument about competition with existing colonies, on the grounds that in fact only limited amounts of indigo and cotton were currently being produced in British colonies. Indigo was produced only in Carolina, whose output was of poor quality and small quantity, so that British demand was currently met by imports from France and Spain. Likewise 'very little' cotton was being cultivated in the British West Indies, because planters there found sugar more profitable, so that Britain was currently obliged to import it from Turkey, 'at great disadvantage'. To the suggestion that the experiment might in time extend to other crops, one Committee member stressed its limitation to indigo and

[93] Inikori, *Africans*, 389.

cotton, though another did suggest that if sugar were to be cultivated in Africa, this would be no bad thing: 'it would enable us to supply all the merchants in Europe', undercutting the French.

Another Committee member did, however, concede the second point, about the security of territorial possession, agreeing that 'the whole of the question rested upon the point whether our property and possessions in Africa were established and secured with respect to the natives, for if our possession was dependant upon the natives, and we were only tenants at will, it was clear that the introducing of culture and produce might prove of bad consequence'.[94] Inikori's own analysis, indeed, concludes by stressing this second point: 'their secure possession of the resources of the Americas made the latter the preferred choice of West European entrepreneurs ... Being in complete possession of the Americas meant that the European entrepreneurs could ensure that the colonial state in the Americas would make rules and regulations [in their favour] ... They could not trust the African states to do the same.'[95]

Van Dantzig likewise suggested that 'a major problem' for projects of establishing plantations was that 'the Europeans had no territorial property on the Gold Coast'.[96] The Director-General of the Dutch WIC in 1717 observed that 'it has ... been shown on many occasions, like in the case of the sugar, indigo and cotton plantations, that the jurisdiction of the Company does not reach beyond the area covered by its cannons'.[97] In particular, the lack of effective control posed problems for protecting the crops against theft: it was noted that one difficulty in the cultivation of sugar was that much of it was stolen by the local people to eat.[98] An official of the British Company of Merchants, asked about the feasibility of European plantations in West Africa in the 1780s, similarly observed that 'plantations would be nowhere secure, except under the guns of our forts', beyond which European enterprises would be vulnerable to harassment 'with disputes with the natives, and depredations on their property'.[99]

This is also consistent with an observation made, with a more general application, by David Eltis, stressing the importance of African

---

[94] All from *Journal of the Commissioners for Trade*, vol. 9, 270–1.
[95] Inikori, *Africans*, 392.
[96] Van Dantzig, *Les Hollandais*, 141.
[97] Van Dantzig, *The Dutch*, no. 223: Resolutions of Council, Elmina, 12 March 1717. The WIC in response suggested that cotton should be cultivated at a site in Elmina which was 'covered by the guns of the forts': ibid., no. 243: Secret Papers of Assembly of Ten, 20 Nov. 1720.
[98] Ibid., no. 138, Minutes of Council, Elmina, 14 March 1707.
[99] *Report of the Committee of Council appointed for consideration all matters relating to Trade and Foreign Plantations ... [concerning] the State of the Trade of Africa* (House of Commons, London, 1789), section on 'Produce', evidence of Archibald Dalzel.

agency and the maintenance of African sovereignty in the operation of West African commerce: Europeans would have preferred to use slaves on plantations (and in mines) in Africa, but this was precluded by 'African resistance', and the slave trade was adopted as 'a second-best alternative'.[100] The argument may be thought persuasive, but its implication is that it was not West Indian competition, but rather difficulties on the ground in West Africa, which ruled out the prospects of African plantations.

An alternative (or supplementary) explanation relates to the commercial calculations of the trading Companies, rather than to pressures upon them from American planters or the metropolitan state – namely, that there was a contradiction between their commitment to the trans-Atlantic slave trade and their projects of establishing plantations in Africa, inasmuch as it was more profitable to export slaves across the Atlantic than to employ them locally in Africa. Thus Daaku observed (of the WIC's plantation efforts): 'It was the slave trade that killed the project', because it starved the African plantations of the labour which they required to be viable.[101] There were certainly instances of problems in the supply of labour to the Gold Coast plantations. For example, the WIC's local factors in 1707 maintained that 'Concerning the sugar, cotton and indigo plantations ... we would be quite able to continue that work, if only the required tools, slaves and other materials were sent, and especially if we were to be allowed to buy as many slaves as [was formerly] proposed'; and again in 1752 they declared that, in order to increase output of cotton, the Company 'must sent them money to buy them more slaves'.[102] The local factors of the RAC in 1723 likewise cited 'want of hands' as an obstacle to the cultivation of cotton, although the Company insisted that, if the existing labour force of 'Castle slaves' was inadequate, additional slaves could be purchased for the work.[103] This, however, begs the question of why it was that the employment of slaves in cultivation in West Africa was less profitable than transporting them to the Americas – if cultivation in Africa had proved viable, it would presumably have made the trans-Atlantic slave trade less attractive, rather than *vice versa*.

At one level, the failure of the West African plantations can be attributed to undercapitalization and mismanagement. Daaku, for example, describes the Dutch and English plantation projects as 'ill thought-out

---

[100] David Eltis, *The Rise of African Slavery in the Americas* (Cambridge, 2000), 140–9.
[101] Daaku, *Trade and Politics*, 45.
[102] Van Dantzig, *The Dutch*, no. 138: Minutes of Council, 14 March 1707; TNA, T70/29, Melvil, CCC, 11 June 1752.
[103] TNA, T70/53. RAC to Jeremiah Tinker, 5 Dec. 1723.

and half-heartedly pursued'.[104] Both the Dutch and British companies undertook these projects, as noted earlier, at a time when they were in financial difficulty, and this is reflected in the frequent complaints of their West African agents of inadequate support – which were not, in fact, restricted to the plantation projects only, but also applied to the supply of goods to prosecute the slave trade. On the other hand, the parent Companies in turn complained equally repeatedly of incompetent management on the part of their agents in Africa. In the 1720s, for example, the RAC's factors on the Gold Coast declared themselves unable to process indigo dye from the leaves, although the Company maintained that the instructions it had sent were 'so plain they could not miscarry'.[105] Likewise, with cotton, the local agents proved consistently unable to operate the cotton-gins which were supplied to them, to the great irritation of the Company in London.[106]

But perhaps there were more basic constraints. One possibly relevant factor related to transport costs. It is easy to assume that West Africa had a comparative advantage over plantations in the Americas, in its 'relative proximity to the home market'.[107] If you look at a map, on the face of it the distance from West Africa to Europe is somewhat less than that from America to Europe, but in the age of sail this was offset by the constraints imposed on movements of shipping by the prevailing patterns of winds and currents in the Atlantic, which made it difficult to sail directly from West Africa to Europe. Ships normally sailed back, not along the African coast, but first out into the open sea, to pick up westward currents and winds, before turning north – what became known in the early days of Portuguese navigation as the *volta da Guiné* (or *volta da Mina*), which in fact involved a voyage of greater distance than that between Europe and America.[108] From the Gold Coast, moreover, by the seventeenth century at least, it was usual for ships bound homeward to head first east to Gabon, before sailing westward along the line of the Equator, extending the voyage even further. These patterns of winds and currents made it convenient to sail home via America, which in turn was a factor which encouraged the transAtlantic trade in slaves. In practice, therefore, West Africa was, in fact, further away, in terms of shipping distances and times, from Europe

---

[104] Daaku, *Trade and Politics*, 44. Inikori likewise says that the Dutch 'never seriously committed themselves' to plantations: *Africans*, 385.
[105] TNA, T70/53, RAC to Chief Merchants, CCC, 9 July 1724.
[106] The Company insisted that the instructions sent with the gins were 'so plain & easy, 'tis hardly possible for a child to err in the putting the same in execution': TNA, T70/53, RAC to Chief Merchants, 5 Dec. 1723.
[107] Whalley, 'Dutch and English Attempts', 3.
[108] P.E.H. Hair, 'Was Columbus' First Very Long Voyage a Voyage from Guinea?', *HA* 22 (1995), 223–37.

than were the Americas. This presumably disadvantaged potential African suppliers in competition with American plantations, to the extent that the longer sea voyage increased not only transport costs,[109] but also the risk of spoilage of crops. This factor might help to explain, in particular, why West African rice, although purchased by Europeans to provision slave-ships, was not exported to Europe, in competition with that cultivated in North America.

But there were perhaps even more fundamental problems, relating to the agricultural potential of West Africa. The economic logic of the slave trade was clearly that the productivity of labour was higher in the Americas than in Africa: otherwise the transportation of labour across the Atlantic would not have been profitable. In part, this higher productivity of African labour in America may have been due to organizational factors, in the form of the plantation system: a more exploitative labour regime could be maintained in America than in Africa, where slaves could more easily run away. But it was also probably in large part due to ecological factors, in the productivity of the soil.

Looking on into the nineteenth century, and further into colonial period, it seems significant that the successful expansion of cash crop agriculture in West Africa, when it occurred, primarily involved crops which were either newly introduced (cocoa, coffee), or for which there was a newly increased demand in Europe (palm-oil, groundnuts). By contrast, the crops which had been the staples of American plantation agriculture in the seventeenth and eighteenth centuries generally did not become successful export crops from West Africa subsequently. In particular, sugar never became a commercially viable export crop in West Africa. It seems clear, in fact, that basic disadvantages of poorly drained, acidic and saline soils over most of coastal West Africa precluded any possibility of successful large-scale cultivation of sugar for export there.[110] The one exception – of a crop produced by slave plantations in America which was subsequently cultivated for export in West Africa – is cotton, which did become a significant export from parts of West Africa in the twentieth century, notably Northern Nigeria and French Soudan (Mali), and the north of the Republic of Bénin. However, all these areas of cotton produc-

---

[109] It might be argued that, since ships returning directly to Europe carried items of low bulk relative to value (especially gold), they would have had excess cargo capacity which was available for other commodities effectively free (apart from the cost of loading and unloading). However, this consideration might equally encourage sailing home via America and carrying slaves, as argued for the Brandenburg case by Adam Jones, *Brandenburg Sources for West African History 1680–1700* (Wiesbaden, 1985), 6.

[110] H.A. Gemery & J.S. Hogendorn, 'Comparative Disadvantage: the Case of Sugar Cultivation in West Africa', *Journal of Interdisciplinary History* 9/3 (1979), 429–49.

tion were situated inland, and it is doubtful whether they could have become economically viable exporters without the introduction of modern mechanised transport, initially in the form of railways, which decisively reduced the cost of transport to the coast.

The conclusion therefore seems inescapable. It is doubtful whether, prior to the development of an enlarged market for different African crops in the nineteenth century, there was any viable prospect of agricultural production for export to European or American markets, which might have served as an alternative to the trans-Atlantic slave trade.

6

The origins of 'legitimate commerce'

CHRISTOPHER LESLIE BROWN

Where did the idea of 'legitimate commerce' come from? At first glance, there would seem to be a simple answer to this question. The Abolitionists, it would appear, invented the idea of legitimate commerce to justify slave trade abolition, to make an economic case for a moral cause. No one championed the transition in the African trade from slaves to staple crops with more ardour. After the formation of the Society for Effecting the Abolition of the Slave Trade in 1787, the economic potential of legitimate commerce became a point of emphasis among the Abolitionists, figuring centrally thereafter in Abolitionist propaganda and testimony in parliament, in the mission of the new Sierra Leone Company at its founding in 1791 and, more broadly, in Danish and French schemes to promote abolition. After British abolition in 1807, Anglican evangelicals organized the African Institution with the aim of encouraging staple crop production in Sierra Leone.[1] The idea of legitimate commerce, therefore, would seem to have been a creature of the Abolitionist movement, and perhaps would not have existed without it.

In this telling, legitimate commerce emerged as a means to an end rather than an end in itself. It took shape first as a claim before it became a conviction; it arose to make the case before it became an

[1] Ralph Austen and Woodruff D. Smith, 'Images of Africa and British Slave-Trade Abolition', *African Historical Studies* 2 (1969), 79–80; Daniel Hopkins, 'Peter Thonning, the Guinea Commission, and Denmark's Postabolition African Colonial Policy, 1803–1850', *William and Mary Quarterly* 3rd series, 66/4 (2009), 781–808; William Cohen, *The French Encounter with Africans: White Response to Blacks, 1530–1880* (Bloomington IN, 1980), 165; Deirdre Coleman, *Romantic Colonization and British Anti-Slavery* (Cambridge, 2005) 28–133; Suzanne Schwarz, 'Commerce, Civilization, and Christianity: The Development of the Sierra Leone Company', *Liverpool and Transatlantic Slavery*, ed. David Richardson, Suzanne Schwarz, and Anthony Tibbles (Liverpool, 2007) 252–76; Wayne Ackerson, *The African Institution (1807–1827) and the Antislavery Movement in Great Britain* (Lewiston NY, 2005), 31–40, 46–8.

article of faith. An abundance of evidence would seem to support this view. The first Abolitionists argued for commercial alternatives to the African trade some time after – not before and not when – they became committed to slave trade abolition. For most, it facilitated the transition from a moral commitment to antislavery principles to a political commitment to Abolitionist action.

This is apparent, for instance, in the work of Thomas Clarkson. His first publication, *An Essay on the Slavery and Commerce of the Human Species* (1785), showed only the most limited interest in commercial alternatives to the Atlantic slave trade. It took several years for Clarkson to decide that abolition could be accomplished only if economic arguments could be made to bolster the moral claims. Once the need to lobby Parliament became the aim, Clarkson began to investigate and then educate others on the commercial benefits that slave trade abolition might yield. By 1787, he had started to gather and share 'specimens of African produce' that British merchants, he contended, too often overlooked. This line of argument culminated with *An Essay on the Impolicy of the African Slave Trade* (1788), the first British antislavery tract to dispense with the moral case for abolition almost entirely and to emphasize, instead, that the Atlantic slave trade diverted the African economy from equally profitable alternatives. Never again, in his later publications, would Clarkson argue for slave trade abolition without also insisting that staple crop trades from Africa could and would take its place.[2]

James Ramsay, Clarkson's predecessor and briefly his mentor, settled on a similar rhetorical strategy for similar reasons. Ramsay spent nearly two decades looking for ways to 'improve' West Indian slavery. A long-time resident of St. Kitts, he knew the injustices and inhumanities of slavery first-hand. Yet it proved impossible, he realized, to attack colonial slavery directly. Parliament lost interest in legislating for the colonies after the American War of Independence and the West Indian elite had no interest in reform of any kind. So, in 1784, Ramsay turned his attention to the British slave trade, which looked to be the more vulnerable target. This shift in strategy led him to examine the African trade for the first time in detail. 'There are many parts of the slave coast', he opined, 'where the inhabitants are sufficiently polished to be capable of carrying on the manufacture of sugar, planting tobacco, and indigo; that they already have rice of a more valuable quality than that of the Carolinas.' Thinking about the subject this way led Ramsay to conclude that Britain needed neither the Atlantic slave trade nor the

---

[2] Marcus Wood, 'Packaging Liberty and Marketing the Gift of Freedom: 1807 and the Legacy of Clarkson's Chest', *The British Slave Trade: Abolition, Parliament, and People,* ed. Stephen Farrell, Melanie Unwin, and James Walvin (Edinburgh, 2007), 218–20.

West Indian colonies to further staple crop production. The Portuguese example of successful (for a time) cultivation in São Tomé suggested that a plantation regime might be established in West Africa under African control. Ramsay knew very little, in fact, about West Africa but this did not stop him from speculating optimistically about the possibilities. If West Africa was 'civilized ... we should open a market, that would fully employ our manufacturers and seamen, morally speaking till the end of time'. In commercial terms, Britain would acquire 'not a colony, but a continent'. Yet, even here, Ramsay promoted commercial alternatives with other and, for him, more important aims in mind – curtailing the flow of captives to the Americas or reducing British dependence on the West Indies altogether. He cared less about profits in the African trade than about abolishing slavery in the West Indies.[3]

For Ramsay, for Clarkson and for others in their cohort, the commitment to antislavery reform came first; the arguments for legitimate commerce arose later. The one possible exception is Henry Smeathman, who emphasized the antislavery potential of his plan to colonize Sierra Leone in order to win financial assistance from Quaker leaders in London just then committing themselves in 1783 to an Abolitionist campaign. Smeathman, though, never showed much interest in the antislavery campaign itself.[4] No one, in the 1780s, became an Abolitionist *because* they wanted to promote legitimate commerce.

Yet, as the example of Smeathman might suggest, the idea of legitimate commerce, like the ideology of free labour, has a history of its own, separate from the history of the Abolitionist movement, although this has been hard to see because legitimate commerce became quickly and closely associated with Abolitionism once the antislavery movement began. The pursuit of commercial alternatives to the Atlantic slave trade could flourish even in the absence of antislavery sentiments. This becomes apparent with a shift of attention to other actors. Henry Smeathman was only one of a number of British speculators in the 1780s who began to look at West Africa 'not merely as it was, but as it might be', in the words of Philip Curtin. Schemes to colonize West Africa and to encourage commercial agriculture there proliferated after the American Revolution, at a time when few could be sure how or where staple crop production in the empire would continue. The authors, with few exceptions, had little interest in antislavery generally,

---

[3] James Ramsay, *An Inquiry into the Effects of Putting a Stop to the African Slave Trade, and of Granting Liberty to the Slaves in the British Sugar Colonies* (London, 1784), 5, 14, 16; Christopher Leslie Brown, *Moral Capital: Foundations of British Abolitionism* (Chapel Hill NC, 2006), 244–53, 325–6.
[4] Brown, *Moral Capital*, 314–16.

or slave trade abolition specifically. Indeed, veterans of the African trade predominated. Liverpool slave-trader Henry Trafford insisted in 1783 that Africa could and should become the source of all tropical commodities produced for the British Empire. In his view, the North American, Caribbean and South Asian colonies, in turn, would become unnecessary. Richard Oswald, who built his fortune in part from the Bence Island slaving station that he owned with several associates, suspected that future fortunes in the African trade might arise through the cultivation of cotton. He directed his agents there in 1783 to 'buy all the Cotton Wool you can get … in this trade with the Natives, and endeavour to learn whether it may not be possible to persuade these people to increase the culture of that article as a commodity of exchange and commerce'. A soldier returned from service on the Gold Coast urged prime minister William Pitt to make Cape Coast Castle the headquarters of a British colony and introduce cotton cultivation there. In 1787, seventeen Manchester textile firms pushed for experiments in cotton production on the Gold Coast and along the Gambia, since 'upon the African continent … it grows … spontaneously'.[5] In the founding year for the London Committee for Effecting the Abolition of the Slave Trade, it still was possible to think of alternatives to the Atlantic slave trade without, necessarily, endorsing Abolitionism.

The Sierra Leone venture, as it happens, represented the second British attempt to establish a trade in staple crops, not the first. The ambition to colonize the West African coast has a longer history in the British Isles than scholars have realized. In 1758, during the Seven Years' War, the British government captured the French West African trading posts of Gorée and Saint-Louis on the Senegambia coast. The 1763 Treaty of Paris confirmed British possession of Saint-Louis and inspired the British government to institute a formal colonial establishment to secure it. From 1765 to 1779, the new British 'province' of Senegambia tried to control trade in and out of the Upper Guinea coast from Portendick, on the Atlantic coast of the Sahara, to the mouth of the Gambia River. The main prize was the trade in gum

[5] Philip Curtin, *The Image of Africa: British Ideas and Action, 1780–1850* (Madison WI, 1964), xii; Bedfordshire Record Office, L29/340/341, Grantham Papers, Henry Trafford to Lord Grantham, March 5; Richard Oswald to Captain Griffiths, July 4, 1783 in Sheila Lambert, ed., *House of Commons Sessional Papers of the Eighteenth Century*, LXVIII, *Minutes of the Evidence on the Slave Trade, 1788–1789* (Wilmington DE, 1975), 283; The National Archives, London (TNA), 30/8/363, The Papers of William Pitt, first Earl of Chatham, 'Lieut. Clarke's Observations on African Affairs submitted to the consideration of the Right Honorable William Pitt, as the groundwork of a Plan for Making Africa a source of Revenue, in Place of an expense to this Country'; TNA, BT 6/140, fols 34–35, William Frodsham to the Lords Committee of the Privy Council of Trade, Nov. 30, 1787. For a good discussion of Trafford, see Coleman, *Romantic Colonization*, 13–16.

arabic, which, over the course of the eighteenth century, quickly had become a crucial component in the manufacture of quality European textiles. The slave trade fulfilled only some of the commercial possibilities of trade with Atlantic Africa, believed those engaged in the new Senegambia venture. London merchant Edward Grace, for example, contracted with an agent at the mouth of the Senegal River in 1767 to send a trading mission upriver to Galam to acquire gold, ivory and cotton from the markets there, but not slaves.[6] From Robin Law's and Colleen Kriger's contributions to this volume, now we know too that plans to diversify the African trade have an even longer lineage, that before 1750 speculators of various kinds encouraged the establishment of staple crop plantations in West Africa to complement and even compete with similar ventures in the Americas. Key elements, then, of legitimate commerce – the hope to cultivate and export staple crops and the ambition to colonize – were in place long before the development of Abolitionist movements anywhere in the Atlantic world. Abolitionism, therefore, facilitated but it did not originate the idea of legitimate commerce.

Indeed, one could denounce the Atlantic slave trade without, also, calling for a new traffic in African commodities, at least for a time. Anthony Benezet presents the most obvious and important example. It took the Quaker pioneer more than a decade to decide that his appeal to end the slave trade would benefit from the identification of a viable commercial alternative. His first concern, instead, was to undermine the cultural prejudices that allowed the Atlantic slave trade to thrive. Justifications of the Atlantic slave trade long had cast the traffic as a kind of rescue. The Atlantic slave trade saved captives either from a brutal death in war or from barbarism, savagery and spiritual darkness, apologists explained. One might regret the Atlantic slave trade in the abstract but, for the victims, it was argued, remaining in Africa likely would have been worse. Benezet insisted that this received picture of Africa was false, that even the European slave-traders resident in Africa recognized that the captives had been taken from prosperous well-ordered societies that possessed an abundance of natural wealth. Benezet described that natural wealth at some length, first in his 1762 *A Short Account of that Part of Africa, Inhabited by the Negroes*, and then in the 1767 *A Caution and Warning to Great Britain and Her Colonies*, but only to vindicate the reputation of Africans and African

[6] James L.A. Webb, Jr., 'The Mid-Eighteenth Century Gum Arabic Trade and the British Conquest of Saint-Louis du Senegal, 1758', *Journal of Imperial and Commonwealth History (JICH)* 25/1 (1997), 37–58; E.C. Martin, *The British West African Settlements, 1750–1821: A Study in Local Administration* (London, 1927), 80–102; J.M. Gray, *A History of the Gambia* (London, 1940) 234–275; Edward Grace to Amable Doct, 23 May 1767, in T.S. Ashton, ed., *Letters of a West African Trader, 1767–1770* London, 1950), 3.

civilization, rather than to encourage commercial alternatives to the Atlantic slave trade.

Only the last and longest of his three publications on Africa, *Some Historical Account of Guinea*, first printed in 1771, sketched the prospective benefits of legitimate commerce. There, in the final paragraph of the text, in a statement that he failed to develop, Benezet suggested that a 'considerable advantage might accrue to the British nation in general, if the slave trade was laid aside by the cultivation of a fair, friendly, and humane commerce with the Africans'. The continent, he said, was 'stored with vast treasures of materials, necessary for the trade and manufactures of Great Britain'. In addition 'most of the commodities which are imported into Europe' from the American colonies could, 'under proper management', be produced in Africa 'in the greatest plenty'. 'The advantages of this trade would soon become so great, that it is evident', he concluded, that 'this subject merits regard and attention of the government'.[7] In this way, Benezet seemed to anticipate, and perhaps even encouraged, later Abolitionist emphasis upon legitimate trade. Yet the claim, the line of argument, never mattered much to him. He made no mention of commercial alternatives to the Atlantic slave trade in later publications or in what now remains of his original correspondence between 1771 and his death in 1784. If arguments for legitimate commerce could be made without reference to antislavery, opposition to the slave trade could take shape in the absence of arguments for legitimate commerce.

That absence, that reluctance to make the economic case, moved later Abolitionists to devise a usable Benezet. The Society for Effecting the Abolition of the Slave Trade republished *Some Historical Account of Guinea* in 1788 to aid the emerging British movement. Since Benezet failed, from their perspective, to make the case for legitimate commerce fully, the publication committee for the Society added an appendix to the original text. This appendix included a series of queries, twelve in all, that told readers why slave trade abolition, from a commercial perspective, might serve Britain well. The first ten are worth listing in full since they delineate what was meant by legitimate commerce as of 1788 and convey, as well as any text, the hopes and assumptions that informed it:

[7] Anthony Benezet, *Some Historical Account of Guinea* (Philadelphia, 1771), 143–144. For Benezet's depiction of African society and history, more generally, see Jonathan D. Sassi, 'Africans in the Quaker Image: Anthony Benezet, African Travel Narratives, and Revolutionary-era Antislavery', *Journal of Early Modern History* 10/1–2 (2006), 95–130; Maurice Jackson, *Let This Voice Be Heard: Anthony Benezet, Father of Atlantic Abolitionism* (Philadelphia PA, 2009), 72–107. The broader debate in Britain on the nature of African society in the eighteenth century is sketched well in George Boulokos, 'Olaudah Equiano and the Eighteenth Century Debate on Africa', *Eighteenth Century Studies* 40/2 (2007), 241–255.

I. Whether so extensive and populous a country as Africa is, will not admit of a far more extensive and profitable trade to Great-Britain than it yet ever has done?

II. Whether the people of this country, notwithstanding their colour, are not capable of being civilized, as well as those of many others have been; and whether the primitive inhabitants of all our countries, so far as we have been able to trace them, were not once as savage and inhumanized as the Negroes of Africa; and whether the ancient Britons themselves of our country were not once upon a level with the Africans?

III. Whether, therefore, there is not a probability that those people might in time, by proper management exercised by the Europeans, become as wise, as industrious, as ingenious and as humane as the people of any other country?

IV. Whether their rational faculties are not, in the general, equal to those of any other human species; and whether they are not, from experience, as capable of mechanical and manufactoral arts and trades, as even the bulk of the Europeans?

V. Whether it would not be more to the interest of all the European nations concerned in the trade to Africa, rather to endeavour to cultivate a friendly, humane, and civilized commerce with those people, into the very center of their extended country, than to content themselves with skimming a trifling portion upon the sea coast of Africa?

VI. Whether the greatest hindrance and obstruction to the Europeans cultivating a humane and Christian-like commerce with those populous countries, has not wholly proceeded from the unjust, inhumane, and unchristian-like traffic called the SLAVE TRADE, which is carried on by the Europeans?

VII. Whether this trade, and this only, was not the primary cause, and still continues to be the chief cause, of those eternal and incessant broils, quarrels and animosities, which subsist between negro princes and chiefs; and consequently, for those eternal wars which subsist among them, and which they are induced to carry on, in order to make prisoners of one another for the sake of the slave trade?

VIII. Whether, if trade was carried on with them for a series of years, as it has been with other countries that have not been less barbarous, and the Europeans gave no encouragement whatever to the slave trade, those cruel wars among the Blacks would not cease, and a fair and honourable commerce in time take place throughout the whole country?

IX. Whether the example of the Dutch in the East Indies, who have civilized innumerable of the natives, and brought them to the European way of cloathing, etc., does not give reasonable hopes that these suggestions are not visionary but founded on experience, as well as on humane and Christian-like principles?

X. Whether commerce in general has not proved the great means of gradually civilizing all nations, even the most savage and brutal; and why not the Africans?[8]

In its assumptions about Africans, about Africa, and about the African trade, these queries marked out the consensus that took shape among nearly all opponents of the Atlantic slave trade in Britain and elsewhere by 1790. All that was missing here was a detailed statement concerning the particular commodities that might be procured from the various parts of the West African coast, the kind of list that would become familiar in the work of Thomas Clarkson and in other Abolitionist literature once the British antislavery movement began.

All the more remarkable then, that that these words, appended to a 1788 Abolitionist text, first appeared in print nearly forty years before, in 1751, in the first edition of the *Universal Dictionary of Trade and Commerce* under the entry for 'English African Company'.[9] Clearly, the idea of legitimate commerce preceded the development of the Abolitionist movements in Britain and America by several decades. The Abolitionists did not invent it. More importantly, the timing of this statement suggests that the ambition to reorganize the African trade arose from circumstances and contexts different from those typically associated with the origins of antislavery ideas. The argument for legitimate commerce, it will appear, developed first out of debates about the management and direction of trade in Africa rather than from concerns about the morality of slaving in the Americas.

Malachy Postlethwayt, the editor of the *Universal Dictionary of Trade and Commerce*, called on the British government to substitute staple crop exports for the slave trade repeatedly between 1751 and his death in 1767. This was a favourite idea of his that ran through almost all of his subsequent publications. He argued the case with increasing emphasis in the later editions of the *Universal Dictionary* in 1758, 1766, and in the posthumous edition of 1774. At the opening of the Seven Years' War he recommended alternatives to the slave trade in his *Great Britain's Commercial Interest, Explained, and Improved* of 1757. The promise of staple crop production in Africa stood at the centre of his 1758 *In Honour to the Administration: The Importance of the African Expedition Considered*, where he recommended that the taking of Gorée and Saint-Louis should lead next to the conquest and settlement of Senegal itself. Then there were the numerous anonymous excerpts published in the periodical literature throughout these years, where Postlethwayt circulated his call for a reorientation of the African trade

---

[8] Benezet, *Some Historical Account*, 123–124. The remaining two queries suggested populating the American colonies with only European settlers.
[9] Malachy Postlethwayt, ed., *Universal Dictionary of Trade and Commerce, Translated from the French of the Celebrated Monsieur Savary* (London, 1751) I, 727.

without, in most instances, attaching his name to the cause.[10] He regarded the good that would come from nurturing alternatives to the Atlantic slave trade as self-evident. Would it not be more beneficial for all the trading European states', he wrote in 1766, 'rather to endeavour to cultivate a friendly, humane, and civilized commerce, with those people, into the very center of their extensive country, than to content themselves only with skimming a trifling portion of trade upon their sea-coasts?' For Postlethwayt, this was a crusade. 'Has not the author of this performance, to no purpose yet, many years since suggested ways and means, whereby this might be done to the immense benefit of the British Empire?'[11] No one in Britain during the third quarter of the eighteenth century promoted the idea of legitimate commerce with greater enthusiasm.

Postlethwayt's publications exercised considerable influence on the early opponents of the British and American slave trades. In the case of Anthony Benezet, the connection was disguised. In the closing paragraphs of his *Some Historical Account of Guinea*, Benezet lifted passages from Postlethwayt's *Universal Dictionary of Trade and Commerce* without attribution. (See the Appendix at the end of this chapter for a comparison of these.) Bostonian James Swan, by contrast, freely acknowledged his dependence on Postlethwayt in his 1772 *Dissuasion to Great Britain and the Colonies, from the Slave Trade to Africa*. From Postlethwayt, Swan learned 'how to put this Trade to Africa on a just and lawful footing'. Portions of the *Dissuasion* cited the *Universal Dictionary* at length and paraphrased its key conclusions. 'It is in the interest of every Merchant in Britain and the Plantations', wrote Swan, 'who are now concerned in traffick to Africa, to cultivate the inland commerce in its utmost extent ... there will be discovered an infinite variety of traficable articles, which the present traders are totally unacquainted.' It closed by adopting Postlethwayt's preferred remedy to the evils of the Atlantic slave trade – the establishment of a chartered corporation charged with funnelling manufactured goods to West Africa in exchange for staple crops.[12] The publications by Postlethwayt equipped the first opponents of the British slave trade to think of commerce with Africa in new ways.

---

[10] Postlethwayt makes reference to these writings for newspapers and periodicals in *The Universal Dictionary of Trade and Commerce* (London, 1751) I, xiv. For an example, see *The Universal Magazine* (March, 1757), 97–105.
[11] Postlethwayt, *Universal Dictionary of Trade and Commerce* (London, 1766 3rd edn) I, vii.
[12] James Swan, *A Dissuasion to Great-Britain and the colonies from the slave trade to Africa* (Boston MA, 1772) xv, 45–46, 50–53, 57–60, quote on page 53. Philip Gould, who also has called attention to the significance of the Postlethwayt queries, emphasizes Postlethwayt's impact on early American Abolitionists: Philip Gould, *Barbaric Traffic: Commerce and Antislavery in the Eighteenth-Century Atlantic World* (Cambridge MA, 2003), 21–24.

Postlethwayt enlarged the sense of possibilities among French critics of the Atlantic slave trade too. From the work of William Cohen on French perceptions of Africa during the eighteenth century, scholars have known for some time that the first public proponent of legitimate commerce in France was the Abbé Pierre-Joseph-André Roubaud, who, during the Seven Years' War suggested that if Africans had never been enslaved in the Americas, but had remained free in their own societies, they would have cultivated, in his words, 'a greater quantity of sugar cane, thicker, and more succulent, and more delicious than what we get from the West Indies'. From the more recent research of Pernille Røge, we know now that the Abbé Roubaud first made this suggestion in 1759, after reading Postlethwayt's *Importance of the African Expedition* of 1758. Roubaud was alarmed at what Postlethwayt proposed – a British Empire in Africa that would bar France from access to slaves there and thereby enable Britain to 'destroy French commerce and navigation, both in Europe and in the Americas'. Roubaud, though, also was taken by Postlethwayt's creative linkage between commercial development and philanthropy. 'It is astonishing', he wrote in the French periodical *Journal de Commerce* in 1759, 'that none of the European nations who have traded along the African coast for so long, and who have recognized an infinity of places advantageously situated and capable of producing the precious commodities of America and Asia, have not [sic] attempted to establish colonies here; especially when it is easy to obtain concessions from African kings in several places; but all have preferred the American colonies, which are infinitely more expensive, more difficult to establish and support, and which are harder to reach.' Twelve years later, in his 1771 *General History of Asia, Africa, and the Americas*, Roubaud developed the argument at greater length and expressed it with more urgency. What do we get from America (he asked)? 'Sugar, coffee, indigo, cotton, syrup, wood, and animal skins. Africa can provide a part of this even without cultivation. It is rich in other kinds of merchandise, has its particular produce, and it even has a majority of the riches of the East Indies. Evidently, it is thus in Europe's interest to introduce culture, the arts, and expertise in this region and dispense with the voyages and ruinous settlements that she maintains sword in hand, in the two Indies.'[13]

With this inspiration from Postlethwayt, the Abbé Roubaud would, in turn, put the rethinking of the African trade on the agenda of the Physiocrats, the French political economists, who favoured agricul-

---

[13] Cohen, *French Encounter with Africans*, 164; Pernille Røge, '"La Clef de Commerce" – The changing role of Africa in France's Atlantic empire ca., 1760–1797', *History of European Ideas* 34 (2008), 433–4.

tural development and the promotion of free trade. In their publications, between 1771 and 1776, the Physiocrats declared repeatedly that the cruel and unjust Atlantic slave trade should be abolished and replaced by a peaceful trade in tropical commodities cultivated by African free labourers in Africa. 'From the very first voyages', wrote the Abbé Badeau in 1771, 'all these slave traders should have done was to have asked for sugar canes instead of human creatures; they would have received big, juicy, delicious ones because Africa has plenty of them.' They should have 'let the Negroes cultivate their sugar cane peacefully in their own countries and to give them in exchange, alcohol, iron, glass and other European merchandise, not for their children and neighbours, but for their raw sugar and indigo'. In the same year, Pierre Dupont de Nemours thought that it would 'suffice to create only a few peaceful establishments on the coast, to send artisans, builders of mills and boilers there, and to say to the Negroes: friends, you see that sugar cane, cut it, pass it through the two rollers we offer you, make the juice boil in the boiler, and we shall pay you for the syrup produced'. This was just talk, it is true, but there was a lot of it among the Physiocrats and their ideas informed the programme of French Abolitionists in the 1790s to an important degree, a programme that would promote the abolition of the French slave trade and the opening of a new French empire in Africa in almost equal measure, just as Postlethwayt first had proposed for Britain many years before.[14]

From one perspective, Postlethwayt would seem an unexpected source for Abolitionist argument. He first came to public notice as a promoter of the Atlantic slave trade, generally, and a defender of the Royal African Company, specifically. Anthony Benezet perhaps refused explicit reference to his debt to Postlethwayt on these grounds. During the 1740s, in the final decade before its dissolution, Postlethwayt served on the Court of Assistants for the company, and served as its chief propagandist. When historians have needed a 'representative' quote from the era attesting to the importance of the slave trade to the British Empire, they often have selected a passage or two from one of his early publications.[15] The series of pamphlets he published in 1745 and 1746 to save the Royal African Company from bankruptcy would seem to make his orientation clear. There was: *The African Trade: The Great Pillar and Support of the British Plantation Trade in America*; *The Importance of Effectually Supporting the Royal African Company of England*; and *The National and Private Advantages of the African Trade Considered*.

[14] Røge, 'La Clef de Commerce', 435–8, Cohen, *French Encounter with Africans*, 164–5.
[15] See, for example, William Darity Jr., 'British Industry and the West Indies Plantations', *The Atlantic Slave Trade: Effects on Economies, Societies, and Peoples in Africa, the Americas, and Europe*, ed. Joseph E. Inikori and Stanley L. Engerman (Durham NC, 1992), 270–73.

Postlethwayt, therefore, would seem a most unlikely candidate to initiate a critique of the Atlantic slave trade, to describe the slave trade as an abomination and a peaceful traffic in export crops as a solution.

Yet there were continuities in Postlethwayt's writings about the African trade, regardless of whether he was defending the slave trade or opposing it. Above all, Postlethwayt worried that the British government, and the British nation more generally, habitually undervalued the African trade, that too frequently it was treated as a peripheral part of British overseas enterprise. Postlethwayt thought, instead, that the African trade made possible much of the rest. The American colonies and, by extension, Atlantic commerce more generally depended on it. Therefore, the state that achieved primacy in the African trade, he argued, soon after would achieve supremacy in the Atlantic World. Yet, in too many sections of the African coast, Britain had seemed to cede influence to European rivals, particularly the French in Senegal and the Dutch along the Gold Coast. By allowing independent merchants to organize and conduct the British slave trade, Parliament had compromised Britain's ability to assert its national interests in Atlantic Africa. For the independent traders cared more about their particular profits than national prestige. If merchants made fortunes for themselves, the nation as a whole steadily lost influence across the Atlantic littoral. Postlethwayt favoured, therefore, a broad public investment in chartered companies that could serve not only as agents of commerce but also of the imperial state. He defended the Royal African Company until its dissolution in 1752 because he thought the company the only body capable of defending and advancing British economic and strategic interests on the West African coast. When, thereafter, he wrote hopefully about commercial alternatives to the Atlantic slave trade, he had in mind a new, chartered corporation that would make the cultivation of staple crops its principal mission. For Postlethwayt, the problem with the Atlantic slave trade was less the slave trade itself than the traffic in staple crops it seemed to discourage.[16]

In making that case, Postlethwayt echoed what the patrons of the Royal African Company had argued for nearly a half century. Once the company proved unable to compete with the independent traders, it began in its public literature to emphasize the ways that its investments on the coast in forts and men enabled the company to accomplish tasks beyond the competence of individual traders. The Royal African Company deserved its annual subsidy from Parliament, they explained, because it kept the trade open to British merchants,

---

[16] Brown, *Moral Capital*, 269–74.

prepared provisions for shippers arriving from Europe, maintained peaceable relations with trading partners along the Gold Coast, and identified new branches of commerce that might benefit British merchants in the future. There was important work in complementing the Atlantic slave trade, the defenders of the company explained in the early eighteenth century.[17] It was Postlethwayt's innovation, after 1750, to suggest that these alternatives perhaps should not only complement the Atlantic slave trade, but, instead, supplant it. His commitment to diversifying the African trade took him from a defence of the Royal African Company to a precocious attack on the Atlantic slave trade as a whole. Those interested in the history of antislavery ideas have overlooked the ways that the dream of reorganizing the African trade emerged first from the failed campaign to preserve chartered companies and extend their coastal establishments.

The broader cultural and intellectual milieu, of course, lent support to these early arguments for diversifying the Africa trade. Postlethwayt and his successors took for granted certain 'truths' about Africa, Africans, and the dynamics of cross-cultural trade. There was first the idea of tropical abundance, that any commodity cultivated in the American tropics could be transplanted and grown in vast quantities in Africa as well. There was, second, the image of the African hinterland as possessing untold riches, as the centre of wealth and, in Emanuel Swedenborg's famous conception, the locus of spiritual power. With these went a theory about the impact of commerce on social development, that the exchange of commodities civilized primitive peoples, facilitated peace between societies, promoted agricultural improvement and, thereby, encouraged social and economic progress. Africans would produce staple crops rather than captives, it was assumed, if given sufficient incentive to do so. When schemes to colonize took shape, a fourth set of assumptions about land in West Africa came into play – that in some places land was unclaimed, underpopulated, or readily available (with initiative and diplomacy) for European settlement. There were assumptions too, in a few instances, about biology and health: that in time, with sufficient disci-

---

[17] *The Case of the Royal African Company in England* (London, 1730); *The Importance of Effectually Supporting the Royal Africa Company of England Impartially Considered* (London, 1745); *A Letter to a Member of Parliament Concerning the African Trade* (London, 1748); *Answers to the Objections Against the Proposals of the Royal African Company for Settling the Trade to Africa* (London, 1748); *The Case of the Royal African Company of England and their Creditors* (London, 1748); *Antidote to Expel the Poison Contained in an Anonymous Pamphlet, Lately Published* (London, 1749); O'Connor, *Considerations on the Trade to Africa Together with a Proposal for securing the Benefits Thereof to this Nation* (London, 1749); *Considerations on the African Trade* (London, 1750). See also Christopher L. Brown, 'The British Government and the Slave Trade, Early Parliamentary Enquiries, 1713–1783', *The British Slave Trade: Abolition, Parliament and People*, ed. Stephen Farrell, Melanie Unwin, and James Walvin (Edinburgh, 2007), 32–35.

pline, and attention to hygiene, European colonists could settle and survive in Africa over many generations. There were others, but these five – about the soil, the interior, about the social effects of commerce, about access to land, and about the possibilities of survival – each figured prominently in the ideology of legitimate commerce.[18]

Yet, it is experience in Africa and with Africa, more than the image of Africa, which we need to know more about. To understand fully the origins of legitimate commerce it helps to pay attention to what might be crudely categorized as experience and event. For European ideas about what should be done in Africa often depended largely upon what people thought was happening along the Atlantic coast at any given moment. An interest in alternatives to the Atlantic slave trade tended to surface when particular groups found it difficult to compete successfully in it. As the case of Malachy Postlethwayt makes clear, a search for alternatives to and then replacements for the Atlantic slave trade emerged from the ultimately unsuccessful attempt to rescue the Royal African Company and then, afterwards, to institute a successor. In chapter 5 of this volume, Robin Law presents examples from the first half of the eighteenth century when both the Dutch West India Company and the Royal African Company began to experiment with staple crop cultivation because of a decline in the gold trade, or a declining share in the Atlantic slave trade.[19]

This connection between a faltering command of the Atlantic slave trade and a nascent interest in alternatives to it recurred across the late eighteenth century. The decade following the Seven Years' War marked the height of French interest in African colonization schemes during the eighteenth century. This had something to do with the writings of the Physiocrats, who pushed plantations schemes with energy and persistence in the early 1770s. But it also resulted from French concerns about the prospects for the nation's African trade after the loss of Senegal during the Seven Years' War. Étienne François duc de Choiseul directed the botanist Michael Adanson in 1763 to write a report 'outlining the agricultural value of Africa', with the aim of preparing the way for a new French settlement on the mainland just south of the Senegal. Seven years later another French official, M. Saget, proposed the establishment of coffee, cotton, cocoa and indigo plantations along the African coast between the island of Gorée and

---

[18] Curtin, *Image of Africa*, 58–87. This is a subject that has yet to receive the extended study it deserves.

[19] Also see Robin Law, 'King Agaja of Dahomey, the Slave Trade and the Question of West African Plantations', *JICH*, 19/2 (1991), 137–63; Joseph E. Inikori, *Africans and the Industrial Revolution in England: A Study in International Trade and Economic Development* (Cambridge, 2002) 389. I suspect that this was the context for the early Danish schemes of the 1760s as well. See Hopkins, 'Peter Thonning', 808.

the Gambia River. It was his view that the difficulty of competing in the Africa trade, as well as its steadily increasing costs, now required France in 1770 to look for new commercial opportunities there. He thought 'the land opposite of Gorée and along the coast … just as good as that of America'. 'The climate is about the same and coffee, sugar, cocoa will grow here just as easily.' In time, a settlement there might render unnecessary the plantation colonies in the French Caribbean.[20] There was a similar dynamic at work among the Danes in the immediate aftermath of the American Revolution. The new Baltic and Guinea Company established in 1781 made inroads into the Gold Coast trade typically dominated by British and Dutch merchants during the last years of the war, shipping more than 2,500 captives to the Americas annually in the years, 1781, 1782, 1784 and 1785. After the peace, though, and with the return of Dutch traders in particular, the Danish share of the Gold Coast trade returned again to relative insignificance in the early 1790s with the embarkation of less than 500 each year from 1790 to 1795. It was at this time that the Danish government took notice of Paul Isert's scheme to transform the Danish establishments along the Gold Coast into profitable plantations.[21] Schemes to promote legitimate trade appealed particularly to those who saw their stake in the Atlantic slave trade reduced.

The loss of colonies in the Americas also could stimulate a fresh look at the potential for colonies in Africa. In the British Isles, the sudden interest in transforming enterprise on the West African coast in the early 1780s was directly related to the independence of the thirteen colonies in North America. There were more than a dozen schemes – Henry Smeathman's was only one – suggesting that new settlements in Africa could compensate for the loss of provinces in America. This becomes clear in how the need for these projects sometimes was described. The British government might send felons to settle in the Gambia River, since it no longer was possible to ship those sentenced to transportation off to North America. It was suggested, in another instance, that a Gambia colony could serve as a breadbasket for the Caribbean Islands, providing 'corn, and other Necessaries which heretofore they received from America'. Another proposed that the great quantity and fine quality of Gambia cotton could serve as a substitute for the loss of Tobago to France at the peace of 1783.[22] The pattern repeated itself during the French Revolution.

[20] Røge, 'La Clef de Commerce', 438–9.
[21] Serena Axelrod Wisnes, trans., *Letters on West Africa and the Slave Trade: Paul Erdman Isert's Journey to Guinea and the Caribbean Islands in Columbia* (1788), (Oxford, 1992) 12–13.
[22] TNA, HO7/1, 'Minutes of the Committee of the House of Commons Respecting a Plan for the Transporting of Felons to the Island of Le Maine in the River Gambia'; CO267/8, Edward

The slave insurrection in Saint-Domingue and the subsequent emancipation of slaves in French possessions promoted fundamental questions about the future of staple crop production in French dominions. The new foreign minister in 1797, Charles Talleyrand, thought that France should establish new colonies 'whose connexion with ourselves', in his words, 'may be more natural, more useful, and more durable'. These he thought should be established on the West African coast.[23]

The seizure of rival trading posts also could provoke a search for new commercial opportunities. Conquest in time of war inspired fantasies of imperial dominion, not merely commercial influence, after the peace across the late eighteenth century. Charles O'Hara, the first British governor of the new province of Senegambia, thought the nascent colony could become a permanent beachhead for Britain on the African coast. Within a year of arrival, he had devised plans to establish white colonists several hundred miles up the Senegal River, near what he thought to be extensive gold mines and 'prodigious quantities of Rice, Wax, Cotton, Indigo, and Tobacco'. He predicted that, in time, Senegambia would become 'one of the richest colonies, belonging to his Majesty', 'that British colonists would extend over every part of this continent that was worth while to settle'. O'Hara remained committed to this vision of turning Senegambia into an agricultural colony throughout his ten years in power. In 1771, he started construction of a new British fort at Podore, 130 miles up the Senegal, in part because 'the neighborhood … abounds' with a 'variety of different kinds of corn, likewise rice, tobacco, indigo, and cotton', in other words, 'every kind of West India commodity in the greatest perfection'. That idea of what the Senegal trade might become for Europeans persisted after the British left the scene. The French navy destroyed the British forts in the Senegal and Gambia during the American Revolution. The merchants who encouraged that conquest explained that the value of Senegal lay not only in its current exports – slaves, gum, ivory, wax and gold, but also, as Pernille Røge reports, in the value its unlimited terrains offered as a site for the production 'of all the cultural products which enrich the inhabitants of the Americas although at a high cost'.[24]

The importance of these circumstances – a declining position in

---

(contd) Morse, 'A Comparative Statement of the Advantages and Disadvantages to be Expected from the Territory of the River Gambia in the Hands of the African Company or Erected in a Colony'; CO267/20, Daniel Francis Houghton to Thomas Townshend, February 24, 1783.

[23] Røge, 'La Clef de Commerce', 442.

[24] TNA CO267/1, Charles O'Hara to the Earl of Dartmouth and the Board of Trade, July 26, 1765; CO268/4, Charles O'Hara to the Board of Trade, Aug. 1, 1772; Røge, 'La Clef de Commerce', 440.

the Atlantic slave trade, the loss of American colonies, the acquisition of new trading posts in Africa – indicates the principal source for the idea of legitimate commerce before the antislavery movements commenced. Those who conceived schemes for agricultural development in Africa typically were engaged in the African trade already, as employees of the chartered or regulated companies, or as agents of the new imperial centres of scientific data-gathering in the metropolitan capitals.[25] They were positioned perfectly to imagine alternatives to the Atlantic slave trade. They possessed an intimate familiarity with the slave trade but often were weakly committed to it. They had enough knowledge to convey a sense of expertise but not enough to dampen dreams of a radically different future for themselves or for their sponsors. Scholars know the most about Henry Smeathman and Paul Isert because new colonization schemes came to life through their initiative. But a full assessment of projects to recast and diversify the trade with Africa in the late eighteenth century depends first upon knowing more about the Europeans who resided on the West African coast for some length of time. Philip Curtin did not notice, for example, how many of the African projects in the 1780s came from returned veterans from the Province of Senegambia, men like Daniel Houghton and Edward Morse who thought the first failure at colonization should provide lessons for future attempts, particularly with the loss of the American colonies. Ambitions to encourage staple crop production along the Gold Coast become more apparent with more detailed research into the history of forts like Cape Coast Castle. Governor Thomas Melvil introduced cotton and indigo cultivation there in the early 1750s with the aim of complementing, if not competing with, production in the British American colonies, Joseph Inikori reports. Three decades later, Emma Christopher has found, Captain Kenneth McKenzie 'purchased some eighty or ninety African slaves and set them to work clearing the ground ready for planting' at Cape Coast Castle. He aimed 'to grow the same cash crops that slaves produced in the West Indies'.[26]

It is at this level of analysis, through attention to experience, event, and the dynamics of intra-national and international commercial competition that we may locate the early plans for alternatives to the Atlantic slave trade. The idea of legitimate commerce – the idea that a trade in natural commodities should serve as a substitute for the trade in captives – emerged unexpectedly early in European thought, in the 1750s in Britain and with some prominence in France in the 1760s

---

[25] Daniel G. Hopkins makes the same observation in 'Peter Thonning', 784.
[26] Inikori, *Africans and the Industrial Revolution*, 390; Emma Christopher, *A Merciless Place: The Fate of Britain's Convicts After the American Revolution* (Oxford, 2010) 173.

and 1770s. We cannot understand that body of thought fully, of course, without also attending to the intellectual milieu from which it surfaced. This is a subject that needs more extensive study. In the end, though, we need to know not only the sequence of influences that encouraged this set of ideas but also the circumstances that gave them power, that allowed them to exercise remarkable influence over an admittedly small but nonetheless important set of actors concerned to rethink the way Europe interacted with Africa. Their early investment in what later generations would know as 'legitimate commerce' reflected the needs and interests of the individuals in question more than the general intellectual environment of the period.

Elsewhere I have argued that the idea of legitimate commerce not only preceded the Abolitionist movement, but also was an important precondition for it. Without this set of ideas and assumptions, without already existing arguments about the African soil, the climate, the land, the body, the natural resources, the civilizing influence of trade, and the nature of slave-taking in Africa, the Abolitionist movement could not have come into existence. Without a compelling set of alternatives to offer, the discomfort with the slave trade might never have moved beyond the sporadic, occasional protest. Antislavery might never have evolved into an Abolitionist movement.[27]

This brief sketch of the origins of legitimate commerce, though, suggests that its history matters outside of its role in the development of Anglo-American Abolitionism. Both the idea of plantations in Africa and the abolition of the Atlantic slave trade shared what must be described, for the time that they were conceived, as an unwarranted, even absurd optimism. Sceptics regarded them as the products of feverish dreams rather than sober calculation. In coming together, in connecting commercial development schemes with plans to abolish a thriving commercial sector they, in some senses, simply multiplied the absurdity. If neither seemed wise alone, it would be hard to see how they might be a good idea together. But these two fantasies of rearranging the economic order crucially were different in character. The critique of the Atlantic slave trade possessed, by the end of the eighteenth century, indisputable moral credit, even if it was hardly clear that it was a good idea in practice. The idea of staple crop production in Africa, by contrast, even with the insistence on the ways legitimate trade would lead to social and cultural progress, did not possess the same moral standing, at least at first. Its appropriation by the Abolitionist movement rescued these ideas from irrelevancy. That movement granted the ambitions to plant in Africa a legitimacy and an

---

[27] Brown, *Moral Capital*, 314–30.

urgency they might never have had otherwise. At the same time, they extended while also recasting a dissenting tradition in the political history of the British slave trade, a tradition that dated back to at least the early eighteenth century and persisted down to the beginnings of the antislavery movement. If the slave merchants governed British interests in Africa, they did so by containing or silencing the views of others, less fortunate or less interested in the traffic, who thought that the potential of the African trade insufficiently realized. The history of early ideas of legitimate commerce sheds light on that much broader history and suggests that, in important respects, the political history of the British slave trade in the eighteenth century remains to be written.

Appendix

Comparison of extract from Benezet's *Some Historical Account of Guinea*, with passages from Postlethwayt's *Universal Dictionary of Trade and Commerce*

Postlethwayt, *Universal Dictionary of Trade and Commerce* (1751), I, 25 (left unchanged in the second and third editions of the *Universal Dictionary* published in 1758 and 1766):

> The obtaining a competent number of servants to work, as the Negroes at present do, in the colonies belonging to the several European potentates who have settlements in America, does not seem at all impracticable. Europe in general affords numberless poor and distressed objects for that purpose; and if these were not over-worked, as the Negroes particularly are in Maritinico, and in other French colonies, the Europeans would make as good servants for the American planters as the blacks do; and if also all the Europeans were upon a level in regard to the price of labour in their colonies, we cannot but think they would all find their account in Laying absolutely aside the slave-trade, and cultivating a fair, friendly, humane, and civilized commerce with the Africans. Till this is done, it does not seems[sic] possible that the inland trade of this country should ever be extended to the degree it is capable of; for, while the spirit of butchery and making slaves of each other, is promoted by the Europeans among these people, they will never be able to travel with safety into the heart of Africa, or to cement such commercial friendships and alliances with them as will effectually introduce our arts and manufactures among them.

Benezet, *Some Historical Account of Guinea* (1771), 143–4:

> If, under proper regulations, liberty was proclaimed through the colonies, the Negroes, from a dangerous grudging half fed slaves, might a become

able willing minded Labourers. And if there was not a sufficient number of these to do the necessary work, a competent number of labouring people might be procured from Europe, which affords numbers of poor distressed objects, who, if not overworked, with proper usage, might, in several respects, better answer every good purpose in performing the necessary labour in the islands than the slaves do now. A farther considerable advantage might accrue to the British nation in general, if the slave trade was laid aside, by the cultivation of a fair, friendly and humane commerce with the Africans, without which it is not possible the inland trade of that country should ever be extended to the degree it is capable of; for while the spirit of butchery and making slaves of each other is promoted by the Europeans amongst the Negroes, no mutual confidence can take place; nor will the Europeans be able to travel with safety into the heart of their country to form and cement such commercial friendships and alliances as might be necessary to introduce the arts and sciences amongst them, and engage their attention to instruction in the principles of the Christian religion, which is the only sure foundations of every social virtue.

Benezet's debt to Postlethwayt is apparent here, as is his decision to mute the economic benefits of a 'fair, friendly, and humane commerce' and 'manufactures' in favour of science and 'Christian religion'.

7

A Danish experiment in commercial agriculture on the Gold Coast, 1788–93

PER HERNÆS[1]

Denmark was an active player in the Atlantic slave trade and responsible for the export of about 100,000 slaves from West Africa in the period 1660–1806.[2] From the end of the seventeenth century Danish trade centred at Christiansborg Castle in Accra, and by the mid-1780s the Danes could operate from a string of forts along the coast from Accra east to the Volta and beyond: Christiansborg, Fredensborg at Ningo, Kongensten at Ada, and Prindsensten at Keta. The early 1780s had been a boom period for Danish trade, which also sustained geographical expansion of activities on the Coast. By the late 1780s the tide turned, and Danish company trade experienced severe recession. By 1792 the Danish king decreed a ban on Danish trans-Atlantic slave-trading, with effect from 1803.

In 1788, when the slave trade was on the decline and abolition was in the offing, the Danish government supported an initiative by a former 'surgeon' at Christiansborg, Paul Erdmann Isert. He had returned to Copenhagen with plans to establish a Danish agricultural settlement in the hilly area of the Gold Coast hinterland, plans inspired by a wish to create a viable alternative to the slave trade. With government support Isert came back to the Gold Coast and managed to establish a small settlement in Akuapem which he called Friederichsnopel. He died shortly afterwards, and the experiment never became a success although the Danes held on to the place for a number of years.

---

[1] For a more detailed study focussing on Friederichsnopel's social and political history, see Per Hernæs, 'Friederichsnopel: A Danish Settler Colony in Akuapem 1788–93', *Transactions of the Historical Society of Ghana* 13 (2011), 81–133.

[2] About 86,000 slaves were exported on Danish ships; the rest were shipped from Christiansborg on foreign vessels. See Per Hernæs *Slaves, Danes, and African Coast Society* (Trondheim, 1995), 232.

The story of Isert's attempt has been outlined by several authors.³ However, a satisfactory in-depth study and analysis is – to my knowledge – still lacking. My intention, then, is to present an analysis of a number of interesting aspects of Isert's endeavour, based on a scrutiny of primary archival sources. The focus will be on colonial objectives inherent in the enterprise, and an examination of the implementation.

Isert and Schimmelmann

Paul Erdman Isert is well known, mostly due to his book *Reise nach Guinea* published in Copenhagen in 1788.⁴ He had stayed in the Gold Coast from November 1783 until October 1786, when he left with the ship *Christiansborg* for the Danish West Indies, where he remained for another eight months before he returned to Copenhagen. The book was based on his varied experiences on the Coast, in the Atlantic crossing, and in the Caribbean where he witnessed slave auctions and gained first-hand knowledge on the living conditions of plantation slaves. It was evidently his observations which made him a convert to

---

³ Daniel P. Hopkins included a section on 'Paul Isert's Colonial Undertaking' in his article 'The Danish Ban on the Atlantic Slave Trade and Denmark's African Colonial Ambitions, 1787–1807', *Itinerario* 25/3–4 (2002), 154–63. This is the most thorough empirical study to date, and has given much inspiration to the treatment offered in this chapter. Among more extended studies are two MPhil theses: T. Aarsand, 'Abolisjonisme og kolonialisme: P.E. Iserts planer for et dansk koloniprosjekt på Gullkysten 1788–89' (University of Trondheim, 1975); K.L. Berg, 'Danmark-Norges Plantasjeanlegg på Gullkysten 1788–1811' (Norwegian University of Science and Technology, Trondheim, 1997), 38–51. See also G. Nørregård, *Guldkysten*, in the series *Vore gamle tropekolonier*, ed. J. Brøndsted (Copenhagen, 1968), 271–3; Ole Justesen, 'Kolonierne i Afrika', *Kolonierne i Asien og Afrika*, ed. O. Feldbæk and O. Justesen, (Copenhagen, 1980), 403–07; H. Jeppesen, 'Danske Plantageanlæg på Guldkysten 1788–1850', *Geografisk Tidsskrift* 65 (1966), 48–72; R.A. Kea, 'Plantations and Labour in the South-east Gold Coast', *From Slave Trade to 'Legitimate Commerce'*, ed. Robin Law (Cambridge, 1995), 119–43. Isert's project is also mentioned in M.A. Kwamena-Poh, *Government and Politics in the Akuapem State 1730–1850* (London and Evanston IL, 1973); L. Wilson, *The Krobo People of Ghana to 1892* (Athens OH, 1991). For archaeological excavations, see K. Randsborg, 'Fredriksnopel: Denmark's first plantation in Ghana', *Current World Archaeology* 2/8 (2006/7). Finally, I would like to draw attention to Selena A. Winsnes (trans., ed.), *Letters on West Africa and the Slave Trade: Paul Erdmann Isert's Journey to Guinea and the Caribbean Islands in Columbia (1788)*, (Oxford and New York 1992): this includes a section on 'Materials Relating to Frederiksnopel' (227–45) which offers an English translation of a number of central documents regarding Isert's project. The documents were first published under the title 'Dokumenter angaaende de af P. E. Isert foreslaaede Kolonianlæg ved Rio Volta' in *Thaarups Archiv (Archiv for Statistik, Politik og Huushholdnings-Videnskaber)*, III (Copenhagen, 1797–98), 231–68. Unless otherwise indicated, all citations from this body of material in this chapter are from the texts in *Thaarups Archiv* (or from the original documents) and are my own translations into English. However, I acknowledge a great debt of gratitude to Selena Winsnes for having read through the entire manuscript and corrected a number of linguistic errors. Thanks also to Robin Law for his help with the English language.

⁴ Paul Erdmann Isert, *Reise nach Guinea und den Caribäischen Inseln in Columbien*, (Copenhagen, 1788); Winsnes, *Letters*.

Abolitionism. In his view the trade in human beings had been 'to the shame of mankind'.[5] As a botanist and natural historian, Isert was convinced that West Indian crops could be successfully produced in West Africa. In his book he raised the question of why the cultivation of sugar, coffee, cocoa and other produce demanded in Europe had not been established in Africa in the first place, where, he assumed, labour would be obtainable at lower cost and land was easily available. By establishing tropical plantations in Africa, he maintained, it would gradually be possible to stop 'the shameful exportation of Blacks from their happy fatherland'.[6] Such ideas had already surfaced in Abolitionist circles. Contemporary 'colonization' experiments took place in Sierra Leone, and even within the Danish environment the plantation idea had been launched before.[7] The ideas were not new, but Isert apparently developed a rough plan to put the ideas into work, and after his arrival in Copenhagen he was able to gain the attention of the powerful finance minister, Count Ernst Schimmelmann. Further elaboration of the plan apparently took place in collaboration with Schimmelmann's private secretary, Ernst Kirstein,[8] through whom the Count must have influenced the process. The outcome was what might be called the 'Isert-Schimmelmann plan' for a Danish colonial establishment.

Why did the Danish finance minister and leading member of the Council of State support Isert's initiative? Schimmelmann was the major architect behind the Danish Royal Edict of 1792 which banned slave-trading by Danish citizens from 1803.[9] Leading up to the decree were conclusions drawn by the so-called Slave Trade Commission initiated by Schimmelmann in July 1791. Thus, by mid-1791 we know for certain that Schimmelmann wanted to abolish the slave trade, although he adopted a strategy of 'abolition-with-a-delay'. Did he entertain similar visions in 1788 when he supported Isert's scheme? It is hard to believe that he did not, although his ideas were yet to crystallize into a concrete plan of action, and in spite of the fact that he

[5] Winsnes, *Letters*, 148. Isert used the expression 'zur Schande der Menschheit' (*Reise nach Guinea*, 248).
[6] Winsnes, *Letters*, 190.
[7] First by L.F. Rømer, who suggested an experimental enclave on an island in the Volta; then by the Guinea Company in the late 1760s and early 1770s. See L.F. Rømer *Tilforladelig Efterretning om Kysten Guinea* (Copenhagen, 1760), 327–8; for an English translation: *A Reliable Account of the Coast of Guinea (1760) and its Nature*, ed. Selena A. Winsnes (Oxford, 2001). For a discussion of early (Dutch, British and Danish) attempts to establish plantations and gardens in the Gold Coast, see K. Berg, 'Danmark-Norges Plantasjeanlegg', 32–7.
[8] Some key documents carry notes indicating that Isert collaborated with Kirstein. See also Hopkins, 'Danish Ban', 156, 178 n. 17.
[9] For a recent study of the Slave Trade Edict of 1792, including a detailed discussion of Schimmelmann's role, see Erik Gøbel, *Det danske slavehandelsforbud 1792: Studier og kilder til forhistorien, forordningen og følgerne* (Odense, 2008).

owned plantation slaves in the Danish West Indies and had invested capital in Danish slave trade companies in the 1780s.[10]

We can only speculate about Schimmelmann's motives, but apparently there was a shift in attitude in late 1787 or early 1788 when he accepted Isert's proposals. Could it be that Isert served as an eye-opener? Isert's book may have made an impression on the Count, but more importantly Isert offered an initiative whereby plantation agriculture could be transplanted to the Gold Coast. In the long run plantation agriculture in Africa would serve as a strategy of abolition. But there was more to it than abolition as such. In theory, successful commercial production of exportable crops in the Gold Coast would open up new avenues for Danish capital, through investments in trade and production, and would serve the State's interests by supplying the Danish market with tropical crops, and by supporting the Danish economy at large. Probably Schimmelmann saw the initiative in such a perspective, that is as an economic alternative to the slave trade and also as a new basis for the continued existence of the Danish forts on the Coast. He must also have envisioned an important political dimension of Isert's initiative: as it turned out, the idea was not to generate 'African production', but European production in Africa, which required territorial control, in other words colonization. We do not know if Isert initially had such ideas himself or if the initiative came from Schimmelmann. At any rate, the final result of the dialogue between them was a plan to establish nothing less than a Danish settler colony in the Gold Coast. It seems quite likely that Schimmelmann, with Isert as his instrument, saw an opportunity to make an attempt to realize a 'colonial alternative' to the slave trade, which may well have influenced the 'road map' to Danish abolition.[11]

Colonial visions: the 'Isert-Schimmelmann plan'

The obvious model behind the project was the 'plantation system' in the West Indies. However, this system involved more than large-scale export agriculture as such; it represented a special 'social formation' where production was closely tied to and dependent on a particular social and political order. Apparently, Isert and/or Schimmelmann thought it possible to transplant – with certain modifications – more

---

[10] The Baltic and Guinea Company to 1787, and thereafter the private company Pingel, Meyer, Prætorius & Co.
[11] I support Hopkins' view that the 'colonial option', supporting the continuation of the Danish fort establishment in the Gold Coast, played an important role in deliberations regarding the 1792 Edict: 'Danish Ban', 154–5.

or less the whole 'package' from the Americas into Africa. Let us now have a closer look at the 'Isert-Schimmelmann' vision of a 'new Establishment' on the Gold Coast.

First, it is clear that we are dealing with a plan to establish a Danish settler colony; 'white' colonists were to be settled on African land under Danish domination. We are here in fact talking of a 'crown colony' whose inhabitants or 'citizens' were placed under the authority of an administration appointed by the king. Second, the colony was to be based on a 'slave mode of production': the colonists were to be served by African slave labour. Acceptance of slavery was, however, justified by a more humane treatment of the slaves.

Available records reveal the following assumptions regarding major features of the envisaged 'colony':

(1) Agricultural production was to be the main pillar of the colonial economy. Commercial production of tropical export crops such as cotton, coffee, indigo and, in time, sugar was supposed to generate income for the colony as well as the mother country. Colonists were also expected to cultivate food crops to avoid the need for food imports.

(2) The colonial plan also had a distinct cultural dimension in that the Protestant Christian religion would be mandatory for the white colonists.[12] Moreover, it was to be a major goal for the colony 'to propagate the Teachings of the Christian religion among the Blacks'. Also, Protestant morality was to be protected and spread, thus, nudity would be prohibited.[13] It is quite telling that the official designation of the colony was 'The Royal Danish African Mission Institution'.[14] Recruitment of missionaries from the Herrnhuters (Moravian Brethren)[15] was an integral part of the colonial plan.[16] The plan represents an assumption of a direct link between Christian missions and colonial establishment in Africa, and we see here an attempt by the Danish State to 'sell' the colonial project under the guise of a Christian and humanitarian 'civilizing mission'.

(3) The social structure was conceived as strictly hierarchical, with a small elite at the top, a group of commoner colonists in the middle, and the slaves at the bottom of the pyramid. The elite would consist

---

[12] 'Udkast til en Anordning for Kolonien', *Thaarups archiv*, III, 260.
[13] 'Anordning', ibid., 261.
[14] 'Instruktion for Hr. P. E. Isert', ibid., 238.
[15] The Brethren had long run a successful mission among the plantation slaves in the Danish West Indies, and *inter alia* the outcome had been effective pacification of the slaves in prevention of slave riots (Aarsand, 'Abolisjonisme og kolonialisme', 52–3). They also made an attempt to set up a mission at the Danish fort Fredensborg in the late 1760s. Thus the choice of missionary partner was obvious.
[16] 'Udtog af en Skrivelse til Oberstlieutenan og Gouverneur Kiøge paa Christiansborg', *Thaarups Archiv*, III, 239; also, 'Iserts Beretning', ibid., 247.

of a 'Principal', or governor, and his council. This group of officials would be given judicial as well as executive powers. Apart from a separate military corps of officers and soldiers, a militia would be formed by the young men among the colonists 'who still do not own property', lead by senior militia officers selected from household heads, who would also serve as a kind of civil magistrates watching over the general demeanour of their fellow-colonists. Civil and military functions overlapped. Ascribed rank was considered a distinct marker of social status which was expected to be exhibited by military uniform: 'The white Inhabitants of the Colony are to be divided in Classes and distinguished by military Attire which they must not fail to wear on Sundays and Festive Days'.[17]

(4) The 'Isert-Schimmelmann' plan saw slave labour as a pragmatic necessity, but slavery on plantations in Africa was to be a lesser evil than in the West Indies. A number of precautions were taken to ensure humane treatment of slaves in the colony.[18] However, it must be emphasized that, no matter what protective measures were taken, a slave was a slave: slaves were denied any form of 'citizenship' and thus remained outsiders; slave status was conceived as hereditary; slaves were not entitled to landed property, except for small subsistence plots; slaves were considered legal minors in matters concerning economic contracts; and slaves were to be distinguished by dress as a servile social category.

(5) The architects of the colonial plan envisaged colonial territory to be open to free Africans on condition they abide by the laws of the colony, in which case they were allegedly to be considered 'full citizens of this State'.[19] However, they were expected to adopt a European way of life, including the Christian religion, and it is apparent that they were subject to the same kind of paternalism as the slaves. One particular statement underscores that Africans in fact were ranked as 'second class citizens': 'No Property having belonged to a European, and whereto Serfs [slaves] are attached, shall be sold to or can be owned by a Negro'.[20] The master-slave division evidently rubbed off on general relations between Europeans and Africans. The image of a colony planned as a racially segregated society is confirmed by the categorical rule that, 'No marital bonds shall exist ... between Europeans and Negroes'.[21] This was to ensure that European settlers

---

[17] 'Anordning', ibid., 259–60.
[18] Such precautions are listed in 'Instruktion', sections 8–10, ibid., 236, and 'Anordning', ibid., 258–9.
[19] Isert's 'Traktat' (Treaty with the King of Akwuapem), ibid., 249.
[20] Ibid., 258.
[21] Ibid., 261.

formed a community where they always constituted 'a Race, a Nation in themselves'.²²

(6) How did the architects of the plan see future relations with local African societies? One basic assumption was that colonization should happen by agreement or treaty, through friendly persuasion, and 'Purchase, Promise of annual Rent', or other compensation. No 'Violence or Humiliation from white Colonists' was acceptable.²³ Nevertheless, the colonial plan foresaw potential conflict, which required provision for defence, as shown above, and, in time, the colony was expected to become a political actor in the regional African power struggle by entering into alliances with neighbouring 'Nations' and by serving as 'protector' of the weak.²⁴ Behind the mask of benevolence and humanitarianism we may here see a vague intention to establish a future protectorate over African territories.

Summing up, it is obvious that the Isert-Schimmelmann plan went much further than plantation production *per se*. It represents an economic, cultural and political 'colonial programme' which took the shape of an expansive core colony of Danish settlers under the Danish Crown, where Europeans were to be the supreme political and economic class in a racially divided society and whose cultural values were considered superior.

The colony and the 'old establishment'

Isert was given substantial autonomy. The Danish Government granted him the power to conduct negotiations and conclude agreements or treaties with African political leaders to obtain land rights and other services for the colonial settlement, and to buy slaves directly from African traders to establish a labour force. Moreover, he was entitled to deal with all practical matters at his own discretion. He was appointed Captain of the Infantry in order to give him official status, and he was considered as a representative of the Danish government acting to fulfil the 'King's Wish'. He was given the full powers of 'governor' of the new establishment which was to be directly under the Crown, via a Colonial Board in Copenhagen.²⁵

On the other hand the colonial project also depended on Christiansborg, or the 'old establishment' on the Coast. In 1788 the forts were controlled by a private slave-trade company in Copenhagen,

---

[22] 'Instruktion', ibid., 237.
[23] Ibid., 233–4.
[24] 'Anordning', ibid., 262.
[25] See 'Et andet Udkast om Koloniens Bestyrelse', ibid., 262–4.

Pingel, Meyer, Prætorius & Co, or the so-called 'Guinea entrepreneurs'. This raises the question if Schimmelmann in fact had in mind two parallel structures? If so, it is easy to imagine a scenario where two rival authorities were pursuing conflicting interests. However, we should remember that the fort administration was dependent on Government subvention, and the Christiansborg governor was appointed upon approval of the Danish King. This meant that Christiansborg could hardly act in open defiance of government demands. We know that Schimmelmann gave strict orders that Isert was to have full co-operation and assistance from Christiansborg, and since Isert at first would bring no cargo of goods for his undertaking, he was granted a letter of credit for 6,000 rigsdalers, which the governor there was requested to honour.[26]

Still, there would be ample opportunity to work against Isert. Communication between the Coast and Copenhagen was slow; excuses for keeping back assistance could easily be found, and rival trade interests might overrule any concerns about the colonial project. Isert's powers were, in part, a challenge to the existing Company's privileges, and Isert's independent position would also be a challenge to the Christiansborg governor's authority and general standing among African traders and political leaders. The existence of two separate authority structures was problematic. The question remains whether Isert was able to have the full co-operation of the Christiansborg governor and council when he came to the Gold Coast.

The establishment of Friederichsnopel

Isert set out from Copenhagen on the ship *Fredensborg* on 14 July 1788, and arrived at Christiansborg on 14 November. He brought with him a cargo of provisions for the new colony, including miscellaneous tools and equipment and some 'European livestock'.[27] He also brought his young wife and two maids, and a company of thirteen 'colonists' recruited in Copenhagen – two assistants for the colonial administration, and eleven craftsmen, including masons, blacksmiths and carpenters.[28]

The initial plan was to set up the new establishment in the Volta

---

[26] 'Udtog af en Skrivelse til Kiøge', ibid., 240. (Rigsdaler was an element of the Danish currency of the time.)
[27] Isert, 'Beretning', 16 Jan. 1789, ibid., 241, 245.
[28] Rigsarkivet (Danish National Archives, hereafter RA) 399: Finanskollegiet, 1792–94 Papirer og documenter vedr kolonien Frederiksnopel m.m., box 1144: A.R. Biørn, 'Journal for Colonien Friderichs Nopel', Christiansborg, 24 Dec. 1792, 12–20, 39–42, 46.

River area.²⁹ As soon as goods were unloaded and colonists safely ashore at Christiansborg, Isert set off to Fort Kongensteen at Ada and from there continued up the Volta river to a town called 'Malphy' (Mlefi). Here he opened negotiations with local rulers to obtain land for the planned colony. However, for some reason he chose to abort the plan. He returned to Fort Fredensborg, from where he travelled to Akropong in the Akuapem Hills, where he actually founded the settlement of Friederichsnopel at a place called 'Ammano-passo', or Amanopa.³⁰

Why did he choose Akuapem instead of the Volta? Isert himself claimed that the retreat from the Volta was due to its poor soil, unhealthy climate, and lack of building materials. The Akuapem environment had all the virtues which the Volta lacked. He praised the fertility of the soil, the fresh water, the availability of building materials, and the healthy climate, which made Akuapem a pleasant area for newly arrived settlers.³¹ On the other hand, there were also disadvantages. Most important, Amanopa had a transportation problem. The Volta offered water-borne transport, while Amanopa lay inland in hilly terrain and the distance from the coast and from Christiansborg was considerable.³² Isert was aware of the problem but did not find it prohibitive. He had a long-term plan to link the colony to the 'Ponny River' (Kpone lagoon), and to set up a trade depot at 'Sikka', at the estuary of the river, and near Ponny town. His plan also included planting cotton on the Accra Plains, thus expanding plantations towards the coast.³³

Isert probably downplayed the role of political concerns in his abandonment of Malphy. Other evidence suggests this: Governor Kipnasse

[29] Indications of the actual location are rather vague. 'Anordning' uses the phrase 'the most fertile hilly areas' (*Thaarups Arkiv*, III, 255); 'Instruktion' refers to 'the area at Rio Volta in Africa' (Ibid., 233) and 'Autorisation for Hr. P. Erdmand Isert' refers to 'the area of the Danish Possessions on the African Coast, at Rio Volta', as well as 'the fertile hilly Area at Rio Volta' (Ibid., 231).

[30] For 'Amanopa', see Kwamena-Poh, *Government and Politics*, 99. For 'Amannopasso', see RA 371: Generaltoldkammer- og Kommercekollegiet. Indisk kontor, 1833–1848, Den guineiske kommisjon af 9. Januar 1833, box 1040: [P. Thonning?] 'Report on a travel to Akropong and visit to Frederiksnopel', Christiansborg, 29 Aug. 1800. Also, RA, Dept. for udenlandske anliggender, Gruppeordnede sager 1756–1848, Guinea, Litra G. 1775–1847, box 872: ad Guineisk Journal 434/ 1803: P. Thonning, 'Indberetning om det danske Territorium i Guinea fornemmelig med Hensyn til nærværende Kultur af indiske Kolonial Producter eller Beqvem for samme'.

[31] Isert, 'Beretning', *Thaarups Arkiv*, III, 241.

[32] Isert indicated 5 to 6 Danish miles (c. 38–45 km) to the coast (Isert, 'Beretning', *Thaarups Arkiv*, III, 242). Other contemporaries suggested 8 Danish miles (c. 60km) from Christiansborg (RA 365: Generaltoldkammeret (Gtk), 1778–1809 Schimmelmanske papirer, box 412: Kipnasse to Schimmelmann and Brandt, Copenhagen, 7 Dec. 1791).

[33] RA 365: Gtk, 1778–1809 Schimmelmanske papirer, box 412: Kipnasse, 7 Dec. 1791; Flindt to Schimmelmann, 19 March 1791. The plan is not mentioned explicitly in Isert's letter of 16 Jan. 1789, but he gives detailed instructions for the expected ship from Copenhagen to sail 'past Christiansborg about 4 Miles [30km] to a place called Sikka' (*Thaarups Arkiv*, III, 246).

opined that negotiations at Malphy had been unsuccessful.[34] Kipnasse's successor, A.R. Biørn, even intimated that the failure at Malphy was due to machinations by Danish fort officials, Isert's former superiors, who now disliked his attitude and envied his new authority.[35] Confronted by such intrigues Isert might well have felt the need to seek greener pastures elsewhere.

Akuapem was a political choice as much as anything else. Isert knew the importance of political goodwill and support from local rulers, and he had reason to believe that such support would be easier to achieve in Akuapem, which was ruled by 'Duke Attiambo', as the Danes called him, i.e. Okuapenhene (King) Obuobi Atiemo.[36] Atiemo had been a loyal 'friend' of the Danes for more than a decade. His military support of Christiansborg in the Dano-Dutch War of 1777–78 had been of crucial importance.[37] Like Isert, he served on the Danish side in the *Sagbadre* War (Anglo-Danish War) in 1784. Isert had visited his former comrade-in-arms at his residence at 'Kommang' (Kwaman) in 1786, enjoying Atiemo's hospitality and friendship for ten days.[38] Thus, Isert sought the assistance of a personal friend and a loyal ally to the Danes; he went to Akropong, the capital of Akuapem and presented his undertaking to the Okuapenhene.

In Akropong, far away from coastal intrigues, Isert experienced a warm welcome.[39] Evidently the king saw an opportunity to strengthen his 'Danish connection'. As we know, Isert selected Amanopa Hill, a place some distance away from Akuapem settlements, for his prospective colonial capital, Friederichsnopel. Further negotiations resulted in an agreement between the parties, and Atiemo eventually put his mark on a written 'Traktat', i.e. treaty.[40]

Isert's 'Traktat', the text of which might even have been conceived in Copenhagen,[41] had little practical importance, but it is an interesting document, which reveals much about Danish colonial thinking.

[34] RA 365: Gtk, 1778–1809 Schimmelmanske Papirer, box 412: Kipnasse to Schimmelmann and Brandt, Christiansborg, 3 Feb. 1789.
[35] RA 399: Finanskollegiet, Regnskab, box 1149: Pro Memoria, Biørn to Schimmelmann and Brandt, Christiansborg, 20 Sept. 1790.
[36] See Kwamena-Poh, *Government and Politics*, 97.
[37] Hernæs, *Slaves*, 57–63.
[38] See Isert's account of his stay in *Letters on West Africa*, 164–72.
[39] Isert, 'Beretning', *Thaarups Archiv*, III, 242, and (for the German version) RA 365: Gtk, 1778–1809 Schimmelmanske papirer, Guinea, box 412: Pro Memoria, Isert to Schimmelmann and Brandt, Friederichsnopel, 16 Jan. 1789.
[40] See 'Traktat', *Thaarups Archiv*, III, 249–50.
[41] The treaty states that the agreement was concluded between the Danish king and 'Crobbo [Krobo]', but it was signed by Atiemo of Akuapem. Why Krobo? A plausible theory is that Isert brought the document with him from Copenhagen and found no reason to change it when presenting it to Atiemo who was illiterate. References to the 'hilly country near the Volta' as the preferred location for the colony in fact point to Krobo.

We see here an attempt to establish the legality of exclusive rights over territory by reference to the law of property as conceived in Europe. However, land tenure as such was not the real issue. The major point was to establish effective control over territory in the name of the Danish king in order to set up a colony. Therefore, the principle of extraterritorial jurisdiction is also an important element in the treaty.

'Purchase', 'property', 'European jurisdiction', this kind of rhetoric was tailored for a Danish audience, to create the impression that a considerable piece of African territory had been added to the domain of the Danish king. Whether Isert actually believed in the rhetoric of the treaty is an open question. We do know, however, that he used the treaty for all its worth in campaigning for his mission. In his report to Schimmelmann he appears convinced that he had achieved his aim to secure necessary land rights for the planned colony. Based on the treaty he claimed to have obtained 'the right to build and cultivate all over the Dukedom [Akuapem] where no one else had occupied the Land, and this amounts to seven-eighths of a Territory of at least 20 square [Danish] Miles [1,134.6km^2]'.[42] Taken literally, this would mean that Isert laid claim to nearly 80 per cent of Akuapem territory! Absurd, of course, but it served the purpose of making an impression in Copenhagen, and it lived up to the 'Isert-Schimmelmann plan' which required space for expansion.

Apart from the sections concerning land rights and jurisdiction, the 'Traktat' in fact followed a long-established standard pattern for Afro-Danish agreements. The parties swore to 'eternal friendship' and mutual defence;[43] runaway slaves from either side were to be returned to their masters,[44] and no European rivals were allowed in Akuapem unless approved by the colony.[45] Finally, it is worth noting that the treaty offered compensation in the form of 'a voluntary Gift'.[46] No *purchase price* is mentioned, and the treaty also included monthly '*Gage*', i.e. stipends, as well as certain customary allowances, to the Okuapenhene and his Grandees.[47] Thus, the 'Traktat' represents a mix of 'new' and 'old' elements in dealings with African rulers. It is tempting to assume that Isert did not take great care to explain to his counterpart

---

[42] One old Danish mile = 7.532km. Twenty square miles would give 1,134.6 km^2.
[43] 'Traktat', section 4, *Thaarups Archiv*, III, 249. The colony was to assist Akuapem with 'a Quantity of War Materials', and the colony was entitled to military assistance from Akuapem.
[44] The colony offered a commission of 2 rigsdalers per returnee; masters who did not comply were to be fined twice the slave's value ('Traktat' section 6–7, ibid., 250).
[45] 'Traktat', section 5, ibid., 249.
[46] 'Traktat', section 10, ibid. 250: 'For these Rights' Acquisition and Eternal Compliance H.M. the King of Denmark shall pay to the Caboceer of [Akuapem] as a voluntary Gift, —— [sic! No amount is given] in various Goods'.
[47] 'Traktat', section 11, ibid., 250–51.

the implications of the 'new' elements regarding property and jurisdiction.

However, we know that Isert concluded an oral *agreement* with Atiemo, which was confirmed by the African practice of 'eating fetish', that is, by solemn swearing of oaths. The important question is how Atiemo interpreted the agreement? We can safely write off Isert's ownership claims to a large tract of land. The Okuapenhene would certainly not have accepted the surrender of 'seven-eighths' of his territory. At most the Okuapenhene had a customary right to dispose of so-called stool land, which constituted only a part of Akuapem territory, and even then the land remained under the stool until legal claims could be established through long-term occupancy and cultivation. According to Akan ideology the king was considered the custodian of the land and had no right to 'sell' it.[48] Probably, what Atiemo did was to cede certain usufruct rights to land at Isert's choice *within* Akuapem 'unoccupied territory', that is, land to which no one laid any claims. Isert misunderstood this, and probably he included fallow lands in his definition of 'unoccupied'. Fallow lands resulted from the practice of shifting cultivation, and their existence did not mean that claims to the land were relinquished. Thus, the area of 'free' land would have been much more limited than Isert asserted.[49]

Atiemo must have seen the deal with Isert as a 'customary' agreement between an African landlord and a European tenant, in accordance with normal practice when Europeans were given the right to build forts and trade lodges on the Coast. Such agreements were based on expectations of mutual advantage and signified political and commercial alliance. They did not involve exclusive rights over African territory, and stipends and customary allowances paid by the Europeans were invariably considered a form of ground-rent by African rulers. The case of Akuapem did not differ from this pattern, and the Danes were to learn that future plantations in Akuapem could not be established unless further demands for compensation and 'ground-rent' were met.

What advantages did Atiemo expect? Obviously, he had a strategic interest in a Danish settlement in Akuapem. The existing relationship, or alliance, would be strengthened by the physical presence of an 'outstation' flying the Danish flag, and payment of ground rent and other economic assets, such as potential loans, enhanced his status within a kingdom rife with conflict.[50] Most important, it opened a

---

[48] See Kwamena-Poh, *Government and Politics*, 100.
[49] Apart from this, Isert's estimate was far from exact. Europeans had no realistic ideas about distance and square measures of hinterland regions, and Isert's guess was no more than a wild shot.
[50] Cf. the series of internal disputes starting in the 1770s: Kwamena-Poh, *Government and Politics*, 59–61.

channel for military support including access to firearms, which would improve Akuapem's defence and increase the Okuapenhene's power to deal with internal disputes.

Concerns relating to trade might also have played a role. Whether Atiemo envisaged Isert's plans as an opportunity to draw revenue from a 'commercially viable agricultural complex', as speculated by Ray Kea, is an open question.[51] A more sober view would be that Atiemo expected an intensification of exchanges with the Danes and therefore increased trade in Akuapem agricultural produce. But again this remains speculative. Strategic political and military concerns appear as the primary motivation of Atiemo and his leading men.

Whatever his reasons, Atiemo gave full support to Isert. Having selected a site, Isert needed buildings, and people sent by Atiemo built two 'African style' houses, one for Isert and his wife and another for the craftsmen he had brought as colonists. Atiemo also mobilized labour to clear a 1.5 Danish mile (11.3 km) stretch of 'road' through dense forest towards the coast. According to Isert, the total workforce numbered '100 to 200 Negroes'. They worked every day for three weeks, and at a very reasonable cost of 400 rigsdalers, or the approximate value of three male slaves.[52] On 21 December 1788 Isert and his companions took part in a solemn ceremony whereby the houses and the land were formally handed over to him by 'Duke Attiambo and his Council'. Atiemo and two of his most distinguished men planted the Danish flag at the doorstep of Isert's house and swore 'eternal Fidelity and Friendship, in Accordance with the written Treaty concluded earlier between His Majesty the King of Denmark and himself [Atiemo]'.[53] The two buildings, together with the foundations of a stone house, or 'Government House' in Isert's wording, a smithy, a bakery and a small garden, all in a cleared space of 300 *alen* (188.4 m) in circumference, surrounded by bush and forest, constituted Isert's 'city' of Friederichsnopel.[54]

With a few slaves at his disposal Isert started clearing the forest to plant food crops as well as cotton, indigo and tobacco, but managed to establish only a small garden where he planted the seeds of 'most of the European Garden Produce', which appeared to thrive well. So, he claimed, did the small number of European domestic animals which he had brought.[55] Apparently, Isert had also imported a distillery to

[51] See Kea 'Plantations and labour', 127; Berg, 'Danmark-Norges Plantasjeanlegg', 43.
[52] Isert, 'Beretning', *Thaarups Archiv*, III, 243.
[53] Ibid.; see also, RA 399: Finanskollegiet, Papirer og dokumenter vedrørende kolonien Frederiknopel, m.m. 1792, 1794, box 1144: Biørn, 'Journal for Colonien Fridrichs Nopel, 1 December 1788–31 Oct 1789', 24 Dec. 1792, 47.
[54] Ibid.; also RA 399: Finanskollegiet, 1789–1792, Kolonien Frederiksnobels regnskab m.m. (hereafter cited as 'Biørn's Regnskab'), box 1149, 3. One Danish 'alen' = 0.628m.
[55] Isert, 'Beretning', *Thaarups Archiv*, III, 244–5.

produce liquor (rum) from sugar-cane.⁵⁶ Such production did not materialize in his lifetime. He planned to add military barracks and a guardroom to the official 'Government House' already under construction, and to build more private dwellings. Moreover, it was his intention to construct simple defence works consisting of a rampart and moat surrounding the 'city' and fortified by twenty-four cannons. This, he reckoned, would suffice against the forces of any African kingdom.⁵⁷

Isert was preparing for expansion. Overly enthusiastic he encouraged Schimmelmann to increase the number of Danish settlers to 300. He suggested an administration of four civil servants, including the governor, supported by a military corps of 30 soldiers. Further, Isert welcomed missionaries to spread the Christian religion.⁵⁸ When Isert wrote his report, dated Friederichsnopel 16 January 1789, two months after his arrival on the coast, he was convinced that a Danish settler colony in Akuapem would become a reality. As we know his vision never came true. Isert had been ill from the time of arrival on the Coast, and on a journey to Christiansborg illness took its toll and he died at Christiansborg on 21 January 1789.⁵⁹ Isert's dream of a colony in Africa had a tragic personal outcome. In our context the important question now is whether the colonial project died with him.

Reactions at Christiansborg and in Copenhagen

Isert's effort had resulted in the small settlement of Friederichsnopel, which he considered the first step in founding the colony. Apart from the Malphy incident there are no indications that the 'old' slave trading establishment at Christiansborg worked against him. Isert himself did not complain. In his report he expressed gratitude to the Governor and Council for rendering all possible assistance.⁶⁰ After his death, certain conflicts of interest became apparent which the 'Isert-Schimmelmann plan' had not taken into account. Governor Kipnasse pointed out that Isert's authority to buy slaves independently of Christiansborg ran counter to the privileges of the company which he represented and thus put Governor and Council under pressure. He suggested that 'slave colonists' should be provided by Christiansborg at

---

⁵⁶ RA 365: Gtk, Schimmelmanske papirer, box 412: Flindt to Schimmelmann, 19 March 1791.
⁵⁷ Isert, 'Beretning', *Thaarups Archiv*, III, 244.
⁵⁸ Ibid., 247.
⁵⁹ RA 365: Gtk, Schimmelmanske papirer 1778–1809, box 412: Kipnasse to Schimmelmann and Brandt, Christiansborg, 3 Feb. 1789.
⁶⁰ Isert, 'Beretning', *Thaarups Archiv*, III, 247–8.

a set price.⁶¹ Second, Kipnasse disapproved that Friederichsnopel's governor had authority to conclude agreements with African rulers without consulting Christiansborg. He claimed that Isert's treaty had caused some amazement in Akuapem that an agreement had been concluded by Isert alone, without the prior knowledge of the Governor.⁶²

Using the lack of new orders from Copenhagen as justification, the Christiansborg Council decided to put the colonial project on hold until further notice from Schimmelmann. Kipnasse adopted a policy of maintaining the status quo and retained Friederichsnopel at minimum cost. When Atiemo came to Christiansborg to settle economic affairs relating to Friederichsnopel, the Governor reluctantly agreed to pay the sum of '50 Benda in Goods'⁶³, for the land rights, and '15 Benda Goods' for the clearing of land and building of the two houses, a total of 1,040 rigsdalers in Danish currency. Added to this were '6 Benda Goods' which Isert had borrowed from Atiemo. Kipnasse tried to postpone the payment, but gave in when Atiemo threatened to align himself with other European nations on the Coast.⁶⁴ Next, Kipnasse appointed a provisional caretaker of the settlement. Initially he suggested Isert's brother and assistant, Carl Christopher Isert.⁶⁵ However, ultimately the choice fell on 'Ober Assistent' Jens Flindt, an employee at Christiansborg. By mid-February Flindt was sent to Akuapem with orders to 'supervise all things at and under Friederichsnopel'.⁶⁶

In Copenhagen Isert's death did not stop Schimmelmann from pursuing his colonial goals. A new governor, A.R. Biørn, was appointed to replace Kipnasse, and in July 1789 Schimmelmann instructed Biørn to sustain the colonial effort.⁶⁷ Biørn was to act according to former instructions given to Isert a year before, and Schimmelmann urged him to take immediate action upon arrival. As Isert's successor the colonial undertaking now rested upon his shoul-

---

⁶¹ RA 365: Gtk, Schimmelmanske papirer 1778–1809, box 412: Kipnasse to Schimmelmann and Brandt, Christiansborg, 3 Feb. 1789; Kipnasse and Council to Company Board, Christiansborg, 2 Feb. 1789, in *Thaarups Archiv*, III, 266–7.
⁶² RA 365: Gtk, Schimmelmanske papirer 1778–1809, box 412: Kipnasse to Schimmelmann and Brandt, Christiansborg, 3 Feb. 1789.
⁶³ At the time the Danes reckoned 16 rigsdalers (Danish currency) to one 'benda goods'.
⁶⁴ RA 365: Gtk, Schimmelmanske papirer 1778–1809, box 412: Kipnasse to Schimmelmann and Brandt, Christiansborg, 3 Feb. 1789.
⁶⁵ Ibid.
⁶⁶ RA 365: Gtk, Schimmelmanske papirer 1778–1809, box 412: Kipnasse to Schimmelmann and Brandt. Christiansborg 11 March 1789.
⁶⁷ RA 399, Finanzkollegiet, 1789–92, Kolonien Frederiksnopel regnskab m.m. (Biørn's Regnskab), Protocol, box 1149: Schimmelmann and Brandt to Biørn, Copenhagen, 14 July 1789, RA 399, 17–22.

ders. It was still at a trial stage, but sure to be continued unless unexpected difficulties turned up. Inspired by Isert's enthusiastic report, Schimmelmann promised to send at least 30 to 40 colonists, with their families, with the next Africa-bound ship from Copenhagen. He also recommended Isert's plan to plant cotton in the plains between Akuapem and the coast, and to solve Friederichsnopel's transport problem by constructing a road through the cotton fields. By now the Count had realized that to ensure co-operation and common purpose, the colony would have to be placed under the authority of Christiansborg.[68] One implication was of course that Christiansborg from now on undertook the supply of slaves to the colony.

Governor Biørn came to Christiansborg with the 'dual mandate' to promote trade and to sustain the colonial undertaking in Akuapem. What was his position regarding Isert's Friederichsnopel? In his immediate response to Schimmelmann he assured him that he would live up to expectations concerning the consolidation of the 'colony', although he questioned Isert's choice of location.[69] About a year later he filed an extensive report in which he accused Isert of having 'withheld certain truths and voiced opinions which defied reality'.[70] Biørn now argued that the River Volta was much more suitable for a plantation colony than Akuapem. It turns out that he had supported the private initiative of a former company employee, Peder Meyer, who had set up a 'colony', i.e. plantation, at 'Tubreku' (Togbloku) near Fort Kongensteen at Ada. Allegedly Meyer owned 40 to 50 slaves, which saved the cost of buying extra slave labour. Biørn had awarded him an annual salary of 400 rigsdalers until the plantation could sustain him and his slaves.[71] Clearly, Biørn did not remain loyal to the original aim to concentrate efforts on a consolidation of Friederichsnopel. Instead, he now presented a grand colonial plan which involved plantations in the Volta area *as well as* in Akuapem, recruitment of 100 settlers (preferably West Indian planters) and a labour force of 600 slaves, import of 100 mules from Cabo Verde, and provision of two seagoing vessels to serve the colony. Total costs ran to 164,500 rigsdalers. This, Biørn said, was what it would take to achieve 'immediate progress'. Such expenses were unthinkable in Copenhagen, and Biørn must have known this, for he also presented a smaller-scale alternative plan representing a more gradual approach.

[68] Ibid.
[69] RA 399, Biørn's Regnskab, Protocol, box 1149: Biørn to Schimmelmann and Brandt, Honfleur, 22 Aug. 1789, 23–25.
[70] RA 399, Biørn's Regnskab, Protocol, box 1149: Biørn to Schimmelmann and Brandt, Christiansborg, 20 Sept.1790, 25–30. See also Hopkins, 'Danish Ban', 162.
[71] Originally Biørn called the Meyer plantation 'Friederikslund'. Later on it was given the name of 'Frydenlund'. For a brief account, see Berg, 'Danmark-Norges Plantasjeanlegg', 56.

In Copenhagen none of his plans were accepted. Communication was slow and it took another year until Schimmelmann responded. By this time (September 1791), Copenhagen had received various complaints about Biørn's administration. The Count had apparently lost confidence in Biørn's ability to advance the colonial project, which had gained support in connection with the Slave Trade Commission leading up to the abolition edict of 1792. Schimmelmann now had alternative ideas for the colonial endeavour.[72] He told Biørn that until a new 'master plan' had been developed, expenses must be cut down to a necessary minimum to retain Friederichsnopel. No plantations in the Volta area would be supported, and Biørn's payment of salary to Peder Meyer was denied approbation.[73] Biørn's diversion of efforts towards the Volta had led to a half-hearted support of Friederichsnopel since his arrival on the Coast in November 1789. No settlers had been sent to the coast from Copenhagen, as promised, and now his 'colonial mandate' was suspended. All this necessarily influenced the state of affairs at Friederichsnopel. Let us now take a closer look at what happened on the spot.

Friederichsnopel after Isert's death

The trajectory of Friederichsnopel's history reveals that Isert's visions were never followed up. From January 1789 until its abandonment four years later, Danish governors at Christiansborg were in charge of the colony. Their first priority was slave-trading. The 'dual mandate' vested in Biørn remained unfulfilled, partly due to non-compliance with directives from Copenhagen, but mostly because no more colonists were ever sent out from Denmark and no additional material or financial support provided. The colonial effort came to a standstill, and Friederichsnopel faded out. A small community of a handful of Danes and a group of slaves survived until January 1793, when the place was abandoned.

As we have seen, Governor Kipnasse sent Senior Assistant Jens Flindt to Akuapem. He was supposed to work together with Carl Isert and the group of craftsmen who had remained at Friederichsnopel after Paul Isert's death. Upon arrival Flindt learnt that most of the remaining six craftsmen refused to obey orders to clear the land or carry out unskilled labour.[74] Kipnasse ordered four of them to Christiansborg

---

[72] See Hopkins, 'Danish Ban', 163–4.
[73] RA 399, Biørn's Regnskab, Protocol, box 1149: Schimmelmann to Biørn, Copenhagen, 15 Sept. 1791, 30–32. Also Hopkins, 'Danish Ban', 163–4.
[74] Isert brought 11 craftsmen. Two died before Flindt took over; three masons had been employed

where he employed them as craftsmen. In replacement he promised to send '10 to 12 Donco slaves [i.e. slaves purchased from the north] who in that place can be of more use in the work currently carried out there'.[75] Thus, already by March 1789 Flindt had lost most of the craftsmen. The slaves promised by Kipnasse did not appear until July 1789, when five men and five women slaves arrived.[76] However, the labour situation did not improve much. Several slaves tried to escape and Flindt complained that they were so 'old and wretched' that they were useless.[77] On 20 December 1789 they were brought to Christiansborg for examination by newly-arrived governor Biørn. He admitted that the slaves were old and in bad condition. However, further questioning revealed that 'they hardly ever got anything to eat, therefore they were forced to run away to the Akuapems to work for food'. Biørn ruled that with proper care and sufficient food they would recover and sent them back to Friederichsnopel.[78] Flindt was forced to comply, but fell out with the governor. He resigned and left for Denmark on 6 February 1790.[79] Flindt explained his resignation by the lack of support from Kipnasse and particularly Biørn, whom he accused of undermining his work at Friederichsnopel: 'I found it evident that Biørn had pursued his own interest more than that of the King; I therefore requested my demission to go home on the ship *Julianehaab*'[80]

What did Flindt actually accomplish? The two identical houses built under Isert were still there, now described as 'clay', or mud, houses with roofs thatched with grass. So was the baking oven, and the smithy with 'a masonry furnace'. Flindt had been able to add a small frame shack sawmill. Otherwise no buildings had been constructed after Isert, although Flindt had quarried 8,000 stones to finish the 'Government

(contd) by Governor Kipnasse at Christiansborg, and three blacksmiths, two carpenters and one gardener remained at Friederichsnopel: see RA 399: Finanskollegiet. Papirer og documenter vedr. Kolonien Fredriksnopel m.m.: Biørn, 'Journal for Colonien Friderichsnopel 1788–89'; RA 365: Gtk. Schimmelmanske papirer, box 412: Kipnasse to Schimmelmann and Brandt, Christiansborg, 3 Feb. 1789.

[75] RA 365: Gtk, Schimmelmanske papirer, box 412: Flindt to Schimmelmann, Copenhagen, 19 March 1791; RA 365: Gtk, Guineiske uafgj. journal sager 1775–1803, box 1037: Pro Memoria, Kipnasse to Schimmelmann and Brandt, Christiansborg, 25 March 1789 (Document titled 'Gienpart af min Skrivelse af 25 Marty 1788 [sic: = 1789]).

[76] Ibid.; RA 365: Gtk, Schimmelmanske papirer, box 412: Kipnasse to Schimmelmann and Brandt, Christiansborg, 16 July 1789. Flindt stated that the slaves did not arrive until the end of July; Kipnasse claimed that he sent them from Christiansborg on 9 July.

[77] RA 399: Finanskollegiet, 1789–1792 Kolonien Frederiksnobels regnskab m.m., box 1149: Biørn's Regnskab, Journal, 25–54: 'Dag Bog', 10 & 27 Nov. 1789.

[78] Ibid., 'Dag Bog', 20 Dec. 1789.

[79] Ibid., 'Dag Bog', 20 & 28 Jan. 1790; also Journal, p. 13 ('Assistent Flindt. Gage Conto'), 6 Feb. 1790.

[80] RA 365: Gtk, Schimmelmanske papirer, box 412: Flindt to Schimmelmann, Copenhagen, 19 March 1791.

House'. Some additional land had been cleared. The area around the houses had been extended from 300 to 800 *alen* in circumference. For plantations Flindt had cleared a plot measuring 600 by 50 *alen* and 500 to 600 'cotton trees' had been planted.[81] The meagre result is quite telling of the problems he faced.

After Flindt's departure, Biørn put Carl Christopher Isert in charge of Friederichsnopel. The latter had been living there for more than a year and gained some experience. He was assisted by gardener Peder Nielsen, a West Indian freed slave and carpenter called Johannes, and the blacksmith Ole Fynberg. In addition, there was the small group of slaves. Carl Isert remained in office only a few months because he died in Akuapem on 28 July 1790.[82] His successor was 'Ober Chirurgus' (chief surgeon) Johan Herman Grotrian, whose knowledge of cultivation and agriculture was questioned by other staff at Christiansborg.[83] Grotrian remained in office until the end of 1792, when he repatriated with the ship *Gregers Juul*.[84]

Information on the level of activity at the colony over the three-year period 1790–92 is limited. As noted earlier, during the first couple of years Biørn did not give Friederichsnopel priority.[85] However, he claimed to have supplied Friederichsnopel with goods and slaves to the total value of about 8,591 rigsdalers.[86] We also know that the labour situation improved when Biørn increased the population of 'colonist slaves'. During this period about 20 slaves or more lived and worked at Friederichsnopel, as well as a handful of 'inventory slaves'.[87] And some work was carried out. Biørn gave the following status of the colony at the end of its lifetime. The two houses from Isert's time had been renovated. The original smithy was still there, but had a rotten roof. A new mud house had been set up, and a second extra building was used for grain storage and as a kitchen. Fourteen small huts had been built for the slaves. Friederichsnopel's garden had now been fenced. It contained various fruit trees and vegetables. Furthermore,

---

[81] All information above based on RA 399: Finanskollegiet, 1789–1792 Kolonien Frederiksnobels regnskab m.m., box 1149: Biørn's Regnskab, 'Bygninger & Plantager', 3–4.
[82] Ibid., Biørn's Regnskab, Journal, 13–14, & 'Dag Bog', 34–35.
[83] RA 365: Gtk, Schimmelmanske papirer 1778–1809, box 412 (enclosure titled 'Guinea. Breve til Schimmelmann og Scheel 1791–94'): Pro Memoria, Cortnum and Wrisberg to Gtk, Christiansborg, 21 June 1791.
[84] RA 399: Finanskollegiet, 1789–1792 Kolonien Frederiksnobels regnskab m.m., box 1149: Bjørn's Regnskab, Journal, 16–19.
[85] RA 399: Finanskollegiet, 1789–92, Frederiksnobels regnskab, m.m., box 1149: Biørn's Regnskab, Protocol, Biørn to Schimmelmann and Brandt, Christiansborg, 20 Sept. 1790; Biørn to Schimmelmann, Copenhagen, 30 April 1794 (postscript), 49–50.
[86] RA 399: Finanskollegiet, 1789–1792 Frederiksnobels regnskab, m.m., box 1149: Biørn's Regnskab, Protocol, Biørn to Schimmelmann, Copenhagen, 30 April 1794; RA, p. 47.
[87] RA 399: Finanskollegiet, 1789–1792 Kolonien Frederiksnobels regnskab m.m., box 1149: Biørn's Regnskab, Protocol, passim.

Biørn reported that nearly five square kilometres of forest and bush land had been cleared and that there were four 'Rusahr places' or plots where the slaves cultivated maize, yams and cassava for their own subsistence.[88] No 'West Indian' crop other than cotton was planted, and how much of it is not known. There is no indication that any cotton was actually exported. The Friederichsnopel accounts reveal no income whatsoever from cotton. If any was harvested, the output must have been negligible. Clearly, the plantation was a commercial fiasco.

Biørn closed the accounts on 31 December 1792. He left the Gold Coast in January 1793. By then Grotrian had been discharged from Friederichsnopel together with Johannes, the only remaining craftsman, and the slaves were put under the supervision of the current *okyeame*, or interpreter, called Assiong.[89] This meant the end of Friederichsnopel in spite of the fact that Schimmelmann, as noted, as early as September 1791, had taken new initiatives to resurrect the colonial project. He recruited Lieutenant Colonel von Rohr, a 'plantation expert', to investigate the conditions in Akuapem and make a new 'master plan' for colonization.[90] Subsequently, Jens Flindt, the former manager of Friederichsnopel, became von Rohr's assistant and he joined him in St. Croix to prepare for the voyage to Africa. Von Rohr never made it to the Gold Coast. However, Flindt arrived at Christiansborg together with Gilbert Woodard, his assistant, on 24 February 1793 and a couple of weeks later they travelled to Friederichsnopel. Flindt took possession of 15 of the 19 slaves left there, finding the rest too old and sick. However, Flindt and Woodard decided not to continue work at the old place but to establish a new settlement, Friderichsstæd, at the foot of the Akuapem Hills near Dodowa.[91] The days of Friederichsnopel were definitively over, and its successor did not last either. Friderichsstæd was closed down in 1794.

---

[88] RA 399: Finanskollegiet, 1789–1792 Kolonien Frederiksnobels regnskab m.m., box 1149: Biørn's Regnskab, 'Bygninger og Plantagen', 53–54. Biørn's information here must be treated with some caution. Figures may well have been inflated. Considering the fact that Friederichsnopel employed about 20 slaves for nearly three years, and also bearing in mind that Biørn's information could be checked, I find it hard to believe that he would dare to give a totally unrealistic picture of the situation.
[89] RA 399: Finanskollegiet, 1789–92, Frederiksnobels regnskab m.m., box 1149: Biørn's Regnskab, Protocol, Biørn to Schimmelmann, Copenhagen, 30 April 1794.
[90] For the von Rohr initiative in general, see Hopkins, 'Danish Ban', 164; Berg, 'Danmark-Norges plantasjer', 51.
[91] RA 365: Gtk, 1778–1809 Schimmelmanske papirer, box 412: J. N. Flindt and G. Woodard, 'Colloniens Conto ved Compagniet', Christiansborg, 2 April 1793; Flindt to Hammer, Christiansborg, 20 April 1793; Flindt to Schimmelmann, Colonien Friderichsstæd, 22 July 1793; G. Woodard to Flindt, Christiansborg, 7 April 1793. See also Berg, 'Danmark-Norges Plantasjeanlegg', 51.

Conclusion

When Paul Erdmann Isert founded Friederichsnopel in Akuapem in November 1788, he was about to start implementing an experiment on behalf of the Danish government which was part of a greater vision. To Isert the Abolitionist perspective was vital. Move production to where the labour is and lessen human suffering! This simple dictum represents Isert's conviction. It also appealed to the Danish finance minister Ernst Schimmelmann who supported Isert's adventure. He saw Abolition coming and became the driving force behind the prohibition of Danish slave exports from 1803. However, Schimmelmann also focussed on possible consequences of Abolition for the Danish West Indies – which explains the ten-year moratorium, resulting in a boom period for Danish slave exports in the 1790s[92] – and for the Danish fort establishment on the Gold Coast. Slave exports constituted the very *raison d'être* for Christiansborg. The finance minister must have looked for alternative economic opportunities to sustain Danish interests in Africa, and Isert's idea to develop commercial agriculture was a potential option. The 'Isert-Schimmelmann plan' reveals, however, that the goal of the experiment went beyond plantation agriculture as such; it presented a vision of a Danish settler colony in Africa modelled on West Indian plantation colonies characterized by a slave-based economy under European control. Friederichsnopel signifies a quest for a commercial alternative in the event of abolition. Even more importantly, it epitomized a political will in Denmark at this early date to pursue actual, territorial colonization. Danish colonial ambitions were indisputable – although quite unrealistic.

Friederichsnopel was a failed attempt to establish commercial agriculture. In the longer run, however, its importance as a beacon of further Danish initiatives becomes apparent. We have seen that finance minister Schimmelmann wanted to develop a 'master plan' for plantation agriculture in Akuapem. No such plan materialized, but Flindt's short-lived establishment at Dodowa was a result of this initiative, and at the beginning of the nineteenth century the government – in search of an alternative basis for its so-called 'possessions' on the Gold Coast after abolition – subsidized a number of private slave-worked plantations aiming at the production and export of cotton and coffee.

---

[92] See Hernæs, *Slaves*, 232, Table 5. Probably as many as 30,000 slaves from the whole of West Africa were exported on Danish ships in the period 1793–1806.

However, none of them managed to export any West Indian crops.[93]

So, all Danish attempts to establish export-oriented commercial agriculture failed, due to political upheavals, lack of finance and managerial expertise, and emerging competition from the palm-oil trade. I would also argue that the very concept of American-style plantation production was fundamentally wrong in the Gold Coast. In 1831 one Danish observer stated that plantations run by white masters and African slaves would not work: 'If the cultivation of colonial products is to become of importance it must probably be carried out by the natives themselves and to their direct advantage.'[94] He could not have been more right. Increasing palm-oil exports from the 1830s demonstrated the superiority of local peasant farmer production, as did the later development of cocoa-farming. For many decades Friederichsnopel thus provided the Danes with a misguided model of commercial agriculture.

---

[93] For nineteenth-century developments, see e.g. Berg, 'Danmark-Norges Plantasjeanlegg', 59–123; Kea, 'Plantations and labour', 119–44.

[94] RA: Papers of the Guinea Commission of 1833, box 1038: B.M. Christensen, 'Bemærkninger om de Danske Besiddelser i Guinea' (1831), 133.

8

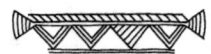

'The Colony has made no progress in agriculture'
Contested perceptions of agriculture in the colonies of Sierra Leone & Liberia

BRONWEN EVERILL

In founding Liberia and Sierra Leone, anti-slavery colonizationists in Britain and America built on the dreams and ambitions of centuries of agricultural planning for Africa. They hoped to establish self-sufficient colonies that would contribute to the production of tropical goods for import into the metropole ('legitimate commerce'), bases from which to operate against the slave trade, and refuges for freed slaves.[1] From the start, however, the many advocates of legitimate commerce in both countries were disappointed by the colonists' apparent lack of enthusiasm for plantation agriculture and the failure of their mission in spreading agriculture to the indigenous Africans. Agriculture was a continual theme in writing by anti-slavery activists interested in these colonies, with new plans for its implementation regularly being formed – from the settlement of new American colonies in Liberia to the 1841 Niger Expedition's model farm. It was also a regular target of both pro-slavery forces and immediate abolitionists, who used reports of the failure of agriculture to lambast the projects. The contested perceptions of West Africa's settler agriculture were carried into the secondary literature also. Most historians of the settler colonies of Freetown and Liberia note that they 'sought wealth through commerce' rather than plantation-style farming.[2] However, in

---

[1] See debates in Robin Law, ed. *From Slave Trade to 'Legitimate' Commerce: The Commercial Transition in Nineteenth-Century West Africa* (Cambridge, 1995); Christopher Leslie Brown, *Moral Capital: Foundations of British Abolitionism* (Chapel Hill NC, 2006), 269–283; Howard Temperley, *British Antislavery 1833–1870* (Columbia SC, 1972), 51; Andrew Porter, "Commerce and Christianity': The Rise and Fall of a Nineteenth-Century Missionary Slogan,' *The Historical Journal* 28/3 (1985), 597–621.

[2] John Iliffe, *Africans: The History of a Continent* (Cambridge, 2007), 161; Philip Curtin, *The Image of Africa: British Ideas and Action, 1780–1850* (Madison WI, 1964), 254; Frankie Hutton, 'Economic Considerations in the American Colonization Society's Early Effort to Emigrate Free Blacks to

both anti-slavery literature and subsequent historiography, the term 'agriculture' hides the multiplicity of expectations the settlers were expected to meet, from combating the slave trade, to establishing self-sufficient utopian settler communities, to spreading the message of civilisation through commerce, to providing a new free-labour arena for plantation agriculture following the abolition of slavery in the West Indies.

This chapter argues that in both Sierra Leone and Liberia, agriculture was portrayed as a 'cure all' that settlers, colonial governments, and metropolitan anti-slavery allies and enemies all expected to take hold in West Africa, but because of a combination of local circumstances, miscommunications and settlers' expectations, it was never perceived as a successful intervention in the slave trade. Like many of the images, stories, and memoirs about Africa in this period, reporting on agriculture in Sierra Leone and Liberia was primarily aimed at making and supporting a particular argument in the metropole, or entertaining European readers. This chapter takes up Philip Curtin's argument that the impressions of Africa and Africans being peddled by metropolitan organisations – both anti- and pro-slavery in this case – were unrelated to the realities on the ground in Africa.[3] With regard to agriculture, competing ideas about the purpose of anti-slavery colonies led to metropolitan confusion about what would constitute 'successful' intervention. Both Sierra Leone and Liberia experienced real practical difficulties in establishing agricultural production. The reasons for these setbacks, however, were almost entirely unrelated to those enumerated by the British and American anti- and pro-slavery presses, who used and reused familiar tropes to promote their arguments in the metropole. As Curtin has argued, 'many affirmations about Africa were made for political, religious, or personal reasons', resulting in reports that were 'sensitive to data that seemed to confirm their European preconceptions and ...were insensitive to contradictory data'.[4]

The first part of this chapter investigates the development of a metropolitan ideal and colonial reality of settler agriculture and anti-slavery intervention in Sierra Leone. British anti-slavery activists used Sierra Leone as a rhetorical tool throughout the early part of the nineteenth century to argue against first the slave trade, and then slavery itself. The metropolitan agricultural development plan was always

Liberia, 1816–36,' *Journal of Negro History* 68/4 (1983), 378; Dwight N. Syfert, 'The Liberian Coasting Trade, 1822–1900,' *Journal of African History (JAH)* 18/2 (1977), 217–235; A.G. Hopkins, *Economic History of West Africa* (London, 1973), 151–3; Martin Lynn, *Commerce and Economic Change in West Africa* (Cambridge, 1997), 23.
[3] Curtin, *Image of Africa*, 318–42.
[4] Ibid., 480, 479.

dependent on local actors – indigenous and colonial. Its expectations, however, were frequently dependent on a de-contextualized model which was unable to recognise small-scale successes or regional differences. Sierra Leone settlers' expectations, their interactions with indigenous Africans, and the miscommunications and misunderstandings created tension between the work against the slave trade in the colony and progress toward a yeoman farmer ideal. Ultimately, the settlers and governors of Sierra Leone achieved a balance between the competing demands of their metropolitan allies, but at the cost of Sierra Leonean settlers' own participation in large-scale commercial agriculture.

The second part examines the situation in Liberia. The American Colonization Society (ACS) responded to perceived developments (or lack of development) based on the networks of communication between settlers, colonial administrators, visitors, speculation, and the propaganda they faced at home. Less concerned with agriculture as a replacement for the slave trade and more concerned with the image of rehabilitated former slaves as free labourers, the ACS and its enemies developed a different set of responses to agricultural development in Liberia. The concluding part of the chapter compares the two examples to show how settler anti-slave trade work and settler agriculture were frequently conflicting goals in West Africa, dictated more by European and American perceptions of Africa's role in the anti-slavery campaign than by the realities of African agriculture. Comparison of the two cases reveals that while many of the stereotypes of the African landscape and environment, black settlers, and indigenous Africans were shared between Britain and America, the colonial realities led to different results both on the ground and in the nature of the contested image of African agricultural production and slave trade prevention.

Sierra Leone

The Freetown peninsula of Sierra Leone was chosen in the 1780s as the site for a new kind of settlement that would combat the slave trade in West Africa. The original plan for the colony proposed that the settlers should be exemplars of a democratic yeomanry, producing their own food and necessities. Despite the early rise of the argument for 'legitimate' commerce to replace the slave trade, the original plan for Sierra Leone was far more concerned with subsistence agriculture, which many of the early settlers participated in and continued to practise.[5]

[5] John Peterson, *Province of Freedom: A History of Sierra Leone, 1787–1870* (Evanston IL and London, 1969), 18–22.

Although Sierra Leone's early commercial advocate, Henry Smeathman, promoted the colony as a chance to replicate the success of the West Indies, the colony's founder, Granville Sharp, was more focused on conducting a constitutional experiment. Sharp, in fact, 'proposed a ban on private landholdings', which suggests that he was not interested in replicating commercial agriculture as it was being practised in the New World.[6] The first settlement of the 'Black Poor' from London in 1787 in Sharp's 'Province of Freedom' embraced this policy. However, problems of disease and local slave-raiding led to the failure of this first experiment.

In order to rescue the project, the Sierra Leone Company was formed, bringing Sharp's yeoman plan into conflict with the Company Chairman Henry Thornton's more ambitious commercial plans.[7] By the time the Sierra Leone Company took control of the colony in 1791, metropolitan ambitions for West Africa had expanded as a result of utopian (and unrealistic) expectations for demonstrating the value of free labour. The arrival of the 'Black Loyalists' from Nova Scotia, who had fought with the British in the American Revolution, spurred the hope that there would be a labour force skilled in cash-crop production. The hope of many anti-slave trade advocates was that the colony would demonstrate that tropical plantation crops of the sort grown in the West Indies could be grown without recourse to enslaved labour.[8] The Company Directors declared that 'all the most valuable productions of the tropical climates seem to grow spontaneously at Sierra Leone; and that nothing but attention and cultivation appear wanting, in order to produce them of every kind, and in sufficient quantities to become articles of trade'.[9]

In 1797, Zachary Macaulay, the Governor of the colony, reported back to the directors that

> The progress of the settlers in cultivating their farms is much more slow than I allowed myself to expect ... they prefer eating casada [i.e cassava] in a miserable way to climbing the Hill, where they may enrich themselves, but when it will no doubt require exertion to do so. But this tho

---

[6] Brown, *Moral Capital*, 316.
[7] Christopher Fyfe, *A History of Sierra Leone* (London, 1962). 27–8.
[8] Suzanne Schwarz, 'Commerce, Civilization and Christianity: The Development of the Sierra Leone Company', *Liverpool and Transatlantic Slavery*, ed. David Richardson, Suzanne Schwarz and Anthony Tibbles (Liverpool, 2007), 257; Suzanne Schwarz, '"Apostolick Warfare": The Reverend Melvill Horne and the Development of Missions in the Late Eighteenth and Early Nineteenth Century', *Bulletin of the John Rylands University Library of Manchester* 85/1 (2003), 65–93.
[9] *Substance of the Report of the Court of Directors of the Sierra Leone Company, 19th October 1791* (London, 1792), 12; Suzanne Schwarz, ed., *Zachary Macaulay and the Development of the Sierra Leone Company, c. 1793–4, Parts I & II* (Institut für Afrikanistik, Leipzig, 2000, 2002); Brown, *Moral Capital*, chapter 5.

a frequent is far from being a universal case. Some twenty or thirty families have fixed themselves in the mountains and are doing well, tho with a little exertion they might do much better.[10]

Despite the exertions of those 'twenty or thirty families', the metropolitan consensus was that the settlers were not interested in cultivation. The perception of the Company's inability to encourage legitimate trade did not hinder later experimenters to pursue the same course. The Sierra Leone Company's failure to make a profit led its directors to petition the British Government to take over the running of the colony. When they agreed, the Sierra Leone Company dissolved and its members formed the African Institution to lobby for the colony. At the time of the government's takeover of the colony in 1808, the African Institution declared that Sierra Leone would be the new British centre for growing cotton in case 'circumstances arise to interrupt our commercial relations with America'.[11] Even with the abolition of the British slave trade, activists maintained an interest in Sierra Leone and its promise for anti-slavery more generally. These competing ideas of what Africa represented for the anti-slavery movement all persisted in various strains in the early governance of Sierra Leone and the gradual development of the colony.

The expectations of the various populations of the Freetown peninsula, their preferences, and their relationships with the constant stream of new Governors in the colony would play an important part in determining the fortunes of commercial agriculture as a replacement for the slave trade. Although the free black settlers from Nova Scotia had bought into the dream of subsistence agriculture, agreeing to migrate to Freetown in 1792 as a result of the promise of free land, the colonial government did not deliver on the free land it had promised them. This weakened the colonial government's position in advocating agriculture amongst the settlers, as rather than pay quit-rents, many gave up farming in favour of trade.[12] However, not only was subsistence still emphasized by missionaries and some humanitarians who preferred Granville Sharp's initial vision of educated yeomanry, it was also the preferred existence of many of those living in the more remote settlements on the Freetown peninsula, or who left the area of British control entirely.

For many in the colony, agriculture was one of a number of co-existing enterprises that were necessary to feed one's family and purchase the necessities of life. By the 1810s, a new group of settlers

[10] Huntington Library: MY 418 Macaulay's Journal, Folder 21, 7 June, 1797, 'Remarks on the Health, Trade, Cultivation, and Civilization of Sierra Leone'.
[11] *Second Report of the Committee of the African Institution* (London, 1808), 6.
[12] J.D. Alie, *A New History of Sierra Leone* (London, 1990), 60.

was arriving in Sierra Leone: those Africans liberated from the slave trade. The so-called Liberated African settlers or 'recaptives' moved regularly in and out of the colony's boundaries to see to their agriculture and conduct trade with both neighbouring groups and long-distance traders. This often upset the authorities, with Governor Turner, for instance, complaining in 1826 about the tendency of the Liberated Africans to 'retrograde in the woods, into a state of nature and barbarism'.[13] However, this practice was typical of the region, where agricultural activities occupied only part of the day and part of the year, with trade or craftsmanship filling out the rest of the labourer's time.[14] Particularly amongst Liberated Africans, a gender imbalance that resulted from the demographic demands of the slave trade, with more male Liberated Africans freed at Sierra Leone than women or children, meant that there was significant intermarriage with the indigenous Temne, Mende and Loko, as well as with existing Liberated African and settler populations, contributing to the expansion of subsistence agriculture as a variety of West African traditional agricultural practices were adapted to the Freetown setting.[15] In the 1832 Census of Liberated Africans in the villages outside of Freetown, 67 per cent of Liberated Africans are recorded as 'Agriculturalists,' with only 12 per cent 'Mechanics' and 20 per cent listed as 'Labourers'. The same census recorded that Liberated Africans in the villages produced 152,746 bushels of cassava and cocoa, 7,390 bushels of Guinea corn, 1,160 bushels of sweet potatoes, and 16,164 bunches of plantains, amongst other produce, including the exportable ginger and arrowroot.[16]

Despite the Liberated Africans' preference for subsistence agriculture, then, export goods were still being produced. Commercial agriculture was developing, even if at a slower pace than anticipated. Exports included 'ship-timber and camwood ... ivory, palm-oil, hides and gold, and a small export of wax, gum, ground-nuts, coffee, arrowroot, dried peppers, starch, ginger'.[17] Although not uncritical of Sierra Leone, even anti-slavery campaigner Thomas Fowell Buxton did

[13] UK Parliamentary Papers, 1826, XXII (389), Governor Turner to Earl Bathurst, 26 Jan. 1826, 4–5.
[14] Hopkins, *Economic History*, 19–21.
[15] Between 1820 and 1861, roughly 66.6% of the slaves disembarked at the Freetown Courts of Mixed Commission were male. Trans-Atlantic Slave Trade Database, ed. David Eltis, Martin Halbert et al. (Emory University, 2008) (TASTD), http://slavevoyages.org/tast/database/search.faces?yearFrom=1820&yearTo=1861&%20slamimpFrom=0&slamimpTo=4000&mjslptimp=60200 (accessed 16 March 2013).
[16] The National Archives, London (TNA): CO267/127 Census and Returns on Liberated Africans, 1832. After 1832 no further census was taken and governors recorded in the annual Blue Books that a proper account was 'impossible'. TNA: CO272/10–38, Blue Books.
[17] Ibid.

acknowledge that: 'The only glimmer of civilization; the only attempt at legitimate commerce; the only prosecution, however faint, of agriculture, are to be found at Sierra Leone ... and there alone the Slave Trade has been in any degree arrested.'[18] While many Abolitionists professed disappointment that the settlers had not taken up plantation-style agriculture, the growing interest in palm-oil revealed that the settlers and anti-slavery activists could still influence the direction of African participation in export agricultural production.

Issues of land allocation and division complicated Sierra Leonean involvement in agriculture. For both the emigrants from the New World and the predominantly Yoruba Liberated Africans, land issues may have been the most significant factor in reducing their inclination to take up commercial agriculture. In the New World, most agriculture took place in isolated and large plantations. In Yorubaland, urbanism was common, with approximately a dozen towns with populations over 20,000. The towns included specialized labour and some leisure class, but most Yoruba people commuted to the countryside to tend their farms for at least part of the day.[19] In Sierra Leone, land presented more of a problem. Unlike the flat landscape of Liberia that would help that colony to create large agricultural plantations, Freetown was pinned between the ocean and steep mountains. This meant that unlike Liberia, where plantations could exist at the edge of the urban areas of Monrovia or Cape Palmas, thereby making elite farming in the model of the American South more feasible, in Sierra Leone commercial farms were unlikely to be present in the capital, and were in fact often established outside the jurisdiction of the colony. Land disputes were also frequent. Despite the emphasis on agriculture in the humanitarian literature, the Governors were responsible for mediating between the needs of the military, the government, the missions, and the settlers. A dispute in 1836 was representative of this ongoing struggle for land. Settlers had established homes and farms along the side of a military barracks. The Governor wrote to the Colonial Office to explain that as land was in short supply and the pursuance of commercial agriculture was an important tenet of the humanitarian efforts in Sierra Leone, the settlers and Liberated Africans living there should be allowed to remain. However, he was in conflict with the military establishment who saw the presence of these Sierra Leoneans as disruptive to military discipline.[20]

Governors played an important role in creating incentives to farm as well. In the 1833 Church Missionary Society report, it was noted

[18] Thomas Fowell Buxton, *The African Slave Trade and Its Remedy* (London, 1839), 365.
[19] Hopkins, *Economic History*, 19.
[20] TNA, CO267 30 April 1836.

that 'A great hindrance to cultivation in this Colony, arises from the difficulty of selling the produce for cash. When the people bring the fruits of their labours to the merchant they are often required to take other articles in exchange for them.'[21] So although trade statistics show that exports of agricultural commodities were not insubstantial, it is likely that by tying producers to barter rather than providing cash, new capital for farm development was not forthcoming. Some Governors attempted to alter this practice in order to support commercial agriculture. In a petition to the British government in support of the former governor, Sir Neil Campbell, Liberated Africans wrote that 'every encouragement to agriculture and industry has been held out to the inhabitants of the several liberated African villages, by Governor Campbell, in the purchase of their arrow-root and other produce'. In fact, the Liberated African petitioners cited an instance that reflects this policy:

> since Governor Campbell has located the young liberated Africans among them, and established large schools, they are enabled to sell their produce on the spot, and the circulation of money which has taken place in consequence, has acted as a stimulus to their industry, and enabled them to clothe themselves, and to purchase those comforts.[22]

However, these policies tended to emanate from the views of individual governors and to change depending on their ideological stance in the fight against slavery and their feelings toward indigenous and British traders and producers. With high mortality rates and high turnover of officials, local government policy frequently changed.

The changes in policy tended to be part of a dialogue with the anti-slavery movement in Britain, where expectations were based on the needs of a changing metropolitan debate. With the re-founding of the Anti-Slavery Society in 1823, aimed at eradicating slavery once and for all, the anti-slavery movement came under attack by the West Indies interests. Sierra Leone, with its many problems, was an easy target in parliamentary debate. In a famous series of letters, James Macqueen wrote to Lord Liverpool arguing that

> The complete failure of every effort which has hitherto been made in and through Sierra Leone, to introduce industry, agriculture and civilization into Africa, leaves the friends and supporters of the place no resource, but to deny boldly that ever any such objects were entertained by those who colonized it, and to assert, that it was merely resorted to as a point

---

[21] Church Missionary Society Archives, University of Birmingham: Section IV, CA 1 M6 Mission Book 1831–34, Report of Mission, for the Year 1833, (J.G. Wilhelm Chr.).
[22] *Addresses, Petitions, &c. from the Kings and Chiefs of Sudan (Africa) and the Inhabitants of Sierra Leone, to his Late Majesty, King William the Fourth* (1838), 8, 11.

from which Christianity, without any reference to industry, commerce and agriculture, might be introduced into Africa.[23]

The colony's struggle either to produce significant cultivated exports – cotton, coffee, sugar, indigo – or to create sizeable new markets of West African consumers for British manufactured exports was deemed a failure by many of the colony's critics. This commercial and agricultural failure was linked to the colony's ineffective suppression of the slave trade and debated in parliament. Although there had been an initial decline in slaving in the Sierra Leone region, between 1820 and 1840 the numbers of slaves embarked in the region soared once again.[24] By making the colony representative of all anti-slavery solutions for all people, the Sierra Leone advocates in the metropole had created for it an impossible measure of success.

This was only one example of the conflicting messages being conveyed to Sierra Leoneans: on the one hand, humanitarians portrayed a utopian vision of free labour plantations; on the other, British commerce demanded a willing market for exports. These competing policies were also expected to produce the means of suppressing the slave trade in the region. As the Sierra Leone authorities would discover, the most effective way of abating the slave trade while simultaneously contributing to the British Empire's commercial expansion was to encourage indigenous West Africans to forgo the slave trade in favour of agriculture or extractive enterprises; it was in fact not through the conversion of settlers into plantation owners and indigenous Africans into their labourers.

In October 1831, Lieutenant Governor Findlay reported on the 'Dissatisfaction of the Chiefs in the neighbourhood of Sierra Leone with the measures taken for seizing the Slaves sent by them thro' the Rivers of the Colony', noting that 'they have stopped the Trade and put the Purrah on the Colony'.[25] Since things had not improved by the end of his time as Lieutenant-Governor, and later Governors were constantly frustrated by the official attempts to end the slave trade through treaties and naval patrols, they instead focused their efforts on one of the two commercial goals – increasing legitimate trade,

[23] James Macqueen, *The Colonial Controversy, Containing a Refutation of the Calumnies of the Anti-colonists; the state of Hayti, Sierra Leone, India, China, Cochin China, Java, &c. &c.; The Production of Sugar, &c. and the State of the Free and Slave Labourers in those Countries Fully Considered in a Series of Letters Addressed to The Earl of Liverpool; with a Supplementary Letter to Mr. Macaulay* (Glasgow, 1825), 88–9. For more on Macqueen, see David Lambert, 'Sierra Leone and Other Sites in the War of Representation over Slavery,' *History Workshop Journal* 64 (2007), 103–32.

[24] TASTD, http://slavevoyages.org/tast/database/search.faces?year%20From=1808&yearTo=1860&mjbyptimp=60200 (accessed 16 March 2013).

[25] TNA: CO714/144, 3 Oct. 1831. 'Purrah' is explained as 'a sort of Ban interdicting trade', referring to a decree of the ruling Poro society.

forgoing their hopes for settler commercial agriculture. They operated in the belief that free and open trade in African produce and British manufactures would quell the trade in slaves. For instance, Governor Henry Dundas Campbell travelled to Magbele, on the Rokelle River, to resolve an ongoing conflict between two factions of the Loko. Instead of insisting that every party to the treaty officially renounce the slave trade, Campbell simply ensured that all routes to the interior from Freetown would be kept open for trade, hoping that the incentive to trade with the colony would itself act as a suppressant to slave-trading.[26]

The fickle nature of both metropolitan and colonial governor's anti-slave trade initiatives created a setback in the development of this policy, however. In the late 1830s, competition from expanding Liberian trade, the arrival of anti-slave trade activist Governor John Jeremie, and vigorous naval and militia attacks on slaving factories along the coast renewed tension between the colony and its hinterland. The trade recession caused by this activity led Thomas Fowell Buxton and other metropolitan activists to conclude that Sierra Leone was unfit for adequately carrying out the multiplicity of anti-slavery goals and activists looked for another solution, this time in the Niger Expedition's model farm and treaty system.[27] As in Sierra Leone, the Niger Expedition represented the confusion of goals and methods of the anti-slavery activists, promoting both plantation agriculture through the model farm, and trade with indigenous communities through the establishment of treaties with local rulers along the Niger. The failure of both of these aspects by 1842 led once again to disillusionment with the anti-slavery mission in West Africa.

An increased focus on trade by Sierra Leone's governors after the disastrous Niger Expedition meant that the colony was able to switch from being a net-importer to a net-exporter and reduce its dependence on grants from the British government. The government's revised position was reflected in the colony's terms of trade during this period. Agricultural exports including palm-oil made up 7.9 per cent of Sierra Leone's overall exports in 1824, and 11.6 per cent in 1837, but the booming palm-oil trade contributed to an overall expansion of agricultural exports and a veritable explosion in Sierra Leonean exports beginning in 1853.[28]

Although palm-oil was produced by Liberated Africans in the villages, the majority of the palm-oil exported from Sierra Leone was

---

[26] TNA, CO267/129 Campbell to Lord Glenelg, 9 Nov. 1835; CO267/132 Campbell to Lord Glenelg, 2 May 1836.
[27] Curtin, *Image of Africa*, 301.
[28] TNA, CO272/1–38, Blue Books.

190 • 'The Colony has made no progress in agriculture'

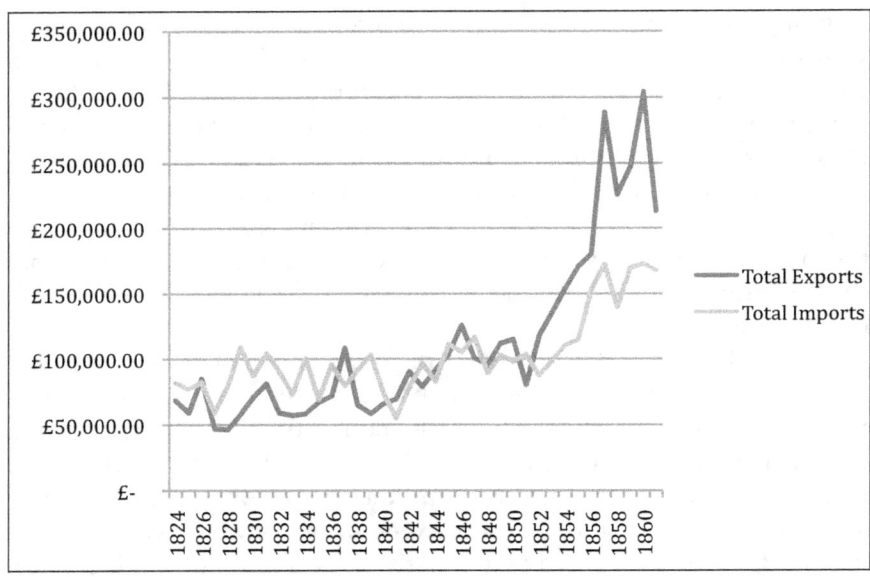

Figure 8.1 Sierra Leone imports and exports in Sterling
(Source: Sierra Leone Blue Books, in the National Archives, London, CO272/1–38)

traded to the colony by surrounding peoples. The colony's shift in focus to middleman trade rather than production of commercial agricultural produce therefore had an impact on both the prosperity of the colony and on the level of slave exports. The period of rising of palm-oil exports finally witnessed a slow overall decline in slave exports from the Sierra Leone region.[29]

However, this turn to trade for agricultural produce, rather than an exclusive focus on a settler yeomanry, reveals that the divided and often utopian metropolitan priorities were not able to be accommodated fully. The indigenous producers relied primarily on domestic slave labour – slaves diverted from the slave trade – in order to produce,

[29] A.G. Hopkins, '"The New International Economic Order" in the Nineteenth Century: Britain's First Development Plan for Africa', *From Slave Trade To 'Legitimate' Commerce: The Commercial Transition in Nineteenth-Century West Africa*, ed. Robin Law (Cambridge, 1995), 247–48; 'Introduction', ibid., 23–6; Martin Lynn, 'The West African Palm Oil Trade in the Nineteenth Century and the "Crisis of Adaptation"', ibid., 57–61; Lynn, *Commerce and Economic Change*; Patrick Manning, 'Slaves, Palm Oil, and Political Power on the West African Coast', *African Historical Studies* 2/2 (1969), 279–88; Paul Lovejoy and David Richardson, 'From slaves to palm oil: Afro-European commercial relations in the Bight of Biafra, 1741–1841', *Maritime Empires: British Imperial Maritime Trade in the Nineteenth Century*, ed. David Killingray, Margarette Lincoln, and Nigel Rigby (Rochester NY and Woodbridge, 2004), 13–29. There were still periods of increased slaving activity, particularly in the late 1830s and early 1840s, and these coincided with conflict between the colony and surrounding areas when these trade/anti-slave trade treaties broke down. TASTD, http://slavevoyages.org/tast/database/search.faces?yearFrom=1807&yearTo%20=1866&mjbyptimp=60200 (accessed 16 March 2013).

extract and especially to transport the cultivated and non-cultivated products being exported by Sierra Leoneans and Britons in the colony. The achievement of some measure of metropolitan anti-slavery success thus resulted from a compromise in the colony's commitment to a free labour ideology and settler commercial agriculture in favour of a mid-nineteenth century emphasis on free trade.

Despite the metropole's vision that commercial agriculture would transform the slave trade from West Africa, it appeared that the most effective way of suppressing the slave trade in reality was the establishment of strong trade networks with the Temne, Loko, and Bulom that depended on Liberated African merchants and itinerant traders from the interior. In 1847, Governor Norman Macdonald wrote to Mohora Suru of Tambaka advising him 'to take an interest in the Trade of the Country and aid and protect strangers and others resorting to the Colony with their produce for the purpose of Trade'.[30] In the same year, a letter to another local ruler, Prince Cain, from Macdonald demonstrated the new sentiments of the anti-slavery settlement: 'I trust that in a very short time your exertions will have completely destroyed the Slave Trade at Cape Mount and its vicinity; and that a good Trade will be established between you and your people and the merchants of this Colony instead.'[31] In Sierra Leone, connections with indigenous groups made it easier (and more friendly to the anti-slavery cause) for Liberated Africans and settlers to act as middlemen for indigenous commercial plantations – particularly those dealing in palm-oil, timber, camwood, bees-wax and other staples of the Sierra Leone trade – than to engage in commercial agriculture themselves.

This had important ramifications as the century progressed. From the late 1840s, Sierra Leonean settlers and Liberated Africans expanded rapidly from their base in Freetown, trading farther into the interior. A group of Yoruba Liberated African merchants left the colony for Badagry (modern Nigeria) at the end of the 1830s and established new settlements, bringing in their wake British anti-slave trade treaties and demand for palm-oil. As a base for both anti-slave trade activity and products such as palm-oil, the Niger Delta was a much richer prospect, and it attracted more Sierra Leonean emigrants over the course of the 1840s and 1850s. The subsequent explosion of the palm-oil trade hints at the role Sierra Leone settlement and anti-slave trade naval presence contributed to the opening of a free port at Lagos.[32] Between 1850 and

---

[30] Sierra Leone Archives: Governor's Local Letters 1846–48, 13 Jan. 1847 to Mohora Suru of Tambacca.
[31] Ibid., 22 May 1847.
[32] Martin Lynn, 'Change and Continuity in the British Palm Oil Trade with West Africa, 1830–55,' *JAH* 22/3 (1981), 331–48, see 335.

1855, imports into Liverpool grew from 30,833 casks of palm-oil to 59,151; imports into London grew from 6,605 casks to 11,898; and imports into Bristol grew from 7,537 casks to 12,121.[33] British West African trade continued to grow over the 1850s. Lagos in particular became increasingly important in the growing palm-oil trade, which reached a peak in 1854–61. The arrival of steam-ships for the palm-oil trade in 1853 contributed to the development of the export industry. The continued growth and expansion of British trade in West Africa contributed to the improved image of Sierra Leone in British public opinion, but also diverted the public's attention away from that original settlement to the potential of other African areas for commercial development. As Lagos grew as a centre for palm-oil exports and anti-slave trade intervention, Sierra Leone was eclipsed, replaced by an interest in the overwhelming success of the legitimate commerce that had replaced the slave trade – the palm-oil trade.[34] In 1861, Palmerston's government supported the annexation of Lagos because the 'occupation of Lagos would be a very useful and important step for the suppression of Slave Trade & for the promotion of Legitimate Commerce'.[35] The experiment in Sierra Leone had demonstrated the potential value of West African produce, but not in the way that its designers had anticipated.

Liberia

Further down the coast of West Africa, Liberia's attempts at commercial agriculture faced similarly conflicting perceptions in America. Liberia's geography, Liberians' relationship with indigenous and family labour, and their system of land use all contributed to the early success of farming in the colony. Despite Liberia's success in promoting commercial agriculture, however, the communication of this success back to the metropolitan population was complicated by assumptions about race as well as discomfort with the labour practices of the colony. After the colony's independence in 1847, these attitudes began to change, helping Liberians to expand their capital investments and promote larger-scale agricultural production.

The American Colonization Society was founded in 1816 with a variety of motives and goals, including the abolition of slavery, the removal of the free black population from America, the spreading of 'civilization', Christian mission and economic expansion.[36] ACS

---

[33] Ibid, 337.
[34] Curtin, *Image of Africa*, 316–17.
[35] TNA: FO2/34, Palmerston Minute Protection of Trade, 22 April 1860.
[36] Bell Irvin Wiley, ed. *Slaves No More: Letters from Liberia 1833–1869* (Lexington KY, 1980), 1.

activists also had a variety of hopes for Liberia: it would be a centre for combating the slave trade; it would provide former slaves with the means to improve their lives and become yeoman famers; it would replace the slave system of the southern plantations with a new form of commercial agriculture; and it would provide a foothold for America's expanding commercial empire in Africa. Many ACS colonizationists believed they were improving the life opportunities of free blacks, giving them a chance for economic self-sufficiency, while at the same time bringing civilization and Christianity to Africa.[37] Liberians were expected to use the land and resources they were given or that they arrived with, just as settlers arriving in America, or migrating to the western territories, did.

However, the argument for agriculture was never as directly tied to the anti-slave trade mission as it was in Sierra Leone. Although the colony was founded with money from the American government for the resettlement of recaptives liberated by the American navy, and between 1827 and 1830 240 recaptives were resettled, by 1835 only 37 more had been sent and by 1839 only nine more.[38] Unlike Sierra Leone, which had the Courts of Mixed Commission to contribute to its intake of recaptives, Liberia had no official court for adjudicating the capture of slave vessels. Compounded with the American government's reluctance to participate wholeheartedly in the anti-slave trade mission, Liberia therefore was slow to get involved in the suppression of the slave trade in its region.

This was not for want of initiative on the part of settlers, many of whom would have gladly participated in militia raids on slaving factories – as they would do at the end of the 1830s – but for want of metropolitan support. ACS members and Liberian settlers were interested primarily in the rehabilitation of former slaves.[39] Liberian settlers saw commercial opportunity in their emigration and took advantage of the availability of land and cheap labour to produce food for local consumption as well as goods for export. ACS supporters hoped to import cotton, coffee and other cash crops from Liberia, with some arguing that this would help reduce America's reliance on slavery in the South. Although this argument changed over time and some ACS supporters and Liberian settlers did explicitly state that their agricultural intervention was helping to hinder the slave trade, there was less

[37] Susan M. Ryan, 'Errand into Africa: Colonization and Nation Building in Sarah J. Hale's Liberia,' *New England Quarterly* 68/4 (1995), 558–83, see 565.
[38] *Tables Showing the Number of Emigrants and Recaptured Africans Sent to the Colony of Liberia by the Government of the United States* (Washington, 1845).
[39] The creation of a domestic ideal reflecting the rise of Victorian family norms and values is dealt with in Marie Tyler-McGraw, *An African Republic: Black & White Virginians in the Making of Liberia* (Chapel Hill NC, 2007), chapter 5.

friction than in the Sierra Leone situation over whether the settlers were supposed to be simple subsistence farmers, commercial producers or traders.

What did affect perceptions of Liberian agriculture in America were networks of communication between Liberians and Americans which helped to exacerbate the pre-existing regional differences between colonists and Abolitionists in America in their attitudes toward the slave trade and agriculture. Two themes dominated the metropolitan discourse surrounding Liberian commercial agriculture. The first was that the settlers were lazy and unmotivated. The second was that the settlers were using slave labour themselves.

The first charge, that Liberians were lazy, demonstrates the important role of information mediation between Liberians and African-Americans by the ACS as well as the ACS's inability in many cases to overcome the propaganda and information 'grapevine' that continued to depict Liberia and Liberians as lazy, unsuccessful and desperate. ACS Agent Joseph Mechlin echoed this view, arguing that 'Another obstacle to the advancement of agriculture arises from the ignorance and indolence of many who are permitted to emigrate. They ... have so long been accustomed to be forced to work, that they will not voluntarily exert themselves'.[40] This was a claim tied to the image of the 'lazy freed slave' that was being circulated around the Atlantic World in the wake of the British abolition of slavery in the West Indies.[41] Some of the colony's officials even complained that the settlers refused to plant in Africa because they preferred imported food.[42] New York's Colonization Society met in 1834 to form their own colony because they felt that 'The colony already established in Africa is more commercial in its character, than [it] is supposed, is most beneficial for the emigrants, or the neighboring population', and it was resolved 'to give their colony a decided agricultural cast, and to make agriculture the controlling, and almost the exclusive occupation of their colonists'.[43]

An investigation of the settlers' attitudes toward land ownership and agriculture by William Allen highlights that Liberians did, in fact, grow and eat indigenous crops.[44] While letters back to America did frequently ask for shipments of wheat or pork products unavailable in Liberia, settlers also wrote back to America about their 'fifteen or

---

[40] *The African Repository and Colonial Journal*, American Colonization Society, 1831 , 260.
[41] Catherine Hall, *Civilising Subjects: Metropole and Colony in the English Imagination, 1830–1867* (Chicago, 2002), 349–51; for a refutation of this depiction of Liberians, see William E. Allen, 'Rethinking the History of Settler Agriculture in Nineteenth-Century Liberia', *International Journal of African Historical Studies* 34 (2004), 435–62.
[42] Allen, 'Rethinking the History of Settler Agriculture', 435–6.
[43] Historical Society of Pennsylvania, *New York Colonization Society* (1834), 16.
[44] Allen, 'Rethinking the History of Settler Agriculture', 435.

twenty acres of land cleared and planted in Potatoes, cassadoes arrow root, corn and about two hundred bushes and about six or seven hundred coffee plants,' or told how they 'have just drawn our five acre lots & are pleased with their situation & soil'.⁴⁵ Even those who complained bitterly of the situation in the country often stated with pride that 'I have got my farm partly cleared down'.⁴⁶ Once they arrived in Africa, it was clear that the majority saw land ownership and cultivation as an important sign of freedom and their connections to the land and to agriculture had not been disrupted by their experiences of slavery.⁴⁷

The charge of slave-holding was more difficult to answer. Thomas Brown, a returned emigrant, submitted himself for examination by the anti-colonizationists in 1833 in order to report on the myths of living in Liberia. When asked 'What are the feelings of the colonists in respect to slavery?' Brown replied: 'I know that some in the colony are disposed to hold slaves. I heard one individual say that the colony would never become anything, that they could never amass wealth without them.'⁴⁸ Settler commercial agriculture was also somewhat more viable than in Sierra Leone, as the self-contained settler population apprenticed indigenous Kru, Gola and Vai, or used them for cheap agricultural work, since they only had to be paid twenty-five cents a day as compared to settler labour at seventy-five cents a day.⁴⁹ Without documentation of the apprenticeships or labourer's contracts, it is difficult to determine to what extent plantation agriculture in Liberia depended on enslaved labour. Large extended families or plantation networks were helpful in providing labour for large commercial plantations.⁵⁰ It is clear, however, that a combination of land policy, environment, agricultural background and labour supply led to something that some settlers resented and others celebrated as being very like the successful commercial plantations of America.⁵¹

---

⁴⁵ George R. Ellis McDonogh to John McDonogh, 14 April 1844 in Wiley, *Slaves No More*, 132; Jacob Gibson to John H. Latrobe and William McKenney, 31 Aug. 1833, ibid., 216. Jacob Gibson was from Maryland. John McDonogh was a slave-holder in Louisiana who gave his slaves the option to work to buy their freedom on the condition that they immigrated to Liberia. Seventy-nine of them chose this route to freedom. See Wiley, *Slaves No More*, 116–17.
⁴⁶ Alexander Hance to William McKenney, 30 Aug. 1835 in Wiley, *Slaves No More*, 218; Paul F. Lansay to John H.B. Latrobe, 16 Jan. 1839, ibid., 219.
⁴⁷ Kimberley K. Smith, *African American Environmental Thought* (Lawrence KS, 2007), 67.
⁴⁸ Historical Society of Pennsylvania, *Examination of Mr. Thomas C. Brown, a Free Colored Citizen of S. Carolina, as to the Actual State Of Things in Liberia in the Years 1833 and 1834* (9 May 1834).
⁴⁹ S.C. Saha, 'Transference of American Values Through Agriculture to Liberia: A Review of Liberian Agriculture During the Nineteenth Century,' *Journal of Negro History* 72/3–4 (1987), 57–65 see 61.
⁵⁰ Allen, 'Rethinking the History of Settler Agriculture', 441.
⁵¹ *Journals of the Rev. James Frederick Schon and Mr. Samuel Crowther, Who with the Sanction of Her Majesty's Government, Accompanied the Expedition up the Niger in 1841 in Behalf of the Church Missionary*

For Liberians, these two images of commercial agriculture in their settlements had a real impact on their ability to convey success to the metropole. The lack of capital to purchase large tracts of land from the government and import the necessary tools and technology for the processing of some cash crops hindered the Liberians' early attempts to grow rich through cash crop production. However, families did try to keep their inheritance intact by forbidding their heirs from dividing landed estates. Isaac Deans, who signed his name with a mark, was in possession of considerable farmland, and directed his heirs that 'neither of them shall sell or cause to be sold any part of the property but the property is to be kept together and inherited from heir to heir'.[52] The distribution of land was one reason why Liberian commercial agriculture was developing successfully. The system of land indenture in Liberia benefited those with large plots who were able to rent them out to other aspiring farmers. However, it also allowed adjacent farmers to expand their plots without requiring them to buy the land outright. A typical indenture lasted anywhere from ten years to the lifetime of the renter.[53] Some merchants used their capital to purchase large plots of land that could then be indentured to aspiring commercial farms, and the ACS itself was willing to sell large plots to those who indicated that they would take up commercial agriculture.[54]

By mid-century, commercial export production was taking place on 224 plantations throughout the Liberian settlements, with more than half of these in the vicinity of Monrovia itself.[55] The *African Repository* optimistically quoted a letter from a settler sent in 1835 that stated that, after the wars of the mid-1830s that resulted in the destruction of the New York and Pennsylvania colonies, 'pressing circumstances have at last, I think, convinced the Colonists of the value of Agriculture. ... Mr. Moore and others of the New Orleans expedition, have cotton growing very beautifully'.[56] An emigrant named Edward Morris from New Orleans wrote that Liberia was 'a healthy colony, well situated for trade, which is greatly on the increase – a good landing place, with a fine river running at the back of the town, with

---

Society (London, 1842), 62–63; Peyton Skipwith to John Hartwell Cocke, 10 Feb. 1834, in Randall M. Miller, *Dear Master: Letters of a Slave Family* (Athens GA, 1990), 58.

[52] Ministry of Foreign Affairs Archives, Monrovia: Last Will and Testament, Isaac Dean, 3 June 1854.

[53] For example, Liberian Archives: Indentures 1833 (2) Joseph Mechlin and Abraham Cheeseman; 1833 (22) Joseph Mechlin and Elijah and Rachel Johnson.

[54] Liberian Archives Indentures; Louis Sheridan for example had 600 acres: Saha, 'Liberian Agriculture', 61.

[55] *African Repository*, XXI (1852), 34; Saha, 'Liberian Agriculture', 61.

[56] *African Repository*, XII (1836), 44.

every accommodation for the landing and shipping of goods'.[57] Indicating the type of commerce that could develop in Liberia, he continued that 'I have planted a farm with three thousand coffee trees, and other produce; my stock of cattle consists of twenty-six head, besides pigs and other animals; my trade with the natives is large for palm oil and other commodities, and upon the whole I am doing very well – thank God for it'. In 1844, Hopkins W. Erskine wrote to ACS Secretary Gurley that:

> Agriculture at the St. Paul River, is becoming quite brisk as far as vegetables is conserned [sic]. The growing of Sugar Cane is becoming quite fashionable, I mean in little quantities … Could requisite sufficient means be preferred to the farmer, your society or some other body of Philanthropic men in America it would be returned to them in the production of our soil in a short time.[58]

After the independence of Liberia in 1847, changing attitudes toward the colony amongst African-Americans helped to promote the image of Liberia as a productive, self-sufficient Republic. This was aided by the introduction of the *Liberia Packet*, a steamship commissioned by the ACS to make regularly scheduled trips between America and the newly independent Liberian Republic. An emigrant wrote in 1849 that he was sending back 'by the *Liberia Packet* one Box containing seventy three pounds of Liberia coffee … the coffee I send is altogether of my own raising'.[59] Randall Kilby, a Liberian emigrant, responded to a letter from John Kilby, writing that 'in Regard to coffe [sic], it is in the Rainy Season and we can not Clean &c; nor gather Coffee. But as soon as the season Returns I will send you Some'.[60]

By the 1850s, renewed interest in the American foothold in Africa grew out of an increasing tone of Anglophobia in the American Congress and recognition that Liberian commercial agriculture was

---

[57] Letter from Sinou, West Coast of Africa, 2 Dec. 1841 in *African Colonization, Slave Trade and Commerce Report of Mr. Kennedy, of Maryland, from the Committee on Commerce of the House of Representatives of the United States* (Washington, 1843), 845–6.

[58] Library of Congress: Records of the ACS, IB29 reel 171 (Hopkins W. Erskine to Gurley 9 Jan. 1844).

[59] Robert Page to Charles W. Andrews, 5 May 1849, in Mary F. Goodwin, ed. 'A Liberian Packet', *The Virginia Magazine of History & Biography* 59/1 (1951), 79. Page refers to 'Liberia coffee,' indicating that this was the native *coffea liberica*, rather than introduced *coffea arabica*. Liberica was more similar to the *robusta* type, which became commercially popular and experienced a production boom in the last quarter of the 19th century, after the variety won the 'Superior Coffee' designation from the 1876 Philadelphia Centennial Exhibition and a prize from Liverpool-based James Irving. See W.E. Allen, 'Historical Methodology and Writing the Liberian Past: The Case of Agriculture in the Nineteenth Century', *History in Africa* 32 (2005), 28–29.

[60] Randall Kilby to John R. Kilby, 26 June 1856, in Wiley, *Slaves No More*.

actually quite successful on a limited scale.[61] Although Liberia was still not recognised as a sovereign country by the American government, individual Congressmen were encouraging of business development in the country. Coffee plantations and palm-oil production also began to develop and by the 1850s exports of coffee, sugar, molasses and pepper had begun. Benjamin Coates, an influential Quaker humanitarian and businessman with ACS connections, began the African Civilization Society, which combined its dedication to 'the destruction of the African Slave trade, by the introduction of lawful commerce and trade into Africa' as well as 'the promotion of the growth of cotton and other products there, whereby the natives may become industrious producers, as well as consumers of articles of commerce'.[62] Coates published a treatise on the cultivation of cotton in Liberia. He argued that since 'the entire yearly consumption of cotton in England alone is upwards of 800,000,000 of lbs., and of this 79 per cent., or more than three-fourths, is raised in the United States, it will be readily perceived how indispensable it is that we should undermine this powerful support of slavery' through the success of cotton production in Liberia.[63]

This interest in cotton production became widespread in the 1850s as the reign of 'King Cotton' in America increasingly worried moderate anti-slavery supporters, who feared that the strength of the South's cotton production would prevent any chance of gradual emancipation. James K. Straw took out a notice in the *Liberia Herald* in 1851 declaring his interest in promoting cotton production in Liberia:

> The undersigned having come to this place for the express purpose of testing the practicability of the culture of cotton, as well as to encourage such persons as may deem it of sufficient interest to try the experiment; begs leave by this notice to inform the people in all the settlements in the Republic of Liberia, that he will award as a premium the sum of fifty dollars to the person who produces the finest five acre plot of cotton; and that he further pledges himself to pay the most liberal price for the proceeds of the same.[64]

---

[61] Both the ACS and the Naval Committee in the House of Representatives played on this idea of Anglophobia, citing Britain's growing strength in the region as a reason for maintaining commercial connections to Liberia and increasing Naval support for the country. *The Thirty-fourth Annual Report of the American Society for Colonizing the Free People of Color of the United States* (Washington DC, 1851), 79; Library of Congress: *H.R. 367* (Report No. 438) 31st Congress, 1st session, 1 Aug. 1850, 1; *Report of the Naval Committee to the House of Representatives, August, 1850, in Favor of the Establishment of a Line of Mail Steamships to the Western Coast of Africa, and thence via the Mediterranean to London; Designed to Promote the Emigration of Free Persons of Color from the United States to Liberia: Also to Increase the Steam Navy, and to Extend the Commerce of the United States. With An Appendix by the American Colonization Society* (Washington DC, 1850), 67.
[62] Benjamin Coates, *Cotton Cultivation in Africa* (Philadelphia PA, 1858), 2.
[63] Ibid., 13.
[64] *Liberia Herald*, 21 May 1851.

Prizes for successful farmers were awarded at annual agricultural fairs in order to encourage the development of cash crops.[65]

Liberian exports were particularly strong to Europe, a fact that did not go unnoticed by American humanitarians. In order to preserve their influence after Liberia's independence in 1847, the commercial agriculture argument began to dominate colonizationist literature in the late 1840s and 1850s.

Liberia's settlers were successfully establishing their country as a base for commercial agriculture in response to the hopes of ACS members and the criticisms of the broader Abolitionist movement. In 1859, Liberia exported $46,515 worth of products more than it imported. Even Sierra Leone was importing goods from Liberia by the early 1860s, although only a limited quantity. In 1862, official statistics show that Sierra Leone imported ten shillings' worth of goods from Liberia; in 1863, that was up to £75 10s., in 1864 £51. and in 1865 £21 18s.[66]

Settlers were clearly not all averse to the idea of commercial agriculture, and in fact, large-scale plantation farming for export in Liberia was moderately successful. Settlers' willingness to adopt the anti-slavery colonizationists' plans for commercial agricultural development were tied into their attitudes toward farming as a respectable middle-class vocation, and were therefore tied into their previous experiences. This made it easier for emigrants to Liberia from southern states or rural areas to adopt commercial farming methods in their new home. Relationships with indigenous groups and availability of land also affected

Table 8.1 Liberian trade returns for 1859

Exports to		
USA	$60,493	
Great Britain	$62,996	
Hamburg	$65,565	
Sierra Leone	$1,315	
Consisting of		
Palm oil	495,194 gal	
Camwood	333 tons	
Ivory	2,335 lbs	(Source: Adapted from Commerce of Liberia – returns for year ending 30 September 1859, Custom House, Port of Monrovia, African Repository, XXXVI (1861), 78–79)
Sugar	19,474 lbs	
Molasses	10, 707 gal	
Coffee	1,007 lbs	
Palm kernels	775 bushels	

[65] American Colonization Society, *Twenty-Fourth Annual Report* (1841), 1; Saha, 'Liberian Agriculture', 63.
[66] TNA: CO272/1–38, Blue Books.

the settlers' attitudes toward commercial agriculture, with those adopting a more dominant role more willing to exploit indigenous labour.

Comparisons

This chapter illustrates two points about the anti-slavery project on the West African coast. Both settlements highlight the environmental and social factors that influenced the type of agriculture that developed; and both highlight the difficulties of aligning anti-slavery ideals with West African realities and of communicating success. In the period from the late eighteenth century to 1861, anti-slavery activists in both Britain and America saw Africa as a blank slate on which to inscribe their plans for a slavery-free world. While attitudes changed after the annexation of Lagos, the subsequent decline in palm-oil revenues, and the outbreak of the American Civil War, in the period leading up to 1861 British and American anti-slavery activists continued to restate and reiterate their hopes for an anti-slavery home base in West Africa. They wanted to abolish the slave trade on the West Coast, and to replace it with 'legitimate' commerce. And finally, they wanted to encourage both the settlers and the indigenous Africans to make something of the supposedly lush tropical abundance in order to prove to the pro-slavery lobby that there was money to be made through free labour.

The debate over the success of commercial agriculture – regardless of the realities on the ground in Sierra Leone and Liberia – is indicative of the wider anxieties which resulted from utopian (and unrealistic) plans, and from miscommunications inherent in the anti-slavery colonization movement that stemmed from the development of imperial and transnational information networks. Rather than adapting their plans and expectations to the evolving situation in West Africa, many pro- and anti-slavery activists simply recycled the familiar images of Africa their supporters expected. With the turn toward the pessimism and social Darwinism of the late nineteenth century, the repeated use of the image of Africans' failures in agriculture moved from a rallying cry for the 'civilising mission' to a darker explanation for white supremacy.

However, through all the fog of metropolitan rhetoric, some subsistence agriculture developed as settlers from agricultural backgrounds translated and adapted their own experiences to their new surroundings. In fact, despite the less than optimum soil and climate conditions for commercial agriculture, both colonies were beginning, by the mid-1830s, to invest in its success. In Sierra Leone, Governors and metropolitan authorities differed on how they would approach the combined goals of abolishing the slave trade, rehabilitating Liberated Africans, and advancing Britain's commercial ambitions for the region. In Liberia, the development of agriculture was almost immediate, but the colony faced considerable trouble convincing its allies in America that this was the case. The 1843 Census of Liberia reveals that many of the stereotypes of Liberian society that proliferated in the Abolitionist press were unfounded. As Tom Shick pointed out of the census findings: 'One hundred and ninety-two farmers were reported which contrasted sharply with only 49 merchants and traders, especially since the literature stresses the settlers' supposed aversion to agriculture and attraction to commerce.'[67] This coincides with local reporting, but contradicts many of the worries of abolitionists and others who hoped to set up an agrarian settlement and were disturbed by the rise of a merchant class.

The different approaches the settlers and their colonial governments took in both Sierra Leone and Liberia were reflective of their understanding of their purposes in migrating to the colonies, their relationships with the indigenous groups, and their ability to communicate effectively with the metropolitan agencies that could offer their support. Differences in environment, land tenure practices, and the types of emigrants and indigenous populations created differences in Sierra Leone and Liberia's approaches to the tensions between anti-slavery, commercial and subsistence aims. Sierra Leone demonstrated that only two out of three of the anti-slavery movement's aims were achievable in conjunction, and the suppression of the slave trade took precedence, particularly because it addressed the needs of the Governors and the settlers, while keeping the colony's critics at bay. As an alternative to the slave trade, settler commercial agriculture never developed in Sierra Leone. However, the metropolitan anti-slavery movement misstated the reasons for this 'failure' and failed to recognise the ways in which Sierra Leone's subsistence agriculture did develop, or the important role Sierra Leonean settlers and Liberated Africans played as middlemen traders in shifting the export economy of West Africa from the slave trade to the palm-oil trade.

In Liberia, commercial agriculture was especially successful starting in the 1840s. Land policies, kinship networks, and the practice of employing cheap indigenous labourers all contributed to the success of coffee, sugar, cotton and rice production in Liberia. However, even here, the communication problems which hinged on the bad press

[67] Tom Shick 'The 1843 Liberian Census: An Analysis of Settler Society before Independence' Indiana University Office of University Archives & Records Management (Bloomington IN), 3.

received by the ACS and the lack of support in the metropole, created by regional experience and communication networks, contributed to the impression in the metropole, and thus in subsequent historiography, that Liberians were either lazy and were not inclined to do 'slave work' or that they were perpetuating practices of the slave society from which they came.[68] This negative impression meant that for many commercial farmers, investment capital for agricultural improvement or production facilities were difficult to come by until the gradual change of American opinion began after Liberian independence in 1847. Changing metropolitan priorities featured in the depiction of Liberian agriculture as well. While in the 1850s growing resentment between the North and South led some to the hope that Liberia might be able to provide an alternative source of cotton, by the early 1860s the focus of metropolitan attention with regard to Liberia was on Britain's aggressive foreign policy under Palmerston and a renewed tension between Britain and America at the start of the Civil War.

The shifting demands of metropolitan expectations, rather than the actual productive capacity, determined in more cases than not, whether Sierra Leone and Liberia were regarded as successful in using commercial agriculture as a replacement for the slave trade. Despite the anti-slavery movement's association with the goal of commercial agriculture in West Africa, in neither Sierra Leone nor Liberia were the metropolitan anti-slavery organizations wholeheartedly behind this approach. While agriculture itself was a consistent theme in writing about the freed slave colonies, Abolitionists and anti-slavery colonizationists were inconsistent in what they envisioned for the colonies. As the needs of the anti-slavery movement changed, so did its proponents' arguments, and Sierra Leone was made to represent the broad canon of anti-slavery solutions. Liberia, representing a narrower strain of colonizationist (rather than abolitionist) thought in America, bore less of that burden. However, the growing debate between African-Americans and the ACS demonstrated the problems of miscommunication and misrepresentation that plagued the colonization movement as it tried to straddle the moderate line between pro-slavery and immediate abolitionist arguments. With a mixture of goals for the colonies, it was inevitable that the settlers would fail in some part.

[68] Tom Shick, *Behold the Promised Land: A History of Afro-American Settler Society in Nineteenth-Century Liberia* (Baltimore MD, 1980), 37; P.J. Staudenraus, *The African Colonization Movement, 1816–1865* (New York, 1961), 153; Eric Burin, *Slavery and the Peculiar Solution: A History of the American Colonization Society* (Gainesville FL, 2005), 151.

9

Church Missionary Society projects of agricultural improvement in nineteenth-century Sierra Leone & Yorubaland

KEHINDE OLABIMTAN

The campaign in Great Britain for the legal abolition of the trans-Atlantic slave trade, from the last quarter of the eighteenth century onwards, always had issues of morality as its central focus. However, during the same period of time that the Abolitionists gradually inched towards their goal of eliminating the trade, the prospect of replacing it with a 'legitimate' one in agricultural produce from Africa also featured prominently in their arguments. Curiously, when the idea of commercial agriculture as a replacement for the trade in human beings became prominent in the struggle it did not excite the same vigorous debate as the morality of the trade. Not even the fact that the tropical environment had proven insalubrious to Europeans tempered the optimism of the advocates of agriculture as an alternative to the slave trade, although the Portuguese, the Danes and the Dutch, as well as the British, had had ample experience of the difficulties of living on the West African coast.

Beyond their casual dismissal of the arguments of the Abolitionists,[1] it appears that opponents of legitimate trade did not seek to draw upon earlier unsuccessful European attempts to create a plantation economy in West Africa, for example the Danish initiative in the Gold Coast from 1788 onwards.[2] Yet it seems unlikely that they could have

---

[1] One such dismissal was 'that even if legitimate trade was more profitable, Africans would prefer the slave trade either because it was "in their blood" or because it involved less physical labour'. Missionaries were also blamed for 'raising a generation of Africans who preferred the pen to the plough and who despised those who worked on the land'. J.B. Webster, 'The Bible and the Plough', *Journal of the Historical Society of Nigeria (JHSN)* 2/4 (1963), 430.

[2] Paul Erdman Isert's venture on the Gold Coast in 1788–89 was one of the early attempts to set up an agricultural plantation in West Africa with the aim to subvert the trans-Atlantic slave trade. His experiences and his death foreshadowed those of others who would later follow in his trail. See Per Hernæs, this volume, chapter 7.

been oblivious of this reality. It is not clear why the advocacy of commercial agriculture could have been sustained over more than half a century, until the disaster of the 1841 Niger Expedition, despite having yielded no lasting results, without provoking a more vigorous debate, which could have drawn from experience of earlier failed agricultural initiatives in West Africa.[3]

Among the earliest advocates of the agricultural potential of Africa as an alternative to the slave trade was Henry Smeathman, a British naturalist who had spent a period on the Banana Islands, off the coast of the future British colony of Sierra Leone, in the 1770s. Smeathman related to a friend in Britain a picture of a romantic paradise in Sierra Leone: 'Pleasant scenes of vernal beauty, tropical luxuriance, where fruit and flowers lavish their fragrance together on the same bough! ... I contemplate the years which I passed in that terrestrial Elysium, as the happiest of my life.'[4] What began as a one-man crusade eventually resulted in the establishment of the colony of Sierra Leone. When the British government took over the administration of the colony in 1808, the slave trade having being declared illegal the previous year, it was obvious that the government had accepted the idea that promoting agriculture in the colony would ensure the death of the now illegitimate trade in human beings.

Smeathman's idea of this agricultural paradise greatly influenced the arguments of the Abolitionists in favour of agriculture as an alternative to the trans-Atlantic slave trade, partly because it seemed to be confirmed by other first-hand testimony. For example, Olaudah Equiano, who was well-known in the circle of the Abolitionists, in his 1789 autobiography painted his home country in Africa (Igboland, in what is today south-eastern Nigeria) as an agricultural paradise, 'a country where nature is prodigal of her favour'.[5] Smeathman's romantic picture and Equiano's personal witness were mutually reinforcing.

Through their political influence in parliament and government, the Abolitionists shaped early British government policy on Sierra Leone. This is evident in the 1808 government memorandum to the

---

[3] The pro-slavery elements in British politics harped mainly on the perceived morality of the trade and the cost of the coastal blockade by the British navy. With the eventual declaration of freedom for the slaves in the British colonies in the New World in 1833, the planters considered the possibility of migrant labour from the West African colony in 1842. Certainly, by then, they were aware of the unhealthiness of the West African climate for the constitution of Europeans, a fact the Niger Expedition finally confirmed.

[4] Quoted in Adam Hochschild, *Bury the Chains – Prophets and Rebels in the Fight to Free an Empire's Slaves* (New York, 2005), 145.

[5] Olaudah Equiano, *The Interesting Narrative of the Life of Olaudah Equiano, or Gustavus Vassa, the African. Written by Himself* (9th edn London, 1794), 12.

first Governor of the crown colony, Thomas P. Thompson, which drew considerably from an earlier memorandum submitted by the Abolitionist and veteran of the colony, Zachary Macaulay, to Lord Castlereagh, the Foreign Secretary.[6] The memorandum to Thompson was loaded with ideas which the Abolitionists cherished and which would be developed further in T.F. Buxton's 1840 publication, *The African Slave Trade and Its Remedy*.[7] Particular notice may be taken in this memorandum of the government's favourable disposition towards immigrants who would set the indigenous people 'an example of profitable cultivation'. It also encouraged the newly appointed Governor to secure treaties with local rulers who would, among other things, give 'privileges and immunities in favour of British planters or traders who may settle among the natives'.[8]

Meanwhile, the emergence of the Church Missionary Society (CMS) in 1799 from among the Abolitionists at Clapham provided a religious approach to realizing the same end of transforming Africa, by taking Christianity to the people and thereby atoning for the 'manifold wrongs' Europeans had committed against them in the slave trade.[9] Beyond the schools they maintained as a means of achieving this religious aim, the CMS home committee and the missionaries who worked in Sierra Leone gave little attention to the dismal economic situation of the colony up till the 1830s. On the other hand, legitimate trade in the colony and its adjoining territories became synonymous with the exploitation of forest resources in the hinterland rather than developing a plantation economy. The role of the colony in this was necessarily restricted to trade, rather than involvement in production. Conversely, the social 'vices' which supposedly attended the trade on the rivers were particularly loathsome to the missionaries who, for many decades, in their West African missions, consequently saw trading as a veritable source of temptation to their converts.

Apart from the negative perception of the missionaries, the 'legitimate trade' in forest resources brought its own problems. It caused complications for the peoples neighbouring the colony, as the interior country became unsettled, power having been diffused beyond the

---

[6] Zachary Macaulay was an eminent Abolitionist and had returned to Britain after his tenure as the Governor of Sierra Leone with fifteen Susu boys to be educated there.

[7] Buxton acknowledged that he drew from earlier arguments offered by others before him in making a case for the *Remedy*. Thomas F. Buxton and Charles Buxton, ed., *Memoirs of Thomas Fowell Buxton, Baronet with Selections from His Correspondence* (Philadelphia, 1849), 369.

[8] The National Archives, London, CO267/24, Memorandum for T.P. Thompson, dated 11 April 1808, quoted in E.A. Ijagbemi, 'The Freetown Colony and the Development of "Legitimate" Commerce in the Adjoining Territories', *JHSN* 5/2 (1970), 249.

[9] Eugene Stock, *The History of the Church Missionary Society* (London, 1899), 99.

traditional rulers through the emergence of 'new men' whose position was based on financial resources, rather than traditional office.[10]

Renewal of the rhetoric of 'legitimate trade'

Two developments in the second half of the 1830s revived among the Abolitionists at home and abroad the idea of promoting agriculture in West Africa as a means of supplanting the trade in slaves. First was the fact that the legal banning of the slave trade did not appear to have reduced the scale of the trade. If anything, it was thought that the slave trade in West Africa had actually increased. Alarm at this led Buxton in 1840 to renew the call for the development of commercial agriculture, alongside mining and forest extractive activities, as a means of supplanting the trade in human beings.[11]

Buxton's *Remedy* appeared at a time when the leadership of the CMS home committee was in transition, bringing the Revd Henry Venn into office as the Honorary Secretary of the Society. Although the Society responded very positively to Buxton's thesis and contributed personnel to the Niger expedition that followed in 1841, it did nothing to realize the socio-economic vision of the expedition in the immediate aftermath of its disastrous result. The vision appeared to have gone into limbo.

What could have been a wake-up call came to the Society in Sierra Leone in the form of the emigration scheme (to recruit workers for the West Indies) promoted by the colonial government in Freetown in 1843. It was evident that various agricultural schemes proposed in the colony had failed. Moreover, planters in West Indies had no motivation to invest in West Africa, where an unhealthy climate decimated European settlers. But Sierra Leone had many mission-educated, energetic young people with bleak prospects of gainful employment in the colony. The attraction of the scheme for the plantations in the West Indies was therefore obvious, and the planters lost no time in encouraging the British government to endorse their immigration scheme. The government responded in 1844 by withdrawing part of the economic support that had sustained the receptive population in Sierra

---

[10] Ijagbemi noted that, 'the series of upheavals that engulfed these territories ruined ... trade, spreading chaos and depredation in the process': 'Freetown Colony', 253.
[11] Buxton's son noted that in *The African Slave Trade and Its Remedy* Buxton 'established the fact, first, that gold, iron, and copper, abound in many districts of the country; secondly, that vast regions are of the most fertile description, and are capable of producing rice, wheat, hemp, indigo, coffee, &c, and above all, the sugarcane and cotton, in any quantities; while the forests contain every kind of timber – mahogany, ebony, dye-woods, the oil palm...': Buxton and Buxton, *Memoirs of Thomas Fowell Buxton,* 371.

Leone.[12] The stir this created in the CMS missionary circle could have been a motivation for the missionaries to take an interest in the economic prospects of their pupils and converts. Regrettably, the controversy the scheme generated among them only cooled off with time.

The second development that brought agriculture to the thought of the CMS mission came in the critique offered by the Reverend Ulrich Graf, the resident missionary in Hastings, Sierra Leone, in 1845. Graf's unsparing critique of the colony-born young people jolted the Sierra Leone mission, exposing the poor results of the cultural transformation the missionaries believed themselves to have been achieving in the colony. But more significant is the fact that, although this had not been his intention, Graf's intervention caused the mission to realize that the need for agricultural development in Africa went beyond using it as an alternative to the slave trade; it was germane also to the realization of the goal of mission, that is raising a transformed people in a transformed society.

In recommending agricultural development as a panacea for what he considered a negative trend in the social values of the colony-born youths in Sierra Leone, Graf drew much from Buxton and earlier arguments advocating agriculture as a means of transforming the colony both socially and economically. Graf's critique was the final call that stirred the CMS to take seriously the economic condition of the colony in which the Society was doing its mission work. It was auspicious that he found in the new Honorary Secretary of the Society, Henry Venn, an ally who shared his vision of the need for the mission to be more involved in the mundane realities impinging on the work of the Society.

Henry Venn followed up on Graf's recommendation of improved agricultural practice in the colony, but he had a broader picture of the situation than his Basel-trained agent in Hastings. His papers between 1845 and 1861 in which he articulated what he considered as the desirable end of mission, for the first time since the emergence of CMS, brought together the visions of the humanitarians and the missionaries. In the process, he not only articulated Buxton's pairing of the Bible and the Plough; he tied the goal of the civilizing mission to the need to appreciate the peculiar cultural contexts of the emerging non-western churches. This was in contrast to the European cultural tastes and values some humanitarians and missionaries were promoting abroad and which sometimes created tension between them and indigenous peoples.[13]

---

[12] Church Missionary Society (CMS) Archives, Cadbury Research Library, University of Birmingham: C/A/M11(1843–1845)/336–7, Lord Stanley to Governor Macdonald, 10 Feb. 1844.
[13] J.F. Ade Ajayi, 'Henry Venn and the Policy of Development', *JHSN* 1/4 (1959), 331–43.

Although Venn's thought matured over the sixteen years from 1845 to 1861, during which he promoted missionary and agricultural initiatives in Sierra Leone and in the fledgling Yoruba mission at Abeokuta, his understanding of how the Bible should connect with the plough in mission work was only vindicated long after he had left the scene and when the CMS had jettisoned his idea. At a time when the connection between indigenous church forms and socio-economic transformation still appeared far-fetched, Venn saw the need to facilitate the process of ameliorating the dire economic situation that dogged the work of the mission in Sierra Leone.

The Sierra Leone agricultural initiative

The perceived need to address the economic realities of Sierra Leone led the CMS home committee and missionaries in the field to push for the formation of the African Improvement Society (AIS) in Sierra Leone in 1849. Its objective was to facilitate the economic transformation of the colony, especially through agricultural development, in partnership with the London-based African Civilization Society. It aimed to achieve this by sponsoring agricultural shows among the farmers in the colony, at which prizes would be awarded to stimulate excellence in production. The AIS also planned to establish a model farm and a museum of modern agricultural implements in Freetown as a means to promote among the local farmers, the cultivation of certain marketable products. This model farm was to be in partnership with the Royal Botanical Garden in Kew (RBGK), London, exchanging plant species between Britain and Africa. The AIS would also facilitate trade with Britain by standardizing and certifying agricultural products from the colony that would be marketed there.

As the peasant farming of their receptive parents was unattractive to colony-born children, the initiative of the AIS held out the prospect of commending agriculture to this rising, anglicized generation as a vocation worthy to be pursued. To effect this, the society also set up two annual prizes in essay-writing among these young people on how to improve the colony. This, it was hoped, would encourage writing skills and stimulate them to do the necessary hard thinking about the future of their society. Obviously, if this succeeded, their ideas would be a vital contribution to realizing the objectives of the AIS.

The London-based partner rightly saw the need for its African counterpart, the AIS, to take the initiative as to what could be done to improve the economic situation of the colony; but it was a mistake to have saddled the missionaries with the responsibility in the bid to ensure that its material contribution to the project was judiciously managed.[14] In the first place, committing such a revolutionary assignment to people who had their own immediate responsibilities was a miscalculation that laid the foundation for the failure of the AIS. Some of the missionaries did not even recognize the problem that caused their colleague to call for the mission's involvement in the material well-being of the colony. Second, the missionaries lacked the expertise required to generate the kind of agricultural activity envisaged. Third, the 'party spirit' of the missionaries in the colony was a major repellent that endangered the scheme.[15] It is no wonder that the AIS was unsteady from the beginning as much of the nurturing of the project was left to Graf and his co-opted assistant, Revd Peyton, as other missionaries were unwilling to add its demands to their existing commitments.

At any rate, the AIS actively functioned for only two years, conducting two agricultural shows in Freetown in February 1849 and February 1850, and organizing periodic lectures. The first agricultural show attracted only fifty farmers in a colony where farming was intended to be the principal engagement of the people. The idea was too novel. The public lectures too did not last long. The plants forwarded to the committee from the RBGK to start the botanical garden in Freetown never reached their intended home. The land for such a venture was not available in Freetown where it was expected to attract the interest of potential subscribers.[16]

It is not clear if the modern agricultural implements which the AIS committee requested ever reached Sierra Leone. The list included 'Cottain's' hand corn mill (for milling maize and rice), a rice-cleaning machine, simple cotton gins, a coffee-cleaning machine, small hoes, hoes for grubbing, spades, shovels, bill-hooks, small scythes, ploughs, iron harrows, machines for grinding arrowroot, 'hair bagging' or sieves for straining the same, chaff-cutters, iron drills, 'Cottain's' portable forge, and wheelbarrows. The committee also requested Sea Island and other cotton seeds and large coffee seeds.[17]

[14] CMS: C/A1/L4(1846–1852)/105, H.Venn to J. Graf, 2 Nov. 1847.
[15] Graf acknowledged that Sierra Leone community was so divided that many would not participate in any project without knowing first those who were involved. Likewise, he feared that the 'exclusive' or 'party' spirit of his own mission community could be an obstacle to the success of his agricultural scheme. CMS: C/A1/M13(1846–1848)/669, J. Graf to H.Venn, 13 May 1848.
[16] Graf wrote that the committee once had in view a land 'where all the Freetown gentry take their evening airing, near the race course, with the great advantage of a streamlet running through it all seasons; but we have not yet been able to induce the proprietor ... to let us have it'. The committee could not secure this or any other land in Freetown before the whole agricultural plan fizzled out the following year, 1851. CMS: C/A1/O105/7a & 9, J. Graf to H.Venn, 19 Jan. 1850, 31 May 1850.
[17] CMS: C/A1/O105/7b, Minute of the Meeting of African Improvement Society, 9 & 10 Jan. 1850.

After two years of activity, Graf, the power behind the intended agro-economic revolution, despaired and wrote with disappointment that:

> [A]s to the Agricultural efforts which have from time to time been made in the colony, it appears certainly very strange that, whether undertaken by Government or by private individuals or by bodies of men here or in England – they should have almost invariably failed and do yet fail, from one cause or another, although there does not appear any cause for such failure inherent to the soil itself! No doubt, it is part of the Providential cloud which has hung for ages over Africa, still reminding us of Ham's Curse.[18]

In spite of the failed mission of the AIS, the African Native Agency Committee in Britain sponsored two young men from the colony, Henry Johnson who went to the RBGK for three months' training in 1853 and Henry Robbin who was sent to Manchester to learn cotton-processing at the factory of the industrialist Thomas Clegg.[19] Henry Johnson returned home to Sierra Leone, while Robbin was sent to manage the CMS project of cotton cultivation at Abeokuta, in Yorubaland (which is discussed further below). Johnson did undertake cultivation on his private farm in Hastings, but the hope that this would become the experimental field for introducing new plants into Africa and for exchanging plants with the RBGK did not materialize. Henry Robbin, en route to take up his appointment at Abeokuta in 1856, set up cotton-presses for both the CMS and the Wesleyan mission in Freetown, though it is not clear whether this was directly connected to Johnson's activities at Hastings.[20] In any case, the project of cotton-growing and processing in Sierra Leone likewise produced no substantial result. Henry Johnson himself left Sierra Leone in 1857, when he was recruited to serve as a Scripture Reader in the CMS mission at Ibadan, in Yorubaland, where he continued to work until his death in 1865.[21]

The failure of the AIS project

Several reasons may be adduced for the failure of the initiatives of the African Improvement Society. In the first place, their enterprise was

[18] CMS: C/A1/O105/57c, J. Graf, Journal Entry, 8 Feb. 1851.
[19] Archives of the Royal Botanical Gardens, Kew: DC 33/425, H. Venn to Sir. W. Hooker, 15 July 1853.
[20] Jean Herskovits Koptyoff, *A Preface to Modern Nigeria: The 'Sierra Leonians' in Nigeria, 1830–1890* (Madison WI, 1965), 118.
[21] Michel R. Doortmont, 'Recapturing the Past: Samuel Johnson and the Construction of the History of the Yoruba' (PhD thesis, Erasmus University, Rotterdam, 1994), 20; J.F. Ade Ajayi, *Christian Missions in Nigeria 1841–1891: The Making of a New Elite* (London, 1965), 164.

beset by cultural problems. The beneficiaries in view, both the recaptives and their colony-born children, did not understand what their European benefactors envisaged. Over many decades of struggle in the British parliament between the Abolitionists and the West Indies planters over whether the British government should continue to fund the colony and maintain the naval squadron patrolling West African waters, the colony was dogged by inconsistent government policies, which sometimes threatened its very existence.

The decision to cut down drastically financial support for the administration of the colony in 1844, in order to induce emigration to the West Indies, was taken as the final evidence to the people that the British government was no longer interested in their welfare. The result of these apparently inconsistent policies was popular scepticism. Rumours that travelled like dry season bush-fires regularly spread across the villages in the late 1840s and early 1850s about Britain's lack of interest in the colony and its plans to disengage from there.[22] The people had thus learnt to distrust permanency in the generosity of Europeans and had convinced themselves that the white man would not give anything in exchange for nothing.[23] It is therefore not surprising that the prizes instituted for the agricultural shows did not stir their indifference. Graf himself noted that the local people could not even differentiate between 'price' and 'prize'.[24]

The younger generation had the additional problem of an identity crisis. Their cultural hybridity was a burden, which caused them to persistently identify civilization with westernization and clerical and administrative jobs. Farming, as they knew it from their parents, was not thought suited for the world supposedly opening before them. This mindset was too deep to be re-orientated within one generation. Members of the AIS appreciated the scale of this problem and knew that the success of their efforts lay in the future.[25] But lack of immediate results could not sustain their efforts. The prospect of future success seemed too far away.

This problem of perception was complicated by the demands involved in raising and nurturing some of the products being introduced into the colony. The intended beneficiaries of the activities of the AIS had no patience for the painstaking efforts required to culti-

---

[22] One rumour which caused panic in 1849 was that the British government had handed over the colony to the French or the Spaniards, claiming that 'all Europeans connected with England (missionaries included) had received orders to leave the place'. The missionary in Hastings found out that the rumour had arisen from 'exaggerated comments' on the 'probable removal' of the British naval squadron from the coast. CMS: C/A1/O105/55a J. Graf, journal, 6 Oct. 1849.
[23] CMS: C/A1/M14(1848–1852)/245, J. Graf to H. Venn, 20 Nov. 1849.
[24] CMS: C/A1/O105/53a, J. Graf, journal, 9 Feb. 1849.
[25] CMS: C/A1/O105/7a & 9, J. Graf to H. Venn, 19 Jan. 1850, 31 May 1850.

vate, process and package them to European standards that would guarantee their acceptance.[26] In the face of the dire economic situation of the colony, Sierra Leoneans of the mid-nineteenth century were not interested in cultivating cotton. It did not take the Reverend David Hinderer long to discover this in 1857 when he visited the colony to recruit agents for his mission in the Yoruba country. He was overwhelmed by applications for employment as scripture readers and teachers, but no one was ready to go and plant cotton in Ibadan.[27]

The implication of these cultural problems is that there was no social angst that would generate in the people a sense of self-definition and determination that would make them affirm their dignity in economic independence. Actually, by the mid-nineteenth century, they were in a state of transition and they were still enchanted with European values and material culture. The growing uncertainty of support from their benefactors was the beginning of a weaning process. When this necessary self-affirmation would come in the nationalism that emerged in the second half of the century, it was not directed towards economic rejuvenation. It was still about claiming equality with their former benefactors, now turned political adversaries.[28]

The second challenge that dogged the AIS at home related to resources. Basic to this was recruitment of a convinced and committed membership. The party spirit of the missionaries aside, mid-nineteenth century Freetown was a divided society. The gentlemen always wanted to know who was involved before they cast their lot in with the initiative being commended to them. Although some eminent persons subscribed to AIS membership, including the Governor, with his lady, as the patron, and the Chief Justice Carr as the president, the much-needed shareholders to invest in the proposed model farm were not forthcoming. When in July 1849 the subcommittee met to raise the sum of £200 through an association of shareholders who were expected to subscribe at the rate of 20 shillings each, it was evident that they were not sure they would meet their target. They knew they were swimming against the tide. Many such agricultural schemes in Sierra Leone had failed in the past, and the idea had come into disrepute among the people. The cost of running them was also too high as high wage rates tend to neutralize whatever could come out as profit to sustain them.[29]

[26] Henry Venn noted that African produce had earned a bad name in Britain because of 'the careless manner in which they are prepared'. To remedy the situation, he furnished the committee in Sierra Leone with 'a Seal, Stamp & Brand (A.I.C.)' to authenticate the quality of the products they would be sending to Britain. CMS: C/A1/L4(1846–1852)/170–1, H. Venn to J. Graf, 1 Nov. 1849.
[27] CMS: C/A2/O49/30, D. Hinderer to H. Venn, 19 Nov. 1857.
[28] Leo Spitzer *The Creoles of Sierra Leone – Responses to Colonialism, 1870–1945* (Madison WI and Ile-Ife, 1975).
[29] CMS: C/A1/M14(1848–1852)/245, J. Graf to H. Venn, 20 Nov. 1849.

Getting land in Freetown for the model farm was also problematic. The society could not contemplate locations outside the city, although it was virtually impossible to acquire one in Freetown; the elite subscribers would not have invested in a scheme to be run outside their own immediate view. This seems to be the reason why Graf could not persuade the committee to consider one of the villages of the colony for establishing the model farm, although he himself seems to have been open to that possibility. With Hastings in mind, he included a personal request to Venn when he sent him the committee's list of agricultural implements for exhibition or for introduction to the people in 1851. Since many of the implements were manually operated, he asked for simple mechanical equipment for the requested mills, wind- or water-powered. His reason was that, 'we have here at Hastings on our own high premises regular winds, sea breeze by day and land breeze by night – the whole machine turning on a pivot can be most easily adjusted to the one or the other. It only requir[es] a set of small vanes that can easily be managed'.[30] Graf might have preferred that the model farm and the museum should be located in Hastings. He might also have thought that mechanizing farming, however small the step, would have intrigued young people and evoked their interest in agriculture. This was not to be, however.

This unavailability of land and local funds to run the activities of the AIS was further compounded by lack of skilled personnel. The committee could not find persons for starting and running the experimental garden. Even versatile lecturers who could give public lectures to stimulate the spirit of enterprise were not available. Graf confessed that there were not many who could adapt to the fancy of the people and sustain their interest.[31]

The third problem that afflicted the agricultural scheme of the AIS was logistical. While dried seeds and specimens were successfully moved across the sea between Sierra Leone and Britain, the challenge of moving live plants by the same means over several weeks could be enormous. Henry Johnson lost some of the plants he was bringing home from Kew Gardens in 1853, although some others survived and took root on his experimental farm in Hastings.[32] When he tried to fulfil his side of the agreement with the RBGK by sending plants to London two years later, they arrived there dead. The vessel that conveyed them from Sierra Leone had some difficulties on the way and was delayed for about three months. Nothing more was heard about Henry Johnson's experimental farm.

[30] CMS: C/A1/O105/7a, J. Graf to H. Venn, 19 Jan. 1850.
[31] CMS: C/A1/O105/7a J. Graf to H. Venn, 19 Jan. 1850.
[32] CMS: C/A1/O122/1, H. Johnson to H. Venn, 7 Dec. 1853.

The luxuriant coastal vegetation that impressed Smeathman and subsequent protagonists of agriculture as a remedy for the African condition in the nineteenth century was deceptive. Even when the AIS scheme had failed, following the failure of earlier ventures, Graf still believed that the problem with such schemes was not inherent in the soil but in the Hamitic curse that was supposedly hanging over Africa. It was easy to be so deceived: Europeans thought that since the gateway to the continent that produced millions of slaves for the New World presented such luxuriant vegetation, certainly the land must be rich. In fact it was, but not at the coastal area with which they were most familiar in the nineteenth century. The tides of the sea, it seems, had washed away the nutrients of the soil and had, over the centuries, deposited arid sands in some parts. For this reason, the soil of the peninsula that constituted the colony was poor and needed improvement for the agricultural success the AIS envisaged.[33] This was further compounded by the fact that the land was rocky and 'did not lend itself even to the simplest forms of nineteenth-century western mechanization'.[34]

Nevertheless, partial success did attend the ideas of the AIS, but not in Sierra Leone. The society's candidate sent for training in Manchester in 1853, Henry Robbin, achieved in Abeokuta what seemed impossible in Sierra Leone. Indeed, the relative success of the Abeokuta venture may itself have been a factor in the failure of the AIS in Sierra Leone, by diverting interest and resources there.[35] Through the activities of Robbin, Abeokuta recorded partial success in growing cotton on a commercial scale for export in the 1850s and proved the possibility of legitimate trade in Africa, although not in the sense European humanitarians conceived it.[36] Yet, the successful exportation of the product was a pointer to the future agriculture could open among the Yoruba as they moved into a new era of British colonial domination at the end of the nineteenth century. This is the subject of the following section.

---

[33] This explanation for the poor quality of the colony's soil is based on a personal discussion with Professor Andrew Walls in April 2010, reflecting his long familiarity with the country.
[34] John Peterson, *Province of Freedom – A History of Sierra Leone 1787–1870* (Evanston IL and London, 1969), 273.
[35] As suggested by Christopher Fyfe, *A History of Sierra Leone* (London, 1962), 253.
[36] The humanitarians had thought that legitimate trade would destroy slavery at its roots in Africa. Whereas the British naval blockade impeded its trans-Atlantic aspect, legitimate trade in Abeokuta did not stop slave-raiding wars in the Yoruba country. Like their Ibadan rivals to the north, Egba people still held persons as slaves and used them to cultivate their fields. The forceful abolition of the trans-Atlantic slave trade in fact boosted domestic slavery, which was part of the culture of the society.

Agricultural development in the Yoruba Mission

Unlike Sierra Leone, where the agricultural initiative of the CMS was motivated by the perceived cultural problem that dogged the colony-born youths, the situation in the Yoruba mission at Abeokuta in the mid-nineteenth century connected with the anti-slavery crusade at the centre of the promotion of agriculture in Africa by British humanitarians. Yet, very much like the interior country beyond the colony of Sierra Leone, which still remained impenetrable to missions, the slave-raiding wars had not abated in the Yoruba country. Warlords were still in control of local politics. If the result of mission in the relatively peaceful colony of Sierra Leone remained controversial and the various attempts to enhance its economy through agriculture had proven impossible, the Egba state of Abeokuta on the face of it should not have been the place where the religious, economic and cultural agenda of Christian missions should have thrived. But the fledgling settlement worked out for missionary agenda, if only for a time.

Basically, the successful thrusts of missionary agenda into the Yoruba country rested on the homecoming of Egba recaptives from Sierra Leone, which had begun from 1838. This unexpected but pleasant reversal of the fortunes of this group in the fratricidal wars that decimated their ancestral homes in the 1820s prepared the ground for missions in Abeokuta. From there they successfully launched their campaign in the entire Yoruba country.

It should be borne in mind that these returnees were essentially recaptives and not colony-born youths who took pride in claiming Sierra Leone as their home. In fact, the natural urge of Egba recaptives to reunite with their kinsmen and women in their new settlement at Abeokuta was not unconnected with the economic and political uncertainties that dogged the colony from the late 1830s. Those in Hastings, who championed this home return, had been persuaded by Graf not to migrate to the West Indies because of the poor organization of the emigration scheme and the possible failure of its prospects. The option available to them was to either remain in the colony or return to their home country. Taking the latter, the first wave of homeward return of these people was essentially made up of those who were born in the indigenous society from where they were torn by the slave-raiding wars that decimated their country.

In orientation, the recaptives were still traditional people, although some of them had embraced Christianity in exile. As beneficiaries of the anti-slavery crusades of Britain, they were the critical mass that prepared their country for the religious, cultural and economic revolutions of the missionaries who followed them home from exile. It is important to emphasize the distinction between the receptives and their colony-born children at this stage of missionary expansion and commercial agricultural initiatives because it is critical to understanding why 'legitimate trade' held out better prospects in Abeokuta in comparison to Sierra Leone. Many of the receptives had always been connected with working on the soil. From the indigenous society to their exile, peasant farming was part of their economic life; they were, therefore, in the position to promote the agricultural agenda of the missionaries in a society that was itself agrarian.

Abeokuta also proved to be an advantageous arena where the agendas of missionaries, eventually including agriculture, could succeed beyond what occurred in Sierra Leone because of the successful resolution of the internal disturbances going on in the town in the mid-nineteenth century. Sodeke, the generalissimo of Abeokuta when the missionaries came on exploratory visits in 1842, was generously disposed to the missionaries and willingly opened his country to them. However, he was dead by the time the agents of the CMS were to begin their work in Abeokuta in 1845. As might be expected, unlike Sodeke, the pro-slave trade war-chiefs and their counterparts in the religious sphere, the priests of the traditional cults in the town, who saw the return home of the former slaves as a threat to their interests, were not favourably disposed to the mission. Sodeke's sudden death provided them the opportunity to attempt to stop the missionaries; but they were not successful.[37] The missionaries not only succeeded in beginning their work, they supported the longing of the people to settle down to civil life rather than live by depredation. Thus Abeokuta, which had emerged from the throes of the Egba holocaust of the 1820s, gradually transformed from a state under the control of warlords to one under a civil administration.

The transformation of the state into a civil system, unlike Ibadan which remained a military state, did not spare the town from the war currents of the age. It only tempered the slave-raiding and expansionist wars of the people. But the growing order that ensued in the town was a basic condition for the positive transformation of society. It combined with the perceived goodwill the missionaries enjoyed from the circumstance of the return home of the exiles and afforded them the necessary foothold and leverage to introduce their religious and economic agenda into the country. When this is borne in mind, together with the fact that the Sierra Leone interior gave missions no

[37] CMS: C/A2/O85/241 & 242, H. Townsend, journal, 2 May 1850, 15 Sept. 1850.

such opportunity, and that the economy of the Yoruba country was traditionally based on agriculture, and not on any illusory clerical and administrative jobs that colonialism might have inadvertently introduced into Sierra Leone, the stage was set for a successful agricultural experimentation in Abeokuta.

The concern to promote agriculture in Abeokuta was incipient from the beginning of the mission in the 1840s, but it became an urgent concern in the aftermath of the abortive attempt of King Gezo of Dahomey to destroy the town in 1851. In consequence of this assault, one of the leaders of the CMS mission in Abeokuta, the Reverend Henry Townsend, saw the need not only for missionary work but also to direct the attention of the people to commercial agriculture as a viable alternative to the slave-raiding wars. In his melancholic memorandum to his fellow-missionaries, he criticized English traders for neglecting to develop the cotton trade in Africa since 1590 when, according to him, the first consignment of that product reached England from Benin. He was of the view that if English people had promoted that trade warfare would not have reached the endemic state it had.[38]

Townsend quickly shook off his disappointment with English cotton-dealers to begin to promote an agricultural product that could prove commercially viable. Tobacco readily came to his mind as holding potential for economic production. It was the major imported luxury article that sustained the slave trade in Abeokuta.[39] Tobacco grew in the streets of the town, but it was not cultivated. The people allowed it to grow because they saw it as capable of keeping away snakes. Townsend at length was able to get a skilled cultivator of tobacco in the American Baptist missionary, Thomas Jefferson Bowen. Bowen assisted Andrew Wilhelm, an African agent of CMS, to set up a model farm for growing tobacco at Ake, which the mission used to promote its cultivation among the Christians in the hope that it would spread to the whole town.[40] From all indications, however, the tobacco-planting campaign did not create the desired breakthrough.

However, against the background of the goodwill which Europeans earned from the return home of the exiles, it did not take much time for the people to respond to the local campaign of the missionaries to

---

[38] Townsend got his information on Benin cotton reaching England in 1590 from Charles Tomlinson, *The Useful Arts and Manufactures of Great Britain* (London, 1848/1861). CMS: C/A2/O85/245, H. Townsend to Missionaries, 3 July 1851. In fact, however, it was cotton cloth, rather than raw cotton, which the early English traders had bought at Benin.

[39] CMS: C/A/A2/O85/240, H. Townsend, journal, 4 March 1848.

[40] CMS: C/A2/O85/246, H. Townsend, journal, 16 Aug. 1851; C/A2/O85/13b, H. Townsend to H. Venn, 29 July 1852.

promote agriculture as against slave-raiding. The immediate success of the campaign can be seen in the changed attitude of the warrior class who began to invest in cash-crop cultivation, a feat that was never achieved in Sierra Leone. Chief Ogunbona of Ikija, an ally of the mission community, was the pace-setter, producing cotton, ginger and pepper on his farm.[41]

But what was their motivation? The ethos of society had changed, and many people wanted to settle down to a peaceful life. In this vein, pillage and depredation had become unpopular, while a new culture was impinging on their society through Christian missions and the growing rapport between Britain and the Egba state. The accompanying potential of deriving new wealth from cash crops had an effective appeal. Moreover, and again unlike Sierra Leone, the Egba saw the rapport as between two equal partners, not between a benefactor and their beneficiary. Hence, national self-determination remained intact at Abeokuta, where agriculture was traditionally regarded as a respectable pursuit. Of course, to this may be added the fact that the land in Abeokuta as in other parts of the Yoruba country was fertile.

In effect, the missionaries' initial work of promoting agriculture as a viable alternative to the trans-Atlantic slave trade in Abeokuta first drew in the old elite, although Henry Venn's vision was to use agriculture to create a new middle class. It was a necessary beginning that prepared the ground for Dr Edward Irving when he arrived in Abeokuta in 1853 and was later succeeded by Henry Robbin from Sierra Leone in 1856 to promote the growing and processing of cotton on a commercial scale. In the second half of the 1850s, the enthusiasm with which the commodity was subsequently grown and processed in the town, which made it one of its principal exports, is attested to by the number of gins and presses that were at work there. Between two and three hundred gins were cleaning cotton across the town while about five to six warehouses were receiving bales ready for export.[42] Nevertheless, cotton-growing on a commercial scale in Abeokuta was only a short-lived success.

To understand what has been referred to as the failure of the cotton business at Abeokuta, the enterprise must be placed in its socio-political context. Webster has attributed the perceived failure not only to the quality of the soil and the climate, which 'permitted cotton to survive but never flourished', but also to the fact that 'labour costs for picking, cleaning and transport were high'.[43] First, it cannot be said

---

[41] CMS: C/A2/O85/11, H. Townsend to H. Straith, 28 Jan. 1851; H. Townsend to H. Venn, 1 May 1852
[42] Webster, 'The Bible', 422.
[43] Ibid.

that such a rapidly developing enterprise as cotton business in Abeokuta in the 1850s had yet reached its full potential within less than five years of introducing new skills in cultivating and processing it. The enterprise was still on the rise when it collapsed.

The perennial feud between Ibadan and Abeokuta, which climaxed in the outbreak of Ijaye war of 1860–62 made it impossible for the Egba to cultivate over half of their land holdings. Much of the land they possessed to the north, east and west of the town could not be cultivated as a result of mutual kidnappings between them and the Ibadans which were a regular feature of warfare. This could only have led to the over-exploitation of the safer land under cultivation, hence diminishing returns. Also disastrous for the cotton trade in Abeokuta was the fact that the war did not end with the destruction of Ijaye in 1862, but shifted to the Ijebu country where it became more complicated and dragged on into the second half of the 1860s. The war situation was a major impediment to the Egbas' cultivation and processing of cotton for export.

At the same time, from the early 1860s, when Abeokuta cotton should have been taking advantage of the fall in the American production because of the Civil War, the relationship between the Egba state and the colonial regime in Lagos began to deteriorate rapidly. The increasing number of the colony-born Egba young people who were now returning to Abeokuta brought to the town the new nationalism that had set them on war path against British government in Sierra Leone. The eventual culmination of this movement came in the face-off between the Egba United Board of Management (EUBM), championed by the volatile Sierra Leonean, G.W. Johnson (no relation to Henry Johnson), which led to the 'housebreaking' episode (of attacks on European properties in Abeokuta) of 1867 and the expulsion of Europeans from the country. This political showdown was distressing for business in general and a rising one like that of cotton trade in Abeokuta could not have survived it when it depended very much on the goodwill of British people. The argument that failure was due to soil and climate cannot be sustained in the face of these socio-political upheavals that undercut the Egba while the cotton business was on the rise. In fact, it was not so much a failure of an already successful enterprise as it was a truncating of a bright prospect.

The argument about labour costs must also be qualified. It is generally true that paid employment was not yet the norm in the Yoruba country in the mid-nineteenth century. The idea was still novel as people thought it below their dignity to be employed for payment. Unless they unavoidably got into debt and had to pawn themselves, the only reason why they would serve another person was if they were slaves.[44] The slaves cultivated the land in nineteenth-century Yorubaland, augmented by the input of the children in the households, hence the prevalence of polygamy. Evidently, therefore, commercial agriculture did not affect slavery on the domestic front despite it being promoted to subvert the transatlantic version. So the agricultural economy of the Yoruba country rested on the institution of domestic slavery in spite of the pretension of the European missionaries who resided there to be opponents of slavery.[45]

The missionaries knew how critical the institution of slavery was to the economy of the country as Townsend amply demonstrated in his alarm at Bishop Weeks' intended proscription of slave-holding among Christians during his visit to the Yoruba mission in 1856.[46] Moreover, the unwillingness of some Egba mission agents, including Sierra Leone recaptives among them, to give up their slaves in 1880 as instructed by the CMS home committee revealed how slave-holding was intricately woven into Egba social system.[47] With the lessening of the slave-raiding wars this human resource was drying up, making labour scarce. But when the available slaves also began to discover a safe haven in Lagos from the late 1850s and were making their escape there, it is understandable that labour costs would be climbing.[48] However, it is possible that this problem impacted upon cotton production at Abeokuta not so much in driving up labour costs but rather in the increasing scarcity of labour itself.

Although by the 1860s, Abeokuta's ambition to be a major cotton

[44] This was a common submission from all the centres of the Yoruba mission – Ibadan, Ondo, Abeokuta and Ota – at the conference that was held on domestic slavery in 1880. CMS: G3/A2/O(1880)/106, 'Domestic Slavery: Minutes of Conference held at Lagos on 16–23 March 1880'.

[45] The same reality was at work in the Gold Coast where, in the mid-19th century, Johannes Zimmermann justified African domestic slavery to his Basel Mission home committee shortly before the forceful attempt to stamp it out among the Akyem people created a protracted feud between the mission and Akyem state. With both domestic slavery and polygamy, home-based mission administrators did not understand how these institutions were critical to the pre-colonial economic production and survival of African societies: Peter Haenger, 'Reindorf and the Basel Mission in the 1860s – A Young Man Stands Up to Mission Pressure,' *The Recovery of the West African Past – African Pastors and African History in the Nineteenth Century – C.C. Reindorf and Samuel Johnson*, ed. Paul Jenkins, (Basel, 1998), 19–20; R. Addo-Fening, *Akyem Abuakwa 1700–1943: From Ofori Panin to Sir Ofori Ata* (Trondheim, 1997), 59–87.

[46] CMS: C/A2/O85/32, H. Townsend to H. Venn, 31 Dec. 1856.

[47] CMS: G3/A2/O(1880)/106/19ff, 'Domestic Slavery', Minutes of Conference held at Lagos, 16–23 March 1880; G3/A2/O(1880)147, S.S. Taylor, et al. to H. Wright, 11 June 1880; G3/A2/O(1880)167, J.B. Wood to E. Hutchinson, 22 Oct. 1880; G3/A2/O(1881)/45, V. Faulkner to C.C. Fenn, 28 Jan. 1881.

[48] The impossibility of retrieving their runaway slaves from the colony was one of the factors that led to the deterioration of the Anglo-Egba relationship from the 1860s. E.A. Oroge, 'The Fugitive Slave Crisis of 1859: A Factor in the Growth of Anti-British Feelings among the Yoruba', *Odu: A Journal of Yoruba, Edo and Related Studies* 12 (1975), 40–54; 'The Fugitive Slave Question in Anglo-Egba Relations, 1861–1886', *JHSN* 8 (1975), 61–80.

exporter had waned, it was nonetheless an eye-opener for the people in realizing the potential agriculture held for trade with Europeans. While cotton-growing continued for domestic purposes and for export, from the closing decade of the nineteenth century the new environment of racial antagonism between Africans and Europeans in church and state would be the perfect storm that would propel Yoruba into an interest in plantation agriculture and export crop cultivation.

The advent of cocoa-farming in Yorubaland – a delayed success

In their reaction to the perceived injustice meted out to Bishop Crowther in the Niger mission, which led to his demise in December 1891, and the growing racial antagonism in the civil service, Henry Venn's idea of the nexus between the national church and national self-determination came alive among Lagos Christians in the last decade of the nineteenth century. The self-weaning that came to these Lagos residents from the mutual antagonism between them and Europeans derived inspiration from the former Honorary Secretary of the CMS and spurred some of them to find their economic niche in plantation agriculture, first at Agbowa,[49] and later at Agege.[50] The former, under the West Indian John Ricketts, turned out to be a successful but short-lived enterprise. The latter, established by Jacob Kehinde Coker, became self-propagating and proved a lasting success for the Yoruba nation.

The lasting success of Coker's cocoa plantation agriculture, in contrast to cotton at Abeokuta, can be attributed, first, to the traumatic circumstances of the age. The social angst forced Coker and his equally disenchanted fellow-members of the Breadfruit Church, Lagos, to resolve their frustration with the missionaries who rather than promote the budding leadership of their Yoruba agents in the church further alienated them. By founding the African Church, independent of European control, they affirmed the independence of the Yoruba church and demonstrated their readiness to set out on their own rather than 'doing the baby for aye'.[51] Hence, unlike Abeokuta and Sierra

---

[49] The Reverend John Ricketts, a West-Indian settler in Nigeria was a Baptist minister who had previously been in Congo. Ricketts successfully combined church planting and agricultural development at Agbowa, north of Lagos Island. Ade Adefuye, 'John, Gershion and Joshua Ricketts: Jamaican Contributions to the Socio-Development of the Colony Province', *Studies in Yoruba History and Culture: Essays in Honour of Professor S. O. Biobaku,* ed. G.O. Olusanya (Ibadan: 1983), 135–52.

[50] Webster, 'The Bible', 427–31; Peter Adebiyi, 'Jacob Kehinde Coker (1866–1945)', *Makers of the Church in Nigeria,* ed. J.A. Omoyajowo (Lagos, 1995), 98–115.

[51] A phrase from Mojola Agbebi, quoted in E.A. Ayandele, *The Missionary Impact on Modern Nigeria 1842–1914: A Political and Social Analysis* (London, 1966), 255.

Leone, the competitive environment of racial pride and social angst were at work in the Agege plantation agriculture. To the extent that Ricketts' coming to Nigeria derived from his association with Mojola Agbebi, a foremost churchman in the racial agitation of the day, his success can also be classified as having emerged from this Yoruba cultural ferment.

On the other hand, Coker's success may also be seen as an extension of Abeokuta's earlier short-lived success in cotton-planting. As a CMS mission-educated Egba man, a son of a successful Abeokuta cotton farmer who profited from its commercial export initiated and promoted by the missionaries, and a member of the Breadfruit Church, Lagos, he had his roots in the CMS. His self-understanding as a part of the fulfilment of Henry Venn's idea of the indigenous church and development was, therefore, not a mere ideological counterpoint. It was a reality that confirmed him as a successful product of the CMS even though neither he nor the mission at the turn of the century would have acknowledged that.

Secondly, Coker's success at Agege drew from the opportunity of the moment. With the cessation of the Yoruba wars in 1893 and the employment of some of the belligerents in the construction of the railroad from Lagos into the country, paid employment became more familiar in the country. Coker's farm in Agege took advantage of this new culture of employment that accompanied the *Pax Britannica* into the country by recruiting workers from the Yoruba hinterland to cultivate and process cocoa for export at Agege. The wide ripple of Coker's success at Agege can be seen in its attracting the educated elite of Lagos as planters and the eventual propagation of the gospel of cocoa-farming in the Yoruba hinterland where the produce became the main export of the country by mid-twentieth century. Coker's material success from cocoa-farming was too attractive to be ignored by the elite of Lagos who, hitherto, had taken pride in grammar school education, clerical and administrative jobs and commerce. The possibility of wealth cocoa-farming opened was their motivation.

This successful propagation of the new agricultural product in the Yoruba country in the first half of the twentieth century is significant in view of the wealth it attracted to the country, making cocoa its economic mainstay well into the 1960s. This success may suggest a more judicious evaluation of the fate that befell Agege in its eventual withering away as the 'nursery of Nigerian cocoa'.[52] Again, as in Abeokuta, the argument here is that the withering away was not so much a failure of the nursery. If Abeokuta was an unrealized poten-

---

[52] Webster, 'The Bible', 431.

tial, Agege represented diffused success that multiplied itself over and again by spreading prosperity throughout the Yoruba country.

Webster has persuasively argued that Coker's adventure with cocoa-farming at Agege was the ultimate vindication of the ideas of Henry Venn. It visibly demolished all the arguments of critics that legitimate trade was impossible in Africa. But much more, this success proved attractive enough to draw the interest of the middle class which Venn saw as the necessary catalyst to realize the new society he and his fellow-humanitarians and mission administrators sought to develop in Africa. Even John Ricketts' short-lived success at Agbowa made its significant contribution to the emergence of this middle class. In this regard, Adefuye mentioned that the most vital contribution of Ricketts' family to the colony of Lagos was the development of waterway transportation, which evolved from the success of Agbowa plantation.[53]

Conclusion

The partial success recorded at Abeokuta and the full one realized at Agege offer an informative comparison with the unsuccessful attempts of the African Improvement Society to promote agriculture for the purpose of cultural renewal in Sierra Leone. The successful attraction into agriculture of members of the educated elite of Lagos may supply vital ingredients to understanding this failure. With the emergence of several local newspapers and the development of a virulent nationalism from the 1880s, the elite of Lagos could have gone the way their kinsmen in Sierra Leone went in the second half of the nineteenth century, frittering away their energy in fruitless contention with the colonial regime in Lagos. But the ingredient that changed their trajectory was the fertility of the soil at Agege, which proved better still as cocoa farming found its way into the country. This advantage was matched by the changing patterns of demand in the international market, making cocoa the most profitable commodity which West Africa could supply in ample quantity to the outside world. The financial success that issued from this combination of good soil and moment of opportunity was lacking in mid-nineteenth century Sierra Leone.

Moreover, even though Henry Johnson and Henry Robbin could have been in Sierra Leone and Abeokuta catalysts like Coker in Lagos, their efforts did not issue from personal eagerness for change and

[53] Adefuye, 'John, Gershion', 142.

success, which could have compelled them to explore and experiment with various products as Coker initially did with coffee and cotton prior to his success with cocoa. They were not entrepreneurs.

Two major ingredients in Coker's agricultural success are distance and time, both in relation to European presence in West Africa. First, there is no evidence that Coker enjoyed an intimate relationship with Europeans like Johnson and Robbin. Second, his active years as an entrepreneur took place in the environment of racial tension in which he was an active participant. Hence he made no pretension to being anglicized. On the contrary, he was conscious of his African roots, the ideological awareness of which propelled his agricultural venture as a response to his struggle against near bankruptcy. However, his two worlds of indigenous Yoruba culture, which kept him close to the soil, and the wider but ideologically tense and business-savvy world of Lagos were the springboards from which he launched the agricultural revolution that changed the perspectives of those who earlier would never have contemplated agriculture as a business pursuit. The dependency that had shaped Sierra Leone over the years, on the other hand, deprived the colony of this compelling quest for success in the mid-nineteenth century.

The AIS also failed because the committee was too restricted in its ideas of calling forth the resources of the soil. Apart from commercial agriculture, which proved unviable, other prospects were not explored, except for Graf's suggestion of establishing a savings bank, which was rejected completely.[54] But in fact, the economic potential of the colony lay in its abundant solid minerals – titanium, bauxite, diamonds and gold – as was eventually demonstrated in the twentieth century. The earliest idea of the presence of these minerals in the colony might have been offered in 1878 by the flamboyant nineteenth-century Sierra Leonean historian, geographer and naturalist, A.B.C. Sibthorpe. But his discovery was dismissed by those in Britain who were asked to examine the supposed precious stones he discovered.[55] It is understandable that missionaries were not enthusiastic about the possible find. They were generally disinclined to trading in products that are merely extracted from nature, such as precious stones. Perhaps if the annual prizes in essay competition had been successful, the course of the adventure the AIS embarked upon to call forth the resources of the soil would have been different. But there were no Sibthorpes in the 1850s who would stir themselves to lead the colony to economic success.

[54] CMS: C/A1/O105/7a, J. Graf to H. Venn, 19 Jan. 1850.
[55] CMS: C/A1/O200/2b & 3, S. Spain to Mr Longman, 22 Nov. 1878; S. Spain to the Secretaries, 21 March 1879.

10

Agricultural enterprise & unfree labour in nineteenth-century Angola

ROQUINALDO FERREIRA

This chapter analyses Portuguese attempts to strengthen colonial ties with Angola by promoting agricultural activities in its African colony between the 1830s and 1860s. Recently, Seymour Drescher has argued that the abolition of the slave trade was not part of British imperialist aspirations towards Africa and that the end of the slave trade did not lay the groundwork for colonial rule over Africa.[1] While this statement might be true in the case of the British, it did not apply to another colonial power with long ties to Africa: Portugal. In the first half of the nineteenth century, as the movement to abolish the trans-Atlantic slave trade gained strength throughout the Atlantic, Portugal sought to reframe its colonial links to Angola by promoting policies aimed at shifting its economy away from the slave trade and toward legal activities such as market-oriented agriculture.

The relationship between the end of the slave trade and Portuguese colonial plans for Africa has long been debated by scholars. In the 1960s, R.J. Hammond argued that Portuguese colonialism was born out of a reflexive ideological attachment to Africa and devoid of economic interest.[2] This stance was challenged by the Portuguese historian Valentim Alexandre, who argued that despite its failure, early

---

[1] Seymour Drescher, 'Emperors of the World: British Abolitionism and Imperialism', *Abolitionism and Imperialism in Britain, Africa, and the Atlantic*, ed. Derek Peterson (Athens OH, 2010), 129–50. Drescher's statement reflects one side of a long debate about how the abolition of the slave trade, the rise of legitimate trade and the beginning of colonialism might have been connected. For the specifics of the debate, see A.G. Hopkins, *An Economic History of West Africa* (London, 1973); Robin Law, 'The Historiography of the Commercial transition in Nineteenth-Century West Africa', *African Historiography: Essays in Honour of Jacob Ade Ajayi*, ed. Toyin Falola (Ikeja Nigeria and Harlow, 1993), 91–115; id. 'Introduction', *From Slave Trade to Legitimate Commerce: The Commercial transition in Nineteenth-Century West Africa*, ed. Robin Law (Cambridge, 1995), 1–31.
[2] R.J. Hammond, *Portugal and Africa 1815–1910: A Study in Uneconomic Imperialism* (Stanford CA, 1966).

Portuguese colonialism was firmly rooted in Portuguese economic interests.³ Recently, Alexandre's view has been rejected by João Pedro Marques, who contends that that Portugal paid little attention to its African colonies before the mid-nineteenth century.⁴ In the most recent assessment, however, Gabriel Paquette states that 'debates about the merits of colonialism did not cease between 1820 and 1850, even though policy was formed haphazardly, applied inconsistently, and generated few lasting results'.⁵

The common thread of these intellectual assessments of imperial policymaking is that while they devote significant attention to the metropolitan milieu in which colonial policies towards Angola were generated, they ignore the execution and impacts of such policies in Angola. Nor do they devote attention to the issue of labour in the context of the implementation of colonial projects to foster agriculture. This chapter places colonial policies towards Angola in the broader context of the Atlantic, illuminating Brazilian investment in coffee production in Angola and the importance of the agricultural colony of Mossamedes, which was founded in southern Angola by Portuguese nationals seeking to escape anti-Portuguese hostility in Brazil. It analyses colonial attempts to engage Africans in agricultural projects, which acknowledged African expertise in farming, and the debate over the type of labour to be used in Angola in the aftermath of the slave trade. It argues that the rise of slavery demonstrates how local conditions shaped colonial policies and undermined early attempts by Portugal to gradually abolish slavery in the colony.

Early agricultural projects

The slave trade was perhaps the element that most profoundly conditioned debates about the viability of agriculture and the execution of agricultural projects in Angola in the first half of the nineteenth century. As early as 1815, exports of slaves were seen by policy-makers as an impediment to the development of an agriculturally-based economy because they depleted Angola of labour.⁶ In 1840, former

---

³ Valentim Alexandre, 'The Portuguese Empire, 1825–90: Ideology and Economies', *From Slave Trade to Empire: Europe and the Colonization of Black Africa, 1780s–1880*, ed. Olivier Pétré-Grenouilleau (New York, 2004), 110–32.
⁴ João Pedro Marques, *The Sounds of Silence: Nineteenth-Century Portugal and the Abolition of the Slave Trade* (New York and Oxford, 2006), 199.
⁵ Gabriel Paquette, 'After Brazil: Portuguese Debates on Empire, c. 1820–1850', *Journal of Colonialism and Colonial History*, 11/2 (2010).
⁶ Diogo Vieira de Tomar e Albuquerque, 'Observações sobre Alguns Importantes Objectos ao Estado da India Portugueza', 31 Aug. 1815, *Annaes do Conselho Ultramarino (Parte não Official), Feve-*

Governor of Cabo Verde, Sebastião Xavier Botelho, wrote that the end of shipments of slaves would inevitably stimulate legitimate trade, thus compensating for the economic impact of Brazilian independence in 1822.[7] In 1846, José Joaquim Lopes de Lima, in a general survey of Portuguese overseas colonies, wrote that 'the slave trade is as infamous as harmful to the true interests of our African possessions, from which it steals hands that could fertilize lands'.[8]

How influential were these ideas? By the mid-1820s, as the prospects of ending imports of slaves into Brazil gained viability, colonial officials in Angola began to seriously consider commercial agriculture as an alternative to the slave trade. In 1827, the Governor of Angola, Nicolau de Abreu Castelo Branco, offered a mixed assessment of commercial agriculture in the colony. While stating that droughts and the arid soil of coastal Angola posed serious obstacles to the development of a sugar industry, he was optimistic about commercial cultivation of cotton and indigo. According to him, cotton was already being cultivated in Golungo, up the River Bengo from Luanda, and wild indigo could be found in the Angolan interior, the *sertões*. Even sugar production, he claimed, could be developed, but it would have to centre on production of cane brandy, a product currently imported from Brazil and widely consumed in Angola.[9]

The short-lived Companhia de Agricultura e Indústria de Angola e Benguela, created in 1835 with the support of the Luanda administration and private investors, was the first attempt to promote commercial agriculture in Angola. The directors of the Company offered a caustic analysis of Angola's dependence on the slave trade which 'had contributed to the neglect of agriculture in Angola', echoing criticism by metropolitan policy-makers. In seeking to gain support from Lisbon administrators, the director proposed that the Company would exist for ten years and be responsible for importing machinery to process sugar, cotton and other crops. While the Luanda administration undertook to allow the Company to use public land

---

reiro de 1854 a Dezembro de 1858 (Lisbon, 1867), 42; Adrien Balbi, 'Variétés Politico-Statistiques sur la Monarchie Portugaise', *Annaes das Sciencias, das Artes, e das Letras* 16 (Paris, 1822), 95–96. See also Alexandre, 'Portuguese Empire', 112. The belief that market-oriented agriculture would provide the basis for colonial relations with Africa informed policies by other European nations as well. See Daniel P. Hopkins, 'Peter Thonning, the Guinea Commission, and Denmark's Postabolition African Colonial Policy, 1803–50, *William and Mary Quarterly (WMQ)* 66/4 (2009), 784.

[7] Sebastião Xavier Botelho, *Escravatura: Beneficios que podem provir às nossas Possessões d'África da proibição daquele Tráfico* (Lisbon, 1840), 20.

[8] José Joaquim Lopes de Lima, *Ensaios sobre a Statistica das Possessões Portuguezas na África Occidental e Oriental; na Asia Occidental; na China e na Oceania* (Lisbon, 1846), p. XXXVIII.

[9] To that end, Castelo Branco said he had recently purchased a sugar mill from Brazil, and another sugar mill had been set up near the Bengo River in 1826. See *Almanak Statistico da Província d'Angola e suas Dependências para o ano de 1852* (Luanda, 1851), 11.

free of charge, the capital to jump-start the company would come from the sale of four hundred shares worth 100,000 *réis* each (approximately £20,000 in total in 1836) to private investors.[10]

Several if not all of the sponsors of the Company were prominent Luanda slave-dealers, who were seeking to develop alternatives to declining exports of slaves from Luanda after the Brazilian law that banned imports of captives into Brazil in 1831.[11] The drop in exports of slaves caused a reduction in prices of captives in Angola:'The abolition of slavery [*sic*: i.e. of the slave trade] has provoked prices of slaves to decline to forty or forty five thousand *réis*, and the prospects are that these prices will drop further.' This situation was seen as an opportunity to develop cash-crop production in Angola and compete against Brazil in the international agricultural market. According to directors of the Company, the end of the slave trade to Brazil would increase labour costs in Brazil and make Brazilian products more expensive in Europe. This would put Angola in a favourable position to sell agricultural products in Europe.[12]

Once Luanda exports of slaves revived in the mid-1830s, however, the investors who had bought shares in the Company abandoned it to return to the slave trade. This development derived from the economic prospects of agriculture vis-à-vis the financial benefits of the slave trade. While profits from exports of slaves were estimated at up to 50 per cent, financial returns to investments in agriculture were slow.[13] As late as 1850, the Governor of Angola, Adrião Acácio da Silveira Pinto, stated that 'one of the great obstacles for the large-scale cultivation of colonial products is the fact that potential investors in agriculture still held up hope that they would be able to ship slaves'.[14]

Yet the resumption of the slave trade in the mid-1830s did not entirely prevent colonial projects for developing commercial agriculture in Angola. The administration of Governor Antonio Manoel de Noronha is a case in point. An old foe of the slave trade, Noronha became Governor of Angola in 1839 and saw the development of agriculture as part of a broader effort to reframe Portuguese colonial links to Angola. In addition to being the first Governor of Angola who seriously acted against the slave trade, Noronha sought to expand colo-

[10] Archivo Histórico Ultramarino, Lisbon (AHU): segunda seção de Angola, pasta 2, 'Regulamentos da Companhia de Agricultura e Indústria de Angola e Benguela', March 21, 1836. For the conversion of *réis* into pounds, see Gervase Clarence-Smith, *The Third Portuguese Empire, 1825–1975: A Study in Economic Imperialism* (Manchester, 1985), 227.
[11] Archivo Histórico de Angola (AHA): cód. 13, fols 5–10v, Governador de Angola, 8 April 1836.
[12] AHU: segunda seção de Angola, pasta 2, 'Primeira Comunicação da Companhia de Agricultura e Indústria de Angola e Benguela', 21 March 1836. See also Marques, *The Sounds of Silence*, 224.
[13] AHA: cód. 13, fols 5–10v, Governador de Angola, 8 April 1836.
[14] AHU: segunda seção de Angola, pasta 16, Governador de Angola, 'Relatório', 22 Feb. 1850.

nial frontiers in the *sertões* (interior), promote mining activities, improve road communication between Luanda and the *sertões*, and develop scientific knowledge about Angola.[15] It was during Noronha's tenure that the Swiss scientist Lang was dispatched to the *sertões* to explore African fauna and flora.

Noronha supported the establishment of a six-hundred household agricultural settlement in the recently conquered outpost, or *presídio*, of Duque de Bragança.[16] To develop this settlement, over one hundred settlers were brought from several parts of Brazil, particularly Rio de Janeiro, Recife and Pará.[17] In addition to waiving import duties on equipment for agricultural projects, the Governor sought to persuade the metropolitan authorities to waive duties on imports of Angolan crops into Portugal.[18] To support Noronha's projects, Lisbon promised to allow merchants to occupy tracts of land in Angola if they invested in agriculture, cattle-herding and commerce.[19] In 1840, the Portuguese investor Francisco Rodrigues Batalha received two tracts of land to set up 'establishments devoted to agriculture and commerce'.[20] More importantly, Noronha also sought to encourage commercial agriculture among Africans and offered exemption from work as porters to 'all [Africans] who cultivate land, planting particularly cotton and rice, while the cultivation of coffee does not take off'.[21]

Noronha's policies achieved mixed results. Two of his main projects were not achieved – the promotion of iron-mining and the construction of a new road between Luanda and Pungo Andongo. Although the Governor argued that Africans were disinclined to engage in agriculture, lack of financial resources and expertise were the real reasons why his plans failed. For example, José Augusto da Silva Neves, who had come to Angola to work for a project to establish sugar mills, sent a petition to Lisbon saying that he had not received any salary since arriving in Luanda five months earlier and wished to return to Brazil.[22] In addition, sugar production was hindered by difficulties in finding

---

[15] *Almanak Statistico da Província d'Angola*, p. XXI. These steps were strikingly similar to measures adopted by French officials to develop legitimate trade in Senegal after the end of the slave trade in the 1810s. See Martin Klein, 'Slaves, Gum, and Peanuts: Adaptation to the End of the Slave Trade in Senegal, 1817– 48', *WMQ* 46/4 (2009), 907.

[16] AHA: cód. 1081, 'Portaria', 17 April 1839.

[17] AHA: cód. 14, fol. 27, Governador de Angola, 24 May 1839.

[18] AHA: cód. 101, fols 90v–91, Conselho de Governo, 'Portaria Circular', 3 April 1839; cód. 19, fols 31v–53v, 'Relatório da Província entre 17 de Agosto de 1848 a 31 de Dezembro de 1849', 20 Jan. 1850.

[19] AHA: cód. 259, fols 107–109v, Ministro da Marinha e Ultramar, 'Portaria', 28 Dec. 1839.

[20] AHA: cód. 259, fols 145v–146v, Ministro da Marinha e Ultramar, 'Portaria', 12 May 1840.

[21] AHA: cód. 14, fols 11v–12v, Governador de Angola, 20 April 1839.

[22] AHU: segunda seção de Angola, pasta 3 C, José Augusto da Silva Neves, 30 Oct. 1840.

adequate land to cultivate sugar.²³ The proposal to exempt Africans from working as porters was never fully implemented because it undermined the commercial interests of Luanda and Benguela merchants, who depended on porters to transport goods to trade for slaves in the interior of Angola.²⁴

State policies in the 1840s

Valentim Alexandre has argued that agricultural projects sponsored by the Portuguese state in the 1830s and 1840s were not able to significantly change production of agricultural goods in Angola.²⁵ In fact, a closer examination suggests that some of the policies pursued during Governor Noronha's tenure were successful in stimulating agricultural production in Angola. African engagement in commercial agriculture is a case in point. Prior to the initiation of Portuguese policies to foster cash-crop production, Africans had already been involved in the production of agricultural goods for subsistence and the domestic market. In the early 1840s, for example, locally-cultivated tobacco was used in a cigar factory in Golungo Alto.²⁶

Most of the tobacco consumed in Angola grew wild, and the Luanda administration sought to promote the sale of this tobacco in Luanda.²⁷ But there were also attempts to promote commercial cultivation of the product among Africans. *Soba* [i.e. chief] Bango Aquitamba from Golungo is a case in point. In 1846, he was reported to be one of the leading farmers of Golungo, prompting the administration to seek to persuade him to grow cotton, coffee and tobacco.²⁸ To encourage him to take up commercial agriculture, the administration offered to facilitate the sale of agricultural products in Luanda. A year later, when returning to Luanda with a cargo of corn, Aquitamba informed the administration that he had begun cultivating coffee, tobacco and cotton.²⁹

There is also evidence that the Luanda administration sought to promote commercial agriculture among white and mixed-race settlers in the Luanda hinterland. In 1846, for example, while pointing out that African-grown tobacco was of excellent quality, Governor Pedro

---

²³ AHU: segunda seção de Angola, pasta 3 C, Governador de Angola, 12 Dec. 1840.
²⁴ AHA: cód. 14, fls. 65v–66, Governador de Angola, 28 Oct. 1839.
²⁵ Alexandre, 'Portuguese Empire', 113.
²⁶ Lopes de Lima, *Ensaios sobre a Statistica*, 13.
²⁷ AHA: cód. 325, fols 42v–43, Secretário Geral da Província de Angola, 12 Sept. 1846; cód. 325, fol. 243, Secretário Geral da Província de Angola, 20 Nov. 1847.
²⁸ AHA: cód. 325, fols 29v–30, Secretário Geral da Provincia de Angola, 26 Aug. 1846.
²⁹ AHA: cód. 325, fols 171–171v, Secretário Geral da Província de Angola, 1 July 1847.

Alexandrino da Cunha remarked that 'the production and export of this product is taking off, in contrast to a few years ago when only black people cultivated it'.[30] To further promote agricultural production, the administration published works devoted to the cultivation of tobacco and distributed seeds of North American tobacco to several parts of Angola.[31] According to the Austrian scientist Frederico Welwitsch, who was hired by Lisbon to study the fauna and flora of Angola in 1852, these efforts had 'very satisfactory results and currently both the cultivation of tobacco from Virginia and cotton of Louisiana are already very extensive in the mountain districts and many fields are cultivated with seeds found in this country.'[32]

By fostering the production of tobacco, the administration sought to strengthen Angola's ties to Portugal. Samples were sent to Lisbon, where the holders of the royal contract to process tobacco committed to purchasing up to 8,000 *arrobas* (1 *arroba* = 32 pounds (lbs) or 15 kg) of Angolan tobacco each year.[33] In 1842, exports to Portugal stood at 200 *arrobas*.[34] Four years later, they were estimated at 340 tons.[35] In 1855, metropolitan authorities ordered the tobacco contractors in Lisbon to purchase 500 *arrobas* of Angolan tobacco.[36] Later, the Lisbon contractors committed to purchasing all tobacco shipped to Lisbon, probably at the request of metropolitan authorities.[37]

Despite official encouragement, tobacco exports amounted to only 118 *arrobas* in 1860, mainly due to the high tobacco consumption in Angola.[38] There are no statistics to document the growth of local tobacco consumption in Angola. However, in Golungo, for example, cigars manufactured with local tobacco were said to be 'good, strong, tasty but poorly made'. In Cazengo, they were described as 'well-done, tasty, made of good tobacco, but very strong'.[39] High consumption of tobacco in Angola is further attested by the production of cigars in Ambaca and Luanda, as well as the

---

[30] AHU, segunda seção de Angola, pasta 10 A, Governador de Angola, 25 June 1846.
[31] Francisco de Salles Ferreira, *Do Tabaco em Angola* (Lisbon, 1877), 42; Frederico Welwitsch, 20 Aug. 1861, cited in Frederico Weltwitsch, *Cultura do Algodão* (Lisbon. 1862), 31–39.
[32] 'Informação do Doutor Frederico Welwitsch', 10 Sept. 1856, *Annaes do Conselho Ultramarino (parte não official), Fevereiro de 1854 a Dezembro de 1858* (Lisbon, 1867), 293–4.
[33] 'Portaria ao Governador de Angola', 6 Feb. 1845, *Annaes Maritimos e Coloniaes* 3 (Feb. 1845), 36.
[34] AHU: pasta 10, Governador de Angola, 25 Oct. 1846.
[35] Most of the product was purchased from Africans by a Luanda merchant, Valentim José Pereira, who even built a cigar factory in the city. See AHA: cód. 325, fols 42v–43, Secretário Geral da Província de Angola, 12 Sept. 1846.
[36] Ferreira, *Do Tabaco em Angola*, 9.
[37] Ibid., 16.
[38] Sebastião Lopes de Calheiros e Menezes, *Relatório do Governador Geral da Província de Angola para o ano de 1861* (Lisbon, 1867), 18.
[39] Ferreira, *Do Tabaco em Angola*, 13.

fact that Angola may have become an importer of tobacco by the early 1860s.[40]

In addition to promoting African engagement in commercial agriculture, the Luanda administration also encouraged private investment in commercial agriculture, generally to complement African production of crops. In the 1840s, for example, an investor named Antonio de Magalhães Mesquita received authorization from the Luanda government to establish two sugar mills in Bengo and Cambambe.[41] Official support for private investors is further illustrated by the investments in sugar production by one of the foremost slave-dealers in Angola, Ana Joaquina dos Santos Silva. In addition to owning one of the largest fleets of slave-ships in Angola, as well as farms around Luanda, Silva actively engaged in projects to develop sugar production, even travelling 'herself to Brazil to witness the process of cultivation and manufacturing sugar and expresses herself pleased and confident as to the result of her present endeavours'.[42] Silva played a crucial role in the development of sugar production in Angola. In 1846, she hired a man from Pernambuco, Pedro Regaire, to assist with the establishment of a sugar mill on one of her farms.[43] Regaire would become part of the state efforts to promote agriculture and was 'very useful due to his knowledge of cultivation and great experience'.[44]

Silva's sugar mill was not the first to be established in Angola, since mills had already been set up along the Bengo River in the 1830s.[45] Others engaged in sugar production as well. In 1841, for example, Governor Manoel Eleutério Malheiros reported that 'there was a resident of Luanda who owned farms [*arimos*] and cattle and who had sought to establish a sugar mill, for which he has the necessary support, despite several obstacles, including lack of wood'.[46]

But none of these investments rivalled those of Ana Joaquina, which dovetailed with the Luanda administration's plan to promote sugar production in areas near Luanda, thus facilitating transport of the crop

---

[40] Bernardino Freire de Figueiredo Abreu e Castro, 'Breve Notícia do Estado da Agricultura nas Nossas Províncias Ultramarinas', *Archivo Rural: Jornal de Agricultura, Artes e Sciencias Correlativas* 4 (1862), 324.
[41] AHA: cód. 15, fol. 11, Governador de Angola, 6 June 1840.
[42] TNA: FO84/671, Edmond Gabriel and George Jackson, 'Report on the Slave Trade', 18 Feb. 1847.
[43] AHA: cód. 325, fols 38–38v, Secretário Geral da Provincia de Angola, 10 Sept. 1846.
[44] AHA: cód. 325, fol. 215v, Secretário Geral da Província de Angola, 5 Oct. 1847; cód. 1178, fol. 22, 'Instruções para o Tenente Paulo Francisco da Silva', 20 Dec. 1847.
[45] AHU: cx. 167, doc. 46, Governador de Angola, 14 Dec. 1830.
[46] AHU: segunda seção de Angola, pasta 4 B, Governador Geral de Angola Manoel Eleutério Malheiros, 'Diário da viagem feita ao Distrito de Icolo e Bengo', 17 Jan. 1841.

to the city for export.[47] Official interest in Silva's sugar farms is confirmed by Governor Adrião Acácio da Silveira Pinto's visit to one of her properties in 1851.[48] Silva's initial investments in sugar production were by no means profitable, with estimated losses of ten million *réis*.[49] By the late 1840s, a British diplomat pointed out that 'the only attempt to give to capital a more wholesome direction ... is now being made by Dona Ana Santos Silva, herself originally a slave dealer and whose name has often figured in the slave trade reports as deeply engaged in the traffic'.[50] Production of sugar on Ana Joaquina Silva's farms became so successful that sugar seeds from these farms were sent to other parts of Angola, including to the newly-established agricultural colony in Mossamedes.[51] In 1852, a Portuguese traveller reported that 'good quality sugar was produced in the margins of Bengo and Dande [rivers] and that blacks daily brought a large number of canes to the market of Quitanda in Luanda and that black people bought it to chew'.[52] By then, 'more than thirty thousand canes are sold [in Luanda] for twenty-five *réis* for two canes'.[53] A great deal of this sugar was produced on Ana Joaquina Silva's ten farms in Icolo, where 1,500 slaves were said to cultivate several types of crops, including sugar.[54]

Silva's investments were so central to sugar production that her death in the late 1850s was an irretrievable setback to the nascent industry. At the beginning of the 1860s, there were only six sugar mills in Angola and the colony had become an importer of sugar.[55] In addition to Ana Joaquina Silva's death, there were two reasons for the failure to develop sugar production in Angola. First, unlike tobacco, sugar production demanded high investment, which kept investors away. Secondly, in the early 1850s, the Luanda administration devoted scarce state resources to support sugar production in Mossamedes, which, as noted before, had been established by Portuguese settlers seeking to escape anti-Portuguese hostility in Brazil. From the onset, production of sugar was central to Mossamedes, and settlers under-

---

[47] AHA: cód. 1178, fols 104v–106v, 'Instruções para o Capitão do Batalhão de Infantaria de Linha de Luanda Joaquim Olavo Gamboa', 28 May 1850.
[48] AHA: cód. 20, fols 62v–63, Governador de Angola, 8 Aug. 1851.
[49] *Almanak Statistico da Província d'Angola*, 11.
[50] TNA: FO84/671, Gabriel and Jackson, 'Report on the Slave Trade', 18 Feb. 1847, fols 99–11.
[51] AHU: segunda seção de Angola, pasta 22 A, Francisco José da Costa Silva, Comentários sobre a 'Memória sobre as imediatas providências que precisa Benguela e Angola', 1856.
[52] Caldeira, *Apontamentos d'uma Viagem*, 211.
[53] *Almanak Statistico da Província d'Angola*, 11.
[54] AHU: papéis de Sá da Bandeira, maço 823, Adrião Acácio da Silveira Pinto, 'Relatório da Visita feita aos distritos do Bengo, Icolo, Barra do Dande e Alto Dande', 28 Aug. 1850; Douglas Wheeler, 'Angolan Woman of Means: D. Ana Joaquina dos Santos e Silva, Mid-Nineteenth Century Luso-Africa Merchant-Capitalist of Luanda', *Santa Barbara Portuguese Studies* 3 (1996), 286.
[55] Menezes, *Relatório do Governo Geral*, 18.

took to bring three mills from Pernambuco in Brazil.⁵⁶ Support for Mossamedes depleted state resources and prevented the Angolan administration from promoting sugar production elsewhere in the colony. Worse still, because of African resistance and several other problems, settlers were unable to develop the Mossamedes economy, including sugar cultivation.⁵⁷

Coffee production

From the early 1820s, coffee became a favourite crop for developing an alternative economic framework to the slave trade in Angola. Even before the end of the slave trade, wild coffee was gathered in several parts of Angola. It was exported to Lisbon by 1830.⁵⁸ Most of the coffee sold in Luanda was wild coffee. In Golungo, 'coffee was found in the lands of *soba* Cabanga Cavalunga, and it has been a long time since it was discovered'. During the 1840s, the sale of wild coffee in Luanda by Africans became a regular activity.⁵⁹ In the 1850s, Governor Adrião Acácio da Silveira Pinto reported that 'a large number of coffee trees' had been found in the district of Dembos.⁶⁰

However, there also were several attempts to promote commercial coffee cultivation. In 1827, Governor Nicolau de Abreu Castelo Branco stated that he had recently promoted the sowing of one thousand coffee seeds in Pungo Andongo.⁶¹ In the early 1830s, the administration sought to further promote the cultivation of coffee in Pungo Andongo, Encoje, Golungo and Cazengo.⁶² The Luanda administration's support for coffee production in this period was clearly influenced by rising coffee prices in the world market.⁶³ However, they were far from evenly successful. In 1840, authorities reported that 'after four and five years [seeking to cultivate coffee in

---

⁵⁶ AHA: cód. 1178, fols 51v–52v, 'Instruções para o Major de Artilharia de Benguela José Francisco Garcia Moreira', 25 Jan. 1848.

⁵⁷ AHA: cód. 18, fols 23–24, Governador de Angola, 30 Sept. 1848.

⁵⁸ Aida Freudenthal, *Arimos e Fazendas: A Transição Agrária em Angola, 1850–1880* (Luanda, 2005), 111; Jill Dias, 'Angola', *Nova História da Expansão Portuguesa: O Império Africano*, ed. Valentim Alexandre and Jill Dias (Lisbon, 1998), X, 374.

⁵⁹ Ferreira, *Do Tabaco em Angola*, 6; AHU: segunda seção de Angola, pasta 4 B, 'Ata da Sessão da Junta da Fazenda', 16 April 1841.

⁶⁰ AHU: segunda seção de Angola, pasta 16, Governador de Angola, 'Relatório', 22 Feb. 1850. For an earlier report of wild coffee in Encoje, see AHA: cód. 325, fols 248v–249, Secretário da Província de Angola, 26 Nov. 1847.

⁶¹ AHU: Angola, cx. 156, Governador de Angola, 10 Oct. 1827.

⁶² AHA: cód. 97, fol. 4, Governador de Angola, 6 March 1829.

⁶³ AHA: cód. 7183, fol. 58v, Antonio Manoel de Noronha, 7 April 1827; AHU: segunda seção de Angola, pasta 4, Antonio Manoel de Noronha, 5 May 1827; segunda seção de Angola, pasta 4, Antonio Manoel de Noronha, 3 July 1827.

Golungo], the coffee was still too small and had not yet blossomed'.[64]

State-sponsored projects to promote cultivation of coffee gained momentum in the second half of the 1840s. In 1846, for instance, authorities 'gave' one hundred workers for Ana Joaquina dos Santos Silva to harvest coffee on a farm that belonged to Joaquim Rodrigues da Graça, her former partner.[65] In Massangano, production thrived with the support of the Luanda administration, which owned farms devoted to coffee cultivation in the Dembos region to the north of Luanda.[66] In fact, commercial production reached 2,400 *arrobas* in Ambaca in 1851.[67] At the time, the total number of coffee trees in Muxima, Massangano, Cambambe, Pungo Andongo, Encoje, Golungo, Zenza, Dembos and Calumbo was estimated at 60,000.[68] By 1856, the authorities reported the growth in coffee production in Golungo Alto as well.[69]

However, the region that came to epitomize Angolan coffee production was Cazengo. As early as 1830, a report commissioned by the administration pointed to a significant amount of wild coffee in territories controlled by African rulers.[70] In 1838, the number of stalks [*pés*] of coffee was estimated at 60,000, and authorities stated that 'it would be very advantageous if workers were brought from Brazil to cultivate [in Angola] the different types of coffee that were cultivated there [in Brazil]'.[71] Indeed, Brazilian participation in Cazengo coffee production had already begun in 1835, when a Brazilian man, João Guilherme Barboza, 'found' wild coffee in the region.[72] As Barboza himself would later admit, 'wild coffee is all over the place, and in some places it is so abundant that one only needs to pick it up and clean it'.[73] Some of the coffee trees were said to grow as high as twenty feet and the mountainous topography of Cazengo was particularly suitable for the crop.[74]

In addition to the commercialization of wild coffee, the Luanda administration also sought to promote the commercial cultivation of

[64] AHU: segunda seção de Angola, pasta 3 A, Chefe do Distrito do Golungo Alto, 3 Jan. 1840.
[65] AHA: cód. 325, fols 30v–31, Secretário de Angola, 29 Aug. 1846.
[66] AHA: cód. 325, fols 199–199v, Secretário Geral da Província de Angola, 3 Sept. 1847; cód. 325, Secretário Geral da Província de Angola, 26 Nov. 1847; cód. 1178, fols 104v–106v, 'Instruções para o Capitão do Batalhão de Infantaria de Linha de Luanda Joaquim Olavo Gamboa', 28 May 1850.
[67] *Almanak Statistico da Província d'Angola*, 12.
[68] Ibid.
[69] AHU: segunda seção de Angola, pasta 22 A, Francisco José da Costa Silva, Comentários sobre a 'Memoria sobre as imediatas providências que precisa Benguela e Angola', 1856.
[70] AHU: Angola, cx. 173, doc. 106, Sargento Cabo, Cazengo, 15 April 1832.
[71] AHA: cód. 259, fols 225–231v, 'Instrução para o Governador de Angola', 4 Nov. 1838.
[72] *Almanak Statistico da Província d'Angola*, 11.
[73] João Guilherme Barboza, 'Descripção do Cazengo', 20 June 1847, *Annaes do Conselho Ultramarino (parte não official), fevereiro de 1854 a dezembro de 1858* (Lisbon, 1867), 472.
[74] Frederico Welwitsch, *Synopse Explicativa das Amostras de Madeiras e Drogas Medicinaes e de outros Objectos mormente Ethnographicos colligidos na Província de Angola* (Lisbon, 1862), 9.

the crop.This is demonstrated by the fact that Barboza received a piece of land from the government to begin commercial cultivation of coffee in 1845.[75] In 1840, Cazengo coffee production reached 170 bags [*sacas*], and the expectation was that it would shortly increase to 1,000 bags.[76] By 1846, Barboza estimated that there were 70,000 coffee trees in Cazengo, and that another 20,000 were being prepared for cultivation.[77] This number was to increase significantly, due to Barboza's investments and organization of commercial production of the crop. By the mid-1850s, there were half a million coffee trees in Cazengo.[78] According to British authorities, coffee production reached 893 hundredweight (cwt) in 1854 and was expected to increase to 2,441 cwt in 1856.[79]

Despite Brazilian investment, Cazengo coffee was not cultivated 'according to the system in practice in Brazil'. The difference derived from the fact that most of the coffee either came from wild tress or was produced according to African farming techniques.[80] The region's success was such that colonial authorities from other parts of Angola tried to acquire coffee seeds from the region, in a bid to replicate its success.[81] Cazengo became so associated with agricultural success that the Luanda administration also promoted the cultivation of other crops there, including ginger, cotton, tobacco and cocoa.[82]

The coffee produced was soon exported to Portugal. Between 1842 and 1844, coffee exports increased from 200 *arrobas* to 380 *arrobas*.[83] In 1846, 345 *arrobas* of coffee were produced in Cazengo.[84] In that year, Governor Pedro Alexandrino da Cunha stated that the 'production and exportation of coffee was increasing significantly'.[85] Indeed,

---

[75] *Almanak Statistico da Província d'Angola*, 11; 'Decreto real', 20 July 1845, *Annaes Maritimos e Coloniaes*. 8 (July 1845), 127.

[76] AHU: segunda seção de Angola, pasta 3 A, 'Ofício do Governador de Angola', 28 Jan. 1840.

[77] João Guilherme Pereira Barboza, 19 July 1846, cited in Luiz Antonio de Figueiredo, *Indice do Boletim Official da Provincia d'Angola: Comprehendendo OS Annos Que Decorrem Desde 13 de Setembro de 1845*, 50.

[78] Jill Dias, 'Changing Patterns of Power in the Luanda Hinterland: The Impact of Trade and Colonization on the Mbundu ca 1845–1920', *Paideuma: Mitteilungen zur Kulturkunde* 32 (1985), 300.

[79] TNA: FO83/1013, Gabriel, 'Report on the Slave Trade', 11 Feb. 1857. For lower estimates, see AHU: segunda seção de Angola, pasta 22 A, 'Petição dos Agricultores do Cazengo', 15 Dec. 1856.

[80] Manuel Alves de Castro Francina, 'Viagem a Cazengo pelo Quanza e Regresso por Terra', 6 Feb. 1847, *Annaes do Conselho Ultramarino (parte não official), fevereiro de 1854 a dezembro de 1858* (Lisbon, 1867), 452–64.

[81] AHA: cód. 110, Carlos Possolo de Sousa, 29 Oct. 1852.

[82] AHA: cód. 1178, fols 48–49v, 'Instruções para o Alferes do Esquadrão de Cavalaria de Luanda', 2 Nov. 1848.

[83] David Birmingham, 'The Coffee Barons of Cazengo', *Journal of African History* 19/4 (1978), 523–38; Lopes de Lima, *Ensaios sobre a Statistica*, 76–77. The estimate for the overall production of coffee comes from João Guilherme Pereira Barboza, 17 Oct. 1845, cited in Luiz Antonio de Figueiredo, *Indice do Boletim Official da Provincia d'Angola (1845–1862)* (Lisbon, 1864), 50.

[84] João Guilherme Pereira Barboza, 25 Feb. 1847, cited in Figueiredo, *Indice do Boletim Official*, 50.

[85] AHA: cód. 16, fols 239–239v, Governador de Angola, 14 Nov. 1846.

exports increased dramatically in the following years. In 1850, Cazengo coffee was exported to Hamburg and two years later, Luanda received 2000 *arrobas*, mostly from Cazengo.[86] Between 1858 and 1859, 15,032 *arrobas* of coffee, mostly cultivated in Cazengo, were exported through the Luanda port.[87] To support the transportation of coffee to Luanda, local authorities built a road to facilitate its delivery by the Lucala River.[88] In addition to building roads, authorities imported camels, which were sent to Cazengo, 'as requested by local authorities since neither cattle nor horses were adaptable to transportation work' in the region.[89]

Unfree labour

Robin Law has noted that the development of commercial agriculture to replace the slave trade led to the rise of a variety of forms of 'social relations of production' in Africa.[90] On the Gold Coast, as demonstrated by Ray Kea, the transition to market-oriented agriculture relied on a combination of free labour, slavery and pawns.[91] In Angola, debates about the nature of labour relations intensified as the slave trade drew to an end in the 1840s and 1850s. In the 1840s, Sebastião Xavier Botelho advocated the gradual ending of slavery and its replacement by free labour.[92] In 1860, however, Governor Sebastião Lopes de Calheiros e Menezes observed that 'free labour in which blacks would receive a salary was a dream, at least for the time being'.[93]

Slavery eventually became the cornerstone of labour relations in commercial agriculture. However, for a short period free labour was seen as an effective way to deal with the organization of labour in Angolan farming, in which women carried out most of the agricultural

---

[86] AHU: segunda seção de Angola, pasta 16, Governador de Angola, 'Relatório', 22 Feb. 1850; Caldeira, *Apontamentos d'uma Viagem*, 212; *Almanak Statistico da Província d'Angola*, 12.
[87] Menezes, *Relatório do Governo Geral*, 17.
[88] Governo Geral de Angola, 10 Feb. 1846, cited in Figueiredo, *Indice do Boletim Official*, 65.
[89] Ibid.; Ministério da Marinha e Ultramar, 'Relatório', 2 March 1846, *Annaes Maritimos e Coloniaes* 8 (July 1845), 157. See also Francisco Valdez, *Africa Occidental: Noticias e Considerações* (Lisbon, 1864), 95. This was not the first time that camels were imported into Angola, as some had been imported in 1839. See Lopes de Lima, *Ensaios sobre a Statistica*, 18. This issue was critical elsewhere in Africa as well. See Law, 'Introduction', 10.
[90] Law, 'Historiography', 109.
[91] Ray A. Kea, 'Plantations and Labour in the South-East Gold Coast from the late Eighteenth to the Mid-Nineteenth Century', *From Slave Trade to 'Legitimate' Commerce*, ed. Robin Law (Cambridge, 1995), 120.
[92] Botelho, *Escravatura*, 20.
[93] Menezes, *Relatório do Governo Geral*, p. 67. For the Gold Coast, see Kea, 'Plantations and Labour', 137.

work.[94] In 1846, for example, the administration sent only women and young male *libertos* (former slaves) to Cazengo, since the local men 'were more difficult to train [in agricultural techniques], to get used to the work of agriculture and could escape'.[95] In 1858, Welwitsch remarked that 'the general custom among the black population is that men do not carry out agriculture, leaving any agricultural, domestic, or rural work to the poor care of their often many women and children'.[96]

The fact that agriculture was women's work led to stereotypical analyses by potential investors in agricultural projects. These claimed that females lacked the physical strength to work in large-scale agriculture and were unsuitable workers because they would have to perform household tasks at home on top of agricultural work.[97] It was in the context of this debate that authorities considered the establishment of free labour, in 1840.[98] This proposal was soon challenged, however, because of claims that the success of commercial agriculture depended on the Brazilian plantation model, which was heavily reliant on male slave labour.[99] This position was soon reflected in debates about the possible abolition of slavery in Angola. By the mid-1840s, Lopes de Lima wrote that 'the number of slaves must increase once the settlers of our cities and *presídios* began to establish agricultural fields like in Brazil'. As Lima pointed out, 'the idea of abolishing slavery right away in the interior of Africa is good in theory and is very philanthropic. However, it is inopportune and impolitic.'[100]

The reliance on slave labour meant that slave-owners had to deal with the constant disruption of agricultural production due to slaves attempting to escape, which even provoked the abandonment of *arimos* in Quifandongo in 1841.[101] Governor Pedro Alexandrino da Cunha, who supported the use of slaves as the primary labour force in Angolan agriculture, stated that slaves who worked in agriculture had a tendency to run away as they feared being shipped to Brazil.[102]

---

[94] João Guilherme Barboza, 'Descripção do Cazengo', 20 June 1847, *Annaes do Conselho Ultramarino (parte não official), fevereiro de 1854 a dezembro de 1858* (Lisbon, 1867), 470.
[95] AHA: cód. 325, fols 87v–88, Secretário Geral da Província de Angola, 8 Dec. 1846.
[96] TNA: FO84/1043, Welwitsch, 15 March 1858.
[97] AHU: segunda seção de Angola, pasta 3 C, Presídio de Pungo Andongo', 14 July 1841.
[98] AHU: segunda seção de Angola, pasta 3 A, Chefe do Distrito do Golungo Alto, 3 Jan. 1840; pasta 3 C, Governador de Angola, 3 March 1840.
[99] AHU, segunda seção de Angola, pasta 3 C, Governador de Benguela', 17 March 1840.
[100] Lopes de Lima, *Ensaios sobre a Statistica*, 7. To defend his view, and ultimately win the debate with Botelho, Lopes de Lima embarked on a public campaign that included articles in Portuguese newspapers. See Douglas Wheeler, 'The Portuguese in Angola, 1836–1891: A Study in Expansion and Administration' (unpublished Ph.D. thesis, Boston University, 1963), 103.
[101] AHA: cód. 15, fols 28–32, 'Diário de uma viagem ao Sertão', 25 July 1840.
[102] AHU: papéis de Sá da Bandeira, maço 825, Pedro Alexandrino da Cunha to Visconde de Sá da Bandeira, 4 June 1839.

Runaway slaves remained a problem for many years. As late as 1861, Governor Sebastião Lopes de Calheiros e Menezes wrote that commercial agriculture would 'never be as productive as in São Tomé, Havana, Brazil or the United States because blacks can easily escape' from slavery.[103]

Despite the problem of fugitive slaves, the slave population of Angola grew significantly. In 1849, the number of slaves in Angola reached nearly 30,000.[104] Seven years later, there were approximately 60,000 enslaved Africans in the colony.[105] This increase was greatly facilitated by the fact that prices of slaves dropped significantly in the wake of the ending of the trans-Atlantic slave trade to Brazil in the 1850s. Already in 1847, the British representative to Luanda stated that 'the low price at which an able bodied negro may now be purchased in this province [is] about one half what would have been given for him two or three years ago'.[106] By 1854, the price of a young slave had dropped from 70–80 to 10–20 US dollars in Luanda.[107] By the early 1860s, prices for slaves ranged from 25,000 to 45,000 *réis* (the dollar being equivalent to c. 2,000 *réis*), depending on age and agricultural skills.[108]

Equally importantly, the growth of slavery can be related to the expansion of commercial agriculture. A report of 1850 about cotton cultivation estimated that each enslaved African would be able to cultivate and process ten *arrobas* of cotton.[109] In Mossamedes, Bernardino Freire de Figueiredo Abreu e Castro, the leader of the local agricultural

---

[103] Menezes, *Relatório do Governo Geral*, 67. For the Gold Coast, see Kea, 'Plantations and Labour', 137. For the Bight of Benin, see Patrick Manning, *Slavery, Colonialism and Economic Growth in Dahomey, 1640–1960* (New York and Cambridge, 1982), 13. For Cameroon, see Ralph Austen and Jonathan Derrick, *Middlemen of the Cameroons Rivers: The Duala and their Hinterland c. 1600–c. 1960* (New York and Cambridge, 1999), 87.

[104] AHU: papéis de Sá da Bandeira, maço 779, 'Mapa da População Escrava de Angola', 18 Aug. 1849. For analyses of the growth of slavery following the end of the slave trade in Africa, see Paul E. Lovejoy, *Transformations in Slavery: A History of Slavery in Africa* (New York and Cambridge, 1983), 135–58; Manning, *Slavery, Colonialism and Economic Growth*, 49; David Gordon, 'The Abolition of the Slave Trade and the Transformation of the South-Central African Interior during the Nineteenth Century', *WMQ* 66/4 (2009), 926.

[105] AHU: papéis de Sá da Bandeira, maço 825, Governador Geral de Angola, 'Relatório', 30 March 1859.

[106] TNA: FO84/671: Edmond Gabriel and George Jackson, 'Report on the Slave Trade', 8 Feb. 1847. For a discussion on prices of slaves elsewhere in Atlantic Africa, see Paul E. Lovejoy and David Richardson, 'British Abolition and its Impacts on Slave Prices along the Atlantic Coast of Africa', *Journal of Economic History* 55/1 (1995), 98–119.

[107] David Livingstone, *Family Letters, 1841–1856* (London, 1959), I, 252–3.

[108] Frederico Welwitsch, 20 Aug. 1861, in Weltwitsch, in *Cultura do Algodão*, 31–39.

[109] AHU: segunda seção de Angola, pasta 18 A, 'Memória sobre a Cultura do Algodão', 28 Feb. 1852. For the connection between agriculture and the growth of slavery in Old Calabar, see A.J.H. Latham, *Old Calabar, 1600–1891: The Impact of the International Economy upon a Traditional Society* (Oxford, 1973), 91.

colony, declared that 'a few settlers had already used money from agricultural activities to purchase slaves'.[110] In 1855, a British traveller who visited Mossamedes stated that the number of slaves was 'considerable'.[111] The relationship between slavery and the expansion of agricultural activities in Angola was also acknowledged by British authorities: 'the existence of slavery in this province and especially the employment of slaves by all the agriculturists in the interior render it doubtful whether any large plantations would be made by them without having recourse to slave labour'.[112]

However, as elsewhere in Africa, slave labour was not the only form of unfree labour in Angola.[113] This was partly due to the failure to enforce fully the first Portuguese law for the gradual abolition of slavery, in 1854. This law decreed that slave-holders had to register their captives within thirty days and that slaves not registered would be considered *libertos*.[114] According to British authorities: 'The immediate result [of this decree] has been the registration of about fifteen thousand slaves belonging to this city [Luanda] and its vicinity.'[115] *Libertos* were obliged to work for the state for seven years before becoming 'free'. Between 1859 and 1863, the number of *libertos* increased by approximately 14,000 to almost 30,000.[116] This growth resulted from the fact that enslaved Africans brought into Portuguese Angola were considered as *libertos* according to the 1854 decree.

However, the 1854 decree was fiercely opposed by settlers and officials in the colony. Instead of providing a path to freedom, the status of *liberto* became another form of unfree labour, barely distinguishable from slavery. A report by colonial authorities in 1856 stated that 'what needs to be done is to tolerate the work of slaves, considering them as free but making them free only when authorities are required to

---

[110] Bernardino Freire de Figueiredo Abreu e Castro, 'Relatório sobre Mossamedes', 5 July 1854, *Annaes do Conselho Ultramarino (parte não official), fevereiro de 1854 a dezembro de 1858* (Lisbon, 1867), 132–4.
[111] William Meseum, 'Notícia de uma Exploração da Costa Occidental da África ao Sul de Benguela', 1855, *Annaes do Conselho Ultramarino (parte não official), fevereiro de 1854 a dezembro de 1858* (Lisbon, 1867), 253–7.
[112] TNA: FO 84/1043, Gabriel, 3 April 1858.
[113] Kristin Mann, 'Owners, Slaves and the Struggle for Labour in the Commercial Transition at Lagos', in *From Slave Trade to Legitimate Commerce*, ed. Robin Law (Cambridge and New York, 1995), 165; id., *Slavery and Birth of An African City: Lagos, 1760–1900* (Bloomington IN, 2007), 160–200.
[114] 'Decreto de 14 de dezembro de 1854', in Ministério da Cultura, *A Abolição do Tráfico e da Escravatura em Angola: Documentos* (Luanda, 1997), 35–38. The expression 'liberto' was also used to refer to Africans released from slave ships in the 1840s.
[115] TNA: FO84/960, Gabriel, 12 Nov. 1855.
[116] AHU: Angola, pasta 26, 'Relação dos libertos registrados na Província d'Angola desde que existem indivíduos (1854) com tal condição até o fim do ano de 1859'; Angola, pasta 34, 'Nota do número de libertos que têm sido registrados na Província de Angola depois do decreto de 14 de dezembro de 1854 até 31 de dezembro de 1863'.

rule over their status'.[117] In 1861, Governor Menezes stated that 'slavery has been abolished by law, but it is believable that it will be revived'.[118] Like slaves, *libertos* were regularly used as workers on agricultural projects, much like enslaved Africans.[119] As stated by the Portuguese-British traveller John Monteiro, 'there are at present in Angola several sugar and cotton plantations worked by slaves, called at present *libertos*'.[120]

Conclusion

As elsewhere in Africa, the Angolan transition to the trade in 'legitimate' goods was characterized by an overlap with the continuing slave trade. Thus, even as thousands of slaves were still shipped from Luanda and Benguela, exports of other goods increased. However, much of this 'legitimate' trade was in non-agricultural commodities. For example, exports of ivory rose from 3,000 lbs to 105,000 lbs between 1832 and 1844.[121] Exports of gum and bees-wax also experienced a significant increase. While 832 *arroba*s of gum were exported in 1828, exports uct reached at least 15,815 *arroba*s in 1833.[122] Likewise, exports of bees-wax rose from approximately 52,000 lbs in 1844 to almost 1,700,000 lbs in 1857.[123] By contrast, agricultural products accounted for only 10 per cent of Angolan exports by the end of the 1850s.[124]

Perhaps the most important aspect of early attempts to develop commercial agriculture in Angola relates to the growth of the internal market for agricultural products, rather than exports. It might even be that the latter was a function of the former, as the production of tobacco seems strongly to suggest. Whatever the case, the most successful formula for developing agriculture was in Cazengo, where a hybrid model combined African know-how and labour with Brazilian investment. This model was not followed in the early

[117] AHA: cód. 857, fols 13–13v, 'Secretário do Governo'. 20 Sept. 1856.
[118] Menezes, *Relatório do Governo Geral*, 82.
[119] For evidence of the use of *libertos* in agricultural projects, see AHA, cód. 9/C-4-10, fol. 11v, Juiz Ordinário, Nov. 1855; cód. 18, fols 130v–137v, Governador de Angola, 'Relatório' 1849; Annaes do Município de Mossamedes (anos de 1838–1849), 31 Dec. 1856, *Annaes do Conselho Ultramarino (parte não official), fevereiro de 1854 a dezembro de 1858* (Lisbon, 1867), 490.
[120] J.J. Monteiro, *Angola and the River Congo* (London, 1875), 75–76.
[121] Jill Dias, 'Changing Patterns', 294; Joseph Miller, 'Cokwe Trade and Conquest in the Nineteenth Century', *Pre-Colonial African Trade: Essays on Trade in Central and Eastern Africa before 1900*, ed. Richard Gray and David Birmingham (London, 1970), 178.
[122] AHU: Angola, cx. 176, 'Certificado da Alfândega de Luanda', 1 Oct. 1833.
[123] Dias, 'Changing Patterns', 294.
[124] Allen Isaacman, 'An Economic History of Angola, 1835–1867' (unpublished MA thesis, University of Wisconsin, 1966, 97; David Eltis, *Economic Growth and the Ending of the Transatlantic Slave Trade* (New York, 1987), 230.

attempts to develop sugar production. In addition to its high costs, sugar production relied on Brazilian techniques and restricted African participation to the provisioning of an enslaved labour force. Sugar production was also hindered by the geopolitical choices of the Portuguese, since the decision to support the agricultural colony of Mossamedes diverted scarce resources that could have been used elsewhere in Angola. More importantly, commercial agriculture became one of the driving forces behind of the rise of slavery in Angola in the wake of the ending of the trans-Atlantic slave trade.

11

Commercial agriculture & the ending of slave-trading and slavery in West Africa, 1780s–1920s

GARETH AUSTIN[1]

This chapter reconsiders two turning-points in the economic and social transition from human to agricultural commodities as the principal exports of West Africa. Discussion of this process usually focuses on the first half of the nineteenth century, featuring British abolition and the growth of 'legitimate commerce' in the form of palm-oil and groundnut exports. It will be argued here that the chronological scope should be lengthened, at both ends of the story. For the beginning, I propose a shift in focus from 1807, the year the most powerful of the slave-trading countries embarked upon abolition, to 1787, when the volume of slave shipments actually began to decline. For the ending, I suggest that the modern social (as opposed to economic) history of African agriculture began not in the early nineteenth but in the early twentieth century.

These changes to the description raise questions about the causality. I will argue first that, putting the evidence on movements in the prices paid for slaves alongside the data on the numbers of slaves shipped, it becomes clear that the 'premature' fall in slave exports from West Africa must be endogenous, in the sense of stemming from the interaction of the Atlantic slave trade and African societies and economies, rather than from events elsewhere in the Atlantic. This has been largely overlooked in the literature, so it will be given relatively detailed attention here. I will go on to consider possible demand-side and supply-side hypotheses, derived from the briefly stated but important ideas of

[1] I am grateful to the editors for their patience, to Robin Law for many excellent suggestions, and to David Eltis and others who made valuable comments on one or other of the four presentations of versions of this paper: the GHIL conference in Sept. 2010; the African Economic History Workshop at the Graduate Institute in Geneva in May 2011; a panel organized by Kazuo Kobayashi at the Asian Association of World Historians in Seoul in April 2012; and at the annual conference of the Société Suisse d'Études Africaines in Bern, Oct. 2012. Any errors are mine.

earlier writers, which will be critically examined through a close reading of already-published statistics. The second argument about causality relates to the prohibition of slavery within West Africa itself, and its relationship with the growth of export agriculture during the early colonial period. I argue that the practical outcome of abolition legislation was largely determined by the socially and spatially uneven distribution of the economic opportunities in the 'cash crop revolution'. Depending on crop and area, some slave-masters were able to make the transition to employing hired labourers. But in other cases, former slaves became free peasants, alongside already free households, producing partly for subsistence and partly for sale, especially for export. This part of the chapter draws partly on my own earlier research and will be expressed briefly because I have presented it elsewhere; what is new here is the attempt to set it in the context of the longer history of abolitionism in West Africa. Finally, I go on to relate both parts of the analysis to the very long-term path(s) of economic development blazed by West Africans, focusing on the interaction between resource endowments and human responses to those endowments.

Was the beginning of the decline of the Atlantic Slave Trade endogenous?

The justified scholarly and public attention to the Abolitionist movement has obscured the need to make an analytical distinction between the decline of the trade, on one hand, and its prohibition and eventual extirpation, on the other. The flow of captives from Atlantic Africa is generally thought of as having first been constricted by legislation during the first decade of the nineteenth century, and then gradually but firmly throttled by diplomatic and naval pressure over several decades, culminating in legislation and enforcement in the remaining shipping and importing countries. There is no doubt that it was the abolition campaign that ultimately killed the Atlantic slave trade. But there has been very little attention to the fact that in West Africa – though not further south, in Angola[2] – the flow had begun to be reduced even before abolition legislation began.

A second reason for inattention is that the Atlantic slave trade is often considered in aggregate, and the overall volume rose in the years immediately preceding the beginning of abolition. Yet in many contexts specialists on African history distinguish sharply between the histories

[2] 'Angola' as a slave-trading 'coast' extended from what is now the Angolan border with Namibia north to the equator.

of West Africa and West-Central Africa.³ If we apply this distinction to the decline of the Atlantic slave trade, we find that the continued growth in shipments was entirely an Angolan phenomenon: whereas the trend, however slightly, was downwards in all six of the slaving 'coasts' (regions) that the Europeans distinguished in West Africa, from the Bight of Biafra to Senegambia. This difference matters analytically, because it implies that there were already downward pressures on the size of the trade before 1807, though they operated in West Africa but not, at least as yet or with any strength, in Angola. The downward pressure in West Africa in turn raises the question of whether the initial decline was endogenous in origin: the result of the interaction between the export of slaves and the wider economies of West Africa.

The opportunity to investigate whether the rate of slave exports began to fall before abolition, and if so, to begin to explore the reasons why, has been created by the energetic and painstaking research on anything quantifiable about the trade that has been carried out in recent decades. The argument in this section is based on reflections on two parts of this data revolution. One concerns the estimates of the volume of the trade presented in the Trans-Atlantic Slave Trade Database compiled by a team headed by David Eltis.⁴ The other is the work on the prices paid for slaves, again by various hands, most notably David Richardson and Paul Lovejoy.⁵ These series are combined in Table 11.1 and Figure 11.1, below.

The volume of the slave trade from West Africa revived in the four years after 1783, recovering from the disruption entailed by the fighting between the slave-shipping countries during the American War of Independence. The annual number of captives embarked reached 54,175 in 1787. Over the remaining twenty years before the British abolition act, the annual totals continued to fluctuate, but the trend was gradually if unevenly downward. Taken in isolation, this overall dwindling might be seen as essentially reflecting events outside Africa, because those two decades included the Haïtian Revolution of 1791–1804 and a general European war. The British entry into the war with France, in 1793, meant the beginning of naval disruption of Euro-African trade, especially when carried by the ships of France and, later, France's continental allies. The war lasted, with an intermission in 1802–3, until several years after British abolition had come into force.

---

[3] Interestingly, though, to my knowledge the only scholar to make this distinction in this context is himself primarily an Atlantic rather than Africa specialist, though he has contributed very valuably to both historiographies: David Richardson, 'Prices of Slaves in West and West-Central Africa: Towards an Annual Series, 1698–1807', *Bulletin of Economic Research* 43/1 (1991), 21–56.
[4] www.slavevoyages.org (2008), hereafter TASTD (accessed 30 May 2009).
[5] Starting with Richardson, 'Prices of Slaves'. See further citations below.

What is remarkable is the trend of slave prices, which was the opposite of what would have been expected from the circumstances on the ocean and in the slaving ports of Europe and the Americas. Of all the slaves shipped across the Atlantic in the forty years before the Haïtian Revolution, more than a fifth (22.2 per cent – 750,000) were headed for Haïti (Saint-Domingue).[6] The drastic reduction in demand from Haïti (just 38,184 slaves were embarked for the island from 1791 to 1800) must have exerted considerable downward pressure on the price of slaves. Similarly, with the Europeans fighting each other, and especially while the British navy blockaded continental Europe, it should have been a buyer's market: with fewer ships turning up on the West African coast, and with many of their captains particularly anxious to buy cheaply because only high profit margins could justify the risk of interception by enemy ships. Had other things been equal, prices would have fallen compared to the years before the slave rising in Saint-Domingue. Evidently, other things were not equal, because the opposite happened: prices rose.

The combination of a downward trend in the number of slaves embarked and an upward trend in the real prices that the ship captains paid for them can be seen in Table 11.1. The price figures come from Richardson's series for the British slave trade.[7] Because British buyers in this period both concentrated on West Africa and dominated the trade there, it is reasonable to follow Lovejoy and Richardson in treating the prices paid by British slavers as representative of prices paid for slaves from West Africa generally; but not of prices paid in Angola, which moved differently, showing that the market(s) there were far from closely integrated with those of West Africa.[8]

One might suppose that higher real prices for slaves – meaning that African sellers received more imported goods per slave – were the result of rising productivity in Britain, as the industrial revolution got underway. But such an effect was not yet visible. On the contrary, Richardson cites Huekel's 1985 index of prices of British industrial goods as showing a rise of up to 15 per cent between 1790 and 1807. Richardson used this index to convert the constant official prices used in UK customs accounts into current or market prices.[9] According to

---

[6] During 1751–90. Calculated from the TASTD.

[7] Richardson, 'Prices of Slaves'.

[8] Paul E. Lovejoy and David Richardson, 'British Abolition and its Impact on Slave Prices along the Atlantic Coast of Africa, 1783–1850', *Journal of Economic History* 55/1 (1995), especially 102, 112–13; Richardson, 'Prices of Slaves', 44 n. For a longer view of West and West-Central Africa as separate slave markets see Joseph C. Miller, 'Slave Prices in the Portuguese Southern Atlantic, 1600–1830', *Africans in Bondage: Studies in Slavery and the Slave Trade*, ed. Paul E. Lovejoy (Madison WI, 1986), 66–9.

[9] Richardson, 'Prices of Slaves', 32–3.

Table 11.1 Real prices of slaves in, and numbers of slaves shipped from, West Africa 1783–1807

Period	Price (GB £)* average, real	Numbers Shipped: average annual
1783–1787	15.6	45,482
1788–1792	19.1	48,023
1793–1797	17.5	29,775
1798–1802	23.3	37,058
1803–1807	25.3	38,574

*Base: mean prices of 1783–87.
(Sources: Prices from Paul E. Lovejoy and David Richardson, 'British Abolition and its Impact on Slave Prices along the Atlantic coast of Africa, 1783–1850', *Journal of Economic History* 55: 1 (1995), Table 3 (p. 113). Numbers shipped are calculated from David Eltis, Martin Halbert et al., *Voyages: The Transatlantic Slave Trade Database* www.slavevoyages.org (2008). Accessed 30 May 2009)

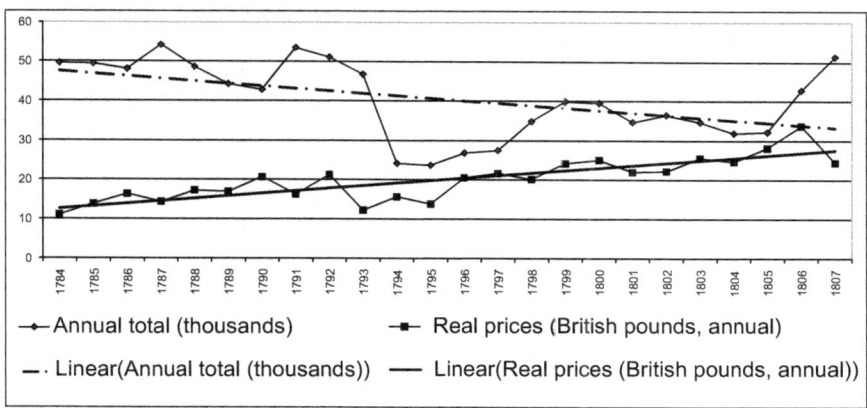

Figure 11.1 Slaves shipped from West Africa, 1784–1807: volume and price
(Sources: numbers shipped, *Voyages*; database in www.slavevoyages.org; prices, D. Richardson, 'Prices of Slaves in West and West-Central Africa: Towards an Annual Series, 1698–1807')

Richardson's five-year moving averages, the price of a slave rose between 1785–89 and 1803–7 by 73.9 per cent in constant terms, and by 94.2 per cent in current terms.[10]

A year-by-year picture of the movement of the numbers and prices of slaves is presented in Figure 11.1. The annual numbers for slaves embarked and for real prices are shown, and linear trend-lines are fitted. The graph shows that the trends of volume and prices were moving in opposite directions during what turned out to be the last

[10] Computed from ibid., appendix, 55–6.

two decades before British abolition. The last year or so of legal slave-trading for British subjects apparently saw a great rush to embark as many captives as possible before the ban came into force on 1 May 1807. Thus the downward trend of the number of people forcibly shipped is much larger still if we exclude the effect of anticipatory transactions by ending the calculation with 1804 and 1805. In those years 31,897 and 32,183 slaves, respectively, were shipped from West Africa, compared to 42,705 in 1806.[11]

If events on and across the ocean would lead one to expect a period of low prices for slaves, we need to look to changes *within* West Africa to explain why prices were actually trending upwards, thus raising the incentive to export slaves, whereas gradually fewer slaves were being shipped. This apparently perverse combination of trends implies that the supply schedule of captives for export was falling: that is, fewer slaves would be available for sale at any given price that the shippers might offer. Indeed, the average prices paid for the declining number of slaves bought from West Africa in this period were much higher than those paid for the increasing number purchased from West Central Africa.[12] Table 11.2 compares Lovejoy and Richardson's price figures for both regions, and calculates the differential.

Thus, not only were prices higher in West Africa than in Angola, but the gap tended to widen. Presumably a process was underway in the former that was not operating, or not as strongly, in the latter.

Table 11.2 Real prices of slaves in West Africa and Angola compared, 1783–1807 (GB £, average)

Period	West Africa	Angola	Ratio (:1)
1783–1787	15.6	12.2	1.28
1788–1792	19.1	13.3	1.44
1793–1797	17.5	14.6	1.20
1798–1802	23.3	13.9	1.68
1803–1807	25.3	13.1	1.93

*Base: mean prices of 1783–87.
(Source: Lovejoy and Richardson, 'British Abolition and its Impact on Slave Prices', 113)

Possible sources of endogenous decline in West Africa

While this phenomenon has not received the detailed research it deserves, two explanations have been offered in the literature. One relates to the supply of slaves, the other to the demand. Let us start with the former.

[11] TASTD.
[12] The price contrast was observed and discussed by Miller, 'Slave Prices', 66–9

In 1970, in the context of her work on the history of cowry shells as currency in Africa, Marion Johnson briefly suggested that as the slave trade went on the costs of acquiring slaves rose, an argument repeated almost *verbatim* in her 1986 book co-written with Jan Hogendorn. This would have been because captives had to be obtained from greater and greater distances from the coast, and perhaps by more expensive means.[13] Though made only partly explicit by the authors, presumably this would be because potential victims nearer the coast were either already enslaved or killed resisting capture, or had succeeded in organizing themselves to evade or defeat slave-raiders or even armies. However, Johnson, and Johnson and Hogendorn, specifically referred to the earlier rise in slave prices, from 'about 100 lb weight of cowries at the end of the seventeenth century to about four times that figure by the 1770s'.[14] But during the early and mid-eighteenth century the annual rate of slave shipments from Western Africa tended to rise, and although the authors referred to 'inherent contradictions' in slave-trading economies,[15] they largely dismissed this as a source of the beginning of the eventual decline of the trade:

> The expansion [of the slave-trading economy] was eventually halted, not by internal strains reaching the breaking-point (though arguably some of the cowry-using states had reached this point by the late eighteenth century), but by the abolition of the slave trade by the very European nations that were bringing the cowries.[16]

However, the implication of the 'inherent contradiction' is that the Europeans and Americans would be charged progressively more for the slaves they bought. This could account simultaneously for the later combination of higher prices and fewer purchases. It was David Richardson made who made the connection, commenting in 1991 that:

> On the evidence of my price data, therefore, the export of over 60,000 slaves a year throughout the half century before 1807 seems to have placed major and possibly unsustainable strains on the slave supply systems of West and West-Central Africa and thus eroded the traditional position of these regions as low-cost suppliers of slave labour to American plantations.[17]

---

[13] Marion Johnson, 'The Cowrie Currencies of West Africa', Part II, *Journal of African History (JAH)*, 11/3 (1970), 348; Jan Hogendorn and Marion Johnson, *The Shell Money of the Slave Trade* (Cambridge, 1986), 111.
[14] As note 13.
[15] As note 13.
[16] Hogendorn and Johnson, *Shell Money*, 112: repeating almost verbatim Johnson, 'Cowrie Currencies', Part II, 349.
[17] Richardson, 'Prices of Slaves', 47.

Thus we have from the literature a plausible supply-side hypothesis, that there was a self-limiting dynamic to the process of seizing and selling captives. Yet, as shown by the West-Central Africa line in Figure 11.2, the number of slaves embarked from Angola continued on an upward trend during the period. If the logic of this hypothesis about the internal dynamics of the slave trade is sound, presumably the slave supply systems of West-Central Africa were not as yet under such severe strain as those of West Africa.

On the other hand, in a 1993 book, James Searing offered a demand-side explanation, albeit specifically for Senegambia, the northernmost of the slaving coasts. He made no comment on whether it might apply elsewhere. He argued that the 'decline of slave exports from Senegambia resulted partly from the growing demand for slave labour in the Senegal River valley'.[18] The latter demand came from a growth of two forms of commercial agriculture. One was the production of grain (millet) in Lower Senegal for sale to markets in both Senegambia (including the provisioning of slaves held awaiting shipment, and of the slave-ships themselves) and the Western Sudan (including merchants supplying the nomads and oases in the western Sahara). The other was the expanding export of gum arabic to Europe:

> Gum extraction required the import of slave labor by desert merchants. The increased desert population and the use of slave labor for gum extraction translated into an expanded demand for savanna grain in the desert. The gum export trade thus increased the demand for slave labor in the desert and in the savanna as well.[19]

Effectively, the African buyers of slaves outbid the Europeans: a growth in effective demand (demand backed by purchasing power) for slaves from within West Africa increased the opportunity cost for African traders of selling captives to the ship captains, rather than to African buyers.

The evidence that Searing provided for his explanation is, by his own admission, 'indirect'.[20] But it is plausible. Moreover, the phenomenon for which he was trying to account was far from confined to Senegambia. On the contrary, as Figure 11.2 shows, it applied more strongly to the three main slave-shipping coasts of West Africa: the Bight

[18] James F. Searing, *West African Slavery and Atlantic Commerce: The Senegal River Valley, 1700–1860* (Cambridge, 1993), 130.
[19] Ibid., 197.
[20] Ibid., 49. On the growth of agricultural slavery in Lower Senegal see ibid., 49–54. On slavery in gum and grain production in the western Sahel see, further, James L.A. Webb, Jr, *Desert Frontier: Ecological and Economic Change Along the Western Sahel 1600–1850* (Madison WI, 1995). Webb argues in general that economic change in the region was not primarily driven by Atlantic influences, though he does not suggest that changes within the region began the decline of the Atlantic slave trade.

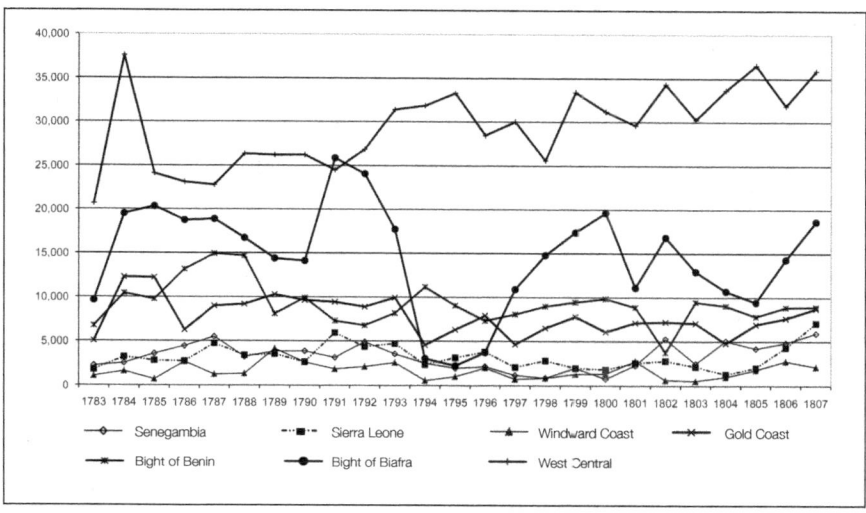

Figure 11.2 Slaves embarked by region, 1783–1807
(Sources: *Voyages* database, www.slavevoyages.org)

of Biafra, the Bight of Benin, and the Gold Coast.[21] For the period from 1787 to 1804 or 1805 it applied to all six slave 'coasts' in West Africa; for 1787 to 1806, the only exception was the smallest, the Windward Coast. Indeed, the share of Senegambia within the decreasing volume of West African slave exports even rose from 10.6 per cent in 1787 to 11.3 per cent in 1806 (reaching 13 per cent the year before). Thus the decline in Senegambia was actually slower than the West African average.[22] In case the end-points of this comparison are not clear in the graph, Table 11.3 gives the numbers for the seven slave coasts for 1785, 1786 and 1787 at one end, and for 1804, 1805 and 1806 at the other.

Which of these hypotheses works better, and should they be combined and supplemented by others? In support of the supply-side approach, there is much evidence of the destruction wrought by slave-raiding and wars that were motivated or facilitated, financially and logistically, by the external slave trade and the access it provided to firearms and ammunition.[23] Indeed, the evidence on the timing of imports of guns and ammunition in relation to the timing of exports of slaves confirms the reality of a 'gun-slave cycle'.[24] This relationship suggests

---

[21] In Figure 11.2, following the TASTD, 'Senegambia' includes 'offshore Atlantic', 'Bight of Biafra' includes 'Gulf of Guinea islands', and 'West-Central Africa' includes St Helena.
[22] Calculated from TASTD.
[23] A notable early study was Kwame Arhin, 'The Financing of the Ashanti Expansion (1700–1820)', *Africa* 37 (1967), 283–91.
[24] A valuable re-examination of the concept is Warren C. Whatley, 'The Guns-Slave Cycle and the 18th Century British Slave Trade in Africa' (paper presented to the 'New Frontiers in African Economic History' Workshop, Graduate Institute, Geneva, Sept. 2012).

Table 11.3 Slaves shipped from the seven 'coasts' of Western Africa, 1785–7 and 1804–6

Year	1785	1786	1787	1804	1805	1806
Senegambia	3561	4435	5741	5052	4195	4829
Sierra Leone	2788	2733	4710	1292	2028	4311
Windward Coast	683	2661	1229	1000	1174	2760
Gold Coast	12,206	6256	8976	4786	6915	7628
Bight of Benin	9743	13,154	14,917	9052	7827	8873
Bight of Biafra	20,327	18745	18,872	10,715	9444	14,304
(West Africa total)	(49,308)	(47,984)	(54,175)	(31,897)	(32,183)	(42,705)
West Central Africa	24,095	23,113	22,971	33,670	36,521	31,888

(Source: *Voyages* database, www.slavevoyages.org)

that the supply of firearms stimulated organized violence that would not have occurred otherwise. This in turn connotes further destruction and the likelihood of some communities making further efforts at more effective resistance or concealment: all of which would raise the costs of obtaining captives for export. Again, recent studies have expanded our information on measures of concealment, defence, organization and negotiation taken by decentralized societies to protect themselves against the aggression of slaving states and raiders.[25] However, the real test of the supply-side proposition would be whether evidence suggests that an increasing proportion of the enslavement that fed the Atlantic trade was done far from the sea; and, bearing in mind the contrast with Angola, that this was more true of West Africa than of West-Central Africa. Pending further research, it seems very likely that increasing costs of capturing and delivering slaves are at least a major part of the explanation for the falling supply schedule of slaves in West Africa. But, to judge from Searing, it is not the full story.

Searing's demand-side argument runs counter to the view, associated with Walter Rodney and most recently revised and restated by Joseph Inikori, that the Atlantic slave trade was destructive of market activity within West Africa.[26] Undoubtedly the slave trade created

---

[25] E.g. James F. Searing, '"No Kings, no Lords, no Slaves": Ethnicity and Religion among the Sereer-Safèn of Western Bawol, 1700–1914', *JAH* 43/3 (2002), 407–30. This is not to imply that all decentralized societies were innocent of slaving themselves. For an overview see Martin A. Klein, 'The Slave Trade and Decentralized Societies', *JAH* 42/1 (2001), 49–65. More generally, see Sylviane A. Diouf (ed.), *Fighting the Slave Trade: West African Strategies* (Athens OH, 2003).

[26] Walter Rodney, 'Gold and Slaves on the Gold Coast', *Transactions of the Historical Society of Ghana*, 10 (1969), 13–28; Joseph E. Inikori, 'Africa and the Globalization Process: Western Africa, 1450–1850', *Journal of Global History* 2/1 (2007), 63–86; Inikori, 'Transatlantic Slavery and Economic Development in the Atlantic world: West Africa, 1450–1850', *The Cambridge World History of Slavery*, vol. 3, *AD 1420–AD 1804*. Ed. David Eltis and Stanley L. Engerman (New York and Cambridge, 2011), 650–73.

economic incentives to all who had the means to engage in slave-raiding and trading to do so, rather than to continue and expand peaceful forms of economic activity. More, it put farmers, artisans and miners at risk of being seized while they worked, if they did not belong to a sufficiently protective state or stateless society. For a demand-side argument to work for West Africa as a whole would require evidence that intra-regional market activity not merely survived the onslaught of the trans-Atlantic slave trade, just as it had survived the long-running trans-Saharan slave trade, but also expanded, perhaps assisted by some stimulus from the Atlantic trade as a whole – the imports, the non-slave exports, and the income received by the elites who controlled the trade on the African (as on the European) side.[27]

One candidate for such a stimulus is institutional reform of the currency system. Philip Curtin noted that pre-colonial West Africa, like much of Eurasia but unlike much of the rest of Sub-Saharan Africa, made much use of metals and cowries in its currency systems.[28] Thus Johnson and Lovejoy described the spread of cowry, and cowry-and-gold currencies across much of West Africa during the Atlantic slave trade.[29] I argue elsewhere that this should be seen as a process of monetary reform, replacing cloth and other commodity currencies with what in monetary terms was the technically superior shell, and creating larger zones within which the same currency operated. All this presumably reduced the cost of transacting, thereby facilitating market activity.[30]

Moreover, the overseas slave trade itself facilitated the acquisition of slaves for use within African economies, especially by owners near the coast or on the slave routes. Slave ports seem to have grown in population as a result of the trade, partly because of retaining some of the slaves that were brought in.[31] The complementarity of the gender preferences of European and African slave owners facilitated this: approximately two-thirds of those embarked in the trans-Atlantic trade were

---

[27] On the last point see A.G. Hopkins, *An Economic History of West Africa* (London 1973), 104–6, 120, 123, 125; E.W. Evans and David Richardson, 'Hunting for Rents: the Economics of Slaving in Pre-colonial Africa', *Economic History Review*, 48/4 (1995), 665–86.
[28] Philip D. Curtin, 'Africa and the Wider Monetary World, 1250–1850', *Precious Metals in the Late Medieval and Early Modern Worlds*, ed. John F. Richards (Durham NC, 1983), 232.
[29] Marion Johnson, 'The Cowrie Currencies of West Africa', Parts I and II, *JAH* 9/1 (1970), 17–49 and 9/3 (1970), 331–53; Johnson, 'The Nineteenth Century Gold "Mithqal" in West and North Africa', *JAH* 9/4 (1968), 547–70; Paul E. Lovejoy, 'Interregional Monetary Flows in the Precolonial Trade of Nigeria', *JAH* 15/4 (1974), 563–85.
[30] Gareth Austin, *Markets, Slaves and States in West Africa* (Cambridge, forthcoming 2014).
[31] Albert van Dantzig, 'Effects of the Atlantic Slave Trade on Some West African Societies', *Revue Française d'Histoire d'Outre-mer* 62: 1–2 (1975), 252–69. On the Dahomian slave 'port' see Robin Law, *Ouidah: The Social History of a West African Slaving 'Port' 1727–1892* (Oxford, 2004), 73–8.

male, whereas it is often argued that the majority of slaves within West Africa were female. Female as well as male slaves laboured to produce commodities.[32] It is likely that the main growth of the slave labour force for agriculture, mining and craft production did not occur until after 1807. But it may well have been prefigured by modest expansions in certain localities, which would have contributed to the reduction in the number of slaves available for the Europeans to buy at any given price. Ironically, the production of provisions to feed slaves and their captors could be part of this.[33] Searing argues convincingly that the 'availability of surplus grain put a cap on the volume of slave exports that could pass through the Senegal valley'.[34] In the context of all this, Searing's Senegambian proposition, that commodity production induced more expenditure on slaves in order to expand output, is worth exploring in other parts of the region.

Inikori has made the crucial point that the 'currencies' imported into West Africa were not accepted back by the Europeans in payment for the goods they sold to Africans.[35] Therefore the transactions they facilitated can only have been internal to Africa. At the point of import, they were currency materials: they only became currencies when Africans used them as such. Inikori went on to formulate an important argument, in two stages. First, he suggested: 'Increases in currency import were, therefore, a measure of growing internal trade requiring corresponding growth in the quantity of money in circulation, while a decline in imports corresponded to contraction of market activity.' Second, he maintained that the Atlantic trade saw a general 'decline in currency imports' after its early years, which can be taken as indicating a 'general decline in the growth of the market economy and inter-regional trade.'[36] Inikori provided detailed examples of currency materials falling as a proportion of British exports to West Africa, while the proportions of manufactured consumer goods and firearms rose (for instance, on the Bight of Benin, summarized here in Table 11.4). As he acknowledged, the evidence for some years comes from just one or two cargos, but taken as a whole it is striking.[37]

Suggestive as relative shares are, it is the absolute level of imported

---

[32] Claire C. Robertson and Martin A. Klein, 'Women's Importance in African Slave Systems', *Women and Slavery in Africa*, ed. Robertson and Klein (Madison WI, 1983), 3–4. See also Law, *Ouidah*, 76–7; Kristin Mann, *Slavery and the Birth of an African City: Lagos, 1760–1900* (Bloomington IN, 2007), 5.
[33] For which see, this volume, David Eltis – chapter 1, and Toby Green – chapter 3.
[34] Searing, *West African Slavery*, 197 (cf. 50).
[35] Inikori, 'Africa and the Globalization Process', 84. The point was made more narrowly, specifically for cowries, by Curtin, 'Africa and the Wider Monetary World', 233 and confirmed by Hogendorn and Johnson, *Shell Money*, 145.
[36] Inikori, 'Africa and the Globalization Process', 84.
[37] Inikori, 'Transatlantic Slavery and Economic Development', 663–72.

Table 11.4 Share of cowries in value of commodities shipped to the Bight of Benin from Britain: Inikori's data for 'select years, 1681–1724'

Year	Number of Cargoes	Cowries (%)	Total (GB £)	Implied value of cowries (£)
1681	2	59.2	3180	1883
1684	1	52.4	1005	527
1690 & 1692	5	38.6	9369	3616
1708	2	20.5	4684	960
1724	2	22.1	3707	819

(Source: Joseph E. Inikori, 'Transatlantic Slavery and Economic Development in the Atlantic World: West Africa, 1450–1850', in David Eltis and Stanley L. Engerman (eds), *The Cambridge World History of Slavery*, Vol. 3, *AD 1420–AD 1804* (New York, 2011), 666)

currency materials that is the better measure of the growth of the money supply within West African economies. Thus I include in Table 4 a column converting the share of cowries in the British cargoes for which Inikori has data into absolute values. This produces a much more uneven trend than the steady decline of cowries in relative terms. It underlines the desirability of achieving a more comprehensive series of absolute values, though the sources severely constrain the extent of comprehensiveness possible. As Inikori notes: 'These cargoes represent only a small part of the actual total for each year.'[38]

'Increases in currency import' were indeed a measure of the growth of internal monetary transactions – their quantity multiplied by their average price, to which we will return – provided that the velocity of circulation remained the same. It does not follow, however, that a 'decline in imports corresponded to contraction of market activity'. If the volume of imports, even if reduced, exceeded the quantity of existing currency that was destroyed or converted to ornamental use (and again assuming the velocity of circulation remained unchanged), the money supply would have continued to expand, albeit more slowly. Inikori's examples suggest that currency materials fell as a proportion of imports, but they do not show that the decline was absolute as well as relative.

Let us pursue this further with cowry shells. Though there is some uncertainty about the overall quantities of imported currency materials, we have rather good information on the biggest single category: cowry shells. According to Johnson, 'Between one-third and one-quarter of the price of slaves was normally paid in cowries.'[39] Hogen-

[38] Ibid., 666 n.
[39] Johnson, 'Cowrie Currencies', Part II, 348. For more details of the variation over time and space see Hogendorn and Johnson, *Shell Money*, 110.

dorn and Johnson compiled an annual series of cowry imports into West Africa during the eighteenth century through the two major channels, the British and Dutch trades, as shown in Table 11.5. This series shows a lot of fluctuation, but no sustained trend over that century as a whole. However, we will return to this.

Johnson and Hogendorn went on to note that, given none of the cowries were re-exported from West Africa, and allowing for a [small] proportion being destroyed or converted to ornamental use, the overall stock of cowries in the cowry-using economies within the region must have continued to rise at least throughout the era when the slave trade was legal. If the volume of goods traded inside West Africa had fallen during the period, the theoretical prediction is that inflation should have occurred: indeed, given the level of imports, 'hyperinflation'.[40]

So far the most detailed and thorough study of prices in pre-colonial West Africa is Robin Law's 1992 account of Dahomey.[41] He collected and rendered commensurable price observations for chickens, other livestock, corn, palm-oil, and a variety of payments for different kinds of labour. The sources yielded observations only for some years, but there were enough to point to a clear overall picture, including in the markets for the foodstuffs most affecting the local African population (corn and palm-oil), that there was indeed a period of 'sustained' inflation from the late seventeenth to the mid-eighteenth

Table 11.5 Cowrie imports to West Africa (lbs avoirdupois) by decade, 1700–1799

Decade	British & Dutch Imports	Decade	British & Dutch Imports
1700–1710	2,292,786	1751–1760	1,453,974
1711–1720	2,900,727	1761–1770	1,361,025
1721–1730	5,150,517	1771–1780	2,400,556
1731–1740	3,103,036	1781–1790	3,131,562
1741–1750	3,417,290	1791–1799	720,268
		Total	25,931,660

(Source: Jan Hogendorn and Marion Johnson, *The Shell Money of the Slave Trade* (Cambridge, 1986), 62–3)

[40] Hogendorn and Johnson, *Shell Money*, 145.
[41] Robin Law, 'Posthumous Questions for Karl Polanyi: Price Inflation in Pre-colonial Dahomey', *JAH* 33/3 (1992), 387–420. Johnson herself collected prices of hens: Johnson, 'Cowrie Currencies', Part II, 347–8; Hogendorn and Johnson, *Shell Money*, 145. But her study is largely overtaken by Law's in terms of the number of observations (a high proportion of hers were also from Dahomey), because she (uncharacteristically) misunderstood one of the units used in the sources, and because his covers a range of goods and services. See Law, 'Posthumous Questions', 403–20.

century.[42] He offered no overall figure but found it comparable in scale to the post-1850 'Great Inflation' when a new surge of cowry imports brought about rapid depreciation. In between, there was a century of price stability, from the mid-eighteenth until the mid-nineteenth century.[43] This periodization matches the rate of importation of cowry shells into West Africa, Dahomey being within the cowry zone. Reverting to Table 11.5, the mid-eighteenth century saw a sharp break between an era of rapidly increasing cowry supply in West Africa, and one of much more gradual increase.

The evidence adduced to date on currency imports and domestic prices is tantalizingly inconclusive with regard to our hypotheses. The relatively high level of cowry imports in the early eighteenth century suggests that the demand for money for transactions within West Africa continued to grow. But if this was accompanied by rapid inflation, it does not necessarily mean that output (slaves for export excepted) rose; it may indeed have fallen.[44] Conversely, in principle, the post-c.1750 combination of reduced cowry imports and relative price stability could reflect a stabilization of the level of output, or a gradual rise in the latter, absorbed without inflation by the continuing but lower rate of cowry imports. A further source of inconclusiveness is that while a more efficient money system, based on the more widespread use of cowries, implies lower transaction costs, this might induce either a rise in the volume of transactions, or an increase in their velocity, or both. Finally, we must be cautious about generalizing from the prices of foodstuffs on the coast, not only because the number of observations is fairly small, but because we cannot assume that foodstuff prices around European forts had any relationship to the prices charged in the distant (or even near) interior. Slaves were made to walk, which made possible a relatively integrated market in captives across West Africa.[45] But in this period, given transport constraints, we must assume food markets were highly localized.

Still, we can surely find out more than we currently know. Conceptually, a starting-point for further research is that our understanding of West African economic history during the Atlantic slave trade, even during its most intense period from the 1680s to 1807, needs to recognize and accommodate two contrary impulses, which are reflected in

[42] Law, 'Posthumous Questions', 403–20 (quote, 418).
[43] Ibid., 403–20.
[44] From the traditional monetary identity, $MV = PQ$, a higher money supply (with the same velocity of circulation) and higher prices are compatible with a lower quantity of output.
[45] Paul E. Lovejoy and David Richardson, 'Competing Markets for Male and Female Slaves: Prices in the Interior of West Africa, 1780–1850', *International Journal of African Historical Studies* 28/2 (1995), 261–94.

the hypotheses we have considered: the destructive effects of the trade on market (as well as non-market) activities within the region; but also a tendency towards the expansion of markets within the region, stimulated in part by a steady reform of currency practices that facilitated domestic and intra-regional commerce. A probable example of the latter is the regional economy of the Central Sudan, following recovery from the mid-eighteenth century drought that afflicted much of West Africa.[46]

Finally, why did slave shipments decline in West Africa while rising in Angola? This question is beyond the scope of this paper. But, evidently, the mechanisms that resulted in the downturn in West African slave exports did not operate as strongly, or were outweighed by other mechanisms, in West-Central Africa. It is worth noting that the drivers of the slave trade from Angola have not always been viewed by historians as the same as those from West Africa. Joseph Miller's emphasis on the contribution of famines in the case of Angola is not replicated, at least as a major explanation, in the historiography of West Africa.[47]

Freedom and cash crops: when did the social organization of agriculture acquire its modern form?

In 1973 A.G. Hopkins famously argued that the adoption of 'legitimate' commerce 'can be seen as the start of the modern economic history of West Africa',[48] because the low capital entry threshold and the absence of economic advantages of scale

> enabled small-scale farmers and traders to play an important part in the overseas exchange economy for the first time. In so far as firms of this type and size are the basis of the export economies of most West African states today, it can be said that modernity dates not from the imposition of colonial rule ... but from the early nineteenth century.[49]

---

[46] Lovejoy, 'Interregional Monetary Flows'.
[47] Joseph C. Miller, 'The Significance of Drought, Disease, and Famine in the Agriculturally Marginal Zones of West-Central Africa', *JAH* 23/1 (1982), 17–61; Miller, *Way of Death: Merchant Capitalism and the Angolan Slave Trade 1730–1830* (London 1988), 143–69. For West Africa, Senegambia is an exception in the importance of famine, but this is for a relatively arid region. In Searing's account famine serves not only as a source of slaves, but also as a constraint upon the number of slaves that could be held (*West African Slavery*, esp. 79–88). See also Boubacar Barry, *Senegambia and the Atlantic Slave Trade* (Cambridge, 1998; transl. from the French edition of 1988 by Ayi Kwei Armah), 108–12, acknowledging an unpublished paper by Charles Becker.
[48] Hopkins, *Economic History*, 124.
[49] Ibid., 125–6.

This thesis provoked much debate and new research.[50] Overall, Hopkins seems to have underestimated the capacity of old mercantile elites, and the political elites with which they overlapped, to retain their dominion. But he was surely right that the groundnut and palm-oil trades brought a mass of small farmers into export production and that this widespread participation in the export market remained a feature of West African economies during and after colonial rule, as it does today.

Hopkins acknowledged that in some cases slaves were used to produce for 'legitimate commerce'.[51] Since 1973, however, an accumulation of evidence has pointed to slaves being more geographically widespread, and present in larger numbers, than is compatible with the idea of family farming that the term 'small-scale' tends to evoke. To be sure, free farming families, producing for export without making use of slave labour, existed.[52] But in the growth of agricultural production for sale, not only for overseas but also for African markets (such as supplying the raw cotton and indigo for the cotton textile industry of the Sokoto Caliphate),[53] the basic social units of commercial agriculture, at least by the mid and late-nineteenth century, were large slave estates combined with 'ordinary' households expanded by the incorporation of captives.[54]

The increasingly widespread and often large-scale use of slaves in agriculture in the post-1807 exchange economy of West Africa was greatly facilitated by the prior history of the Atlantic slave trade, which had stimulated the creation of a massively enlarged slave-supplying system, followed by the abolition campaign. According to Lovejoy and

---

[50] For a survey up to 1995 see Robin Law, 'The Historiography of the Commercial Transition in Nineteenth-century West Africa', *African Historiography: Essays in Honour of Jacob Ade Ajayi*, ed. Toyin Falola (Harlow,1993), 91–115. This was followed by Robin Law (ed.), *From Slave Trade to 'Legitimate' Commerce: the Commercial Transition in Nineteenth-Century West Africa* (Cambridge, 1995), and Martin Lynn: *Commerce and Economic Change in West Africa: the Palm Oil Trade in the Nineteenth Century* (Cambridge, 1997).

[51] It was Hopkins himself who introduced to African historiography the standard economic theory of slavery (*Economic History*, 23–7). This is reconsidered in Gareth Austin, 'La coercition et les marchés : concilier des explications économiques et sociales de l'esclavage en Afrique de l'Ouest précoloniale, c.1450–c1900 ' (translated by Darla Gervais), *Revue d'histoire moderne et contemporaine* (forthcoming, 2013), and in Austin, *Markets, Slaves and States* (forthcoming, 2014).

[52] For a case-study see Susan Martin, 'Slaves, Igbo Women and Palm Oil in the Nineteenth Century', *From Slave Trade to 'Legitimate' Commerce*, ed. Robin Law (Cambridge, 1995), 172–94.

[53] Paul E. Lovejoy, 'Plantations in the economy of the Sokoto Caliphate', *JAH* 19/3 (1978), 341–68; Jan S. Hogendorn, 'The economics of slave use on two "plantations" in the Zaria emirate of the Sokoto Caliphate', *IJAHS* 10/3 (1977), 369–83.

[54] For overviews see Paul E. Lovejoy, *Transformations in Slavery: A History of Slavery in Africa*, 3rd edn (Cambridge, 1983, 2012); Manning, *Slavery and African Life*; also Law, *From Slave Trade to 'Legitimate' Commerce*. More recent case-studies include Gareth Austin, *Labour, Land and Capital in Ghana: From Slavery to Free Labour in Asante, 1807–1956* (Rochester NY, 2005), esp. 122–7, 486–8; Mann, *Slavery and the Birth of an African City*.

Richardson, the average real price of slaves on the coast of West Africa slumped after 1807 until in 1815–20 it averaged only 30 per cent of the average for the five years preceding British abolition (1803–7). Conversely, the substantial recovery of slave prices that followed reflected the increasing demand for slaves to produce palm-oil, kola nuts, raw cotton or other 'legitimate' commodities, whether for maritime or intra-regional markets. The average real price for 1826–30 was back to 67 per cent of the 1803–7 average.[55]

By the late nineteenth century the proportion of slaves in most West African societies was remarkably high. The fullest evidence on the incidence of slavery is from French surveys carried out in 1894 and 1904 (by the latter year, many slaves had left their masters at their own initiative). Having scrutinized this source carefully, Martin Klein estimated that for French West Africa as a whole, the proportion of slaves in the population in 1904 was over 30 per cent.[56] Unlike with the external slave trade, there is no evidence that slave-trading and slave-holding within the region were in decline by the time they began to be the subject of abolitionist legislation, during the early colonial period. For instance, on the contrary, in the kingdom of Asante, the rate of inflow of slaves in the years immediately before the colonial occupation of 1896 was almost certainly a record.[57] It took the prohibition of internal slave-trading in the early colonial period, and the often delayed introduction of at least partial measures against slave-holding, to create the legal conditions for the decline of slavery within West Africa.

By itself, however, the abolition of the legal recognition of the property of slave-owners in slaves, or even the prohibition of slave-holding, weakened but often did not remove the hold of masters over their former slaves, especially when the latter had no known or geographically accessible home to which to return.[58] Former slaves needed to be able to make a living on their own. Conversely, the economic logic of slavery within this relatively land-abundant, labour-scarce region

[55] Lovejoy and Richardson, 'British Abolition'.
[56] Martin A. Klein, *Slavery and Colonial Rule in French West Africa* (Cambridge, 1998), esp. Appendix 1, 252–6. On problems with the source see also Richard L. Roberts, *Warriors, Merchants, and Slaves: The State and the Economy in the Middle Niger Valley, 1700–1914* (Stanford CA, 1987), 118–19. In the case of the Wolof kingdoms of Kajor and Bawol, Searing considers Klein's estimates for 1904 to be too high, because of slave runaways, but he estimated that in 1880 between a quarter and a third of the population would have been slaves in Kajor, and rather less in Bawol. See James F. Searing, *'God Alone is King': Islam and Emancipation in Senegal. The Wolof Kingdoms of Kajoor and Bawol, 1859–1914* (Portsmouth NH, 2002), 166–72, 184–8, 191–3.
[57] Austin, *Labour, Land and Capital*, 123, 486–7.
[58] This and the next paragraph are based on Gareth Austin, 'Cash Crops and Freedom: Export Agriculture and the Decline of Slavery in Colonial West Africa', *International Review of Social History*, 54/1 (2009), 1–37.

had been to lower the cost of labour enough to make it profitable to pay for labour power.[59] Abolition by itself would not have produced a significant growth of hired labour, because the latter would have been unaffordable.

It took the additional ingredient of the rapid further spread of export agriculture in response to the larger and more diverse markets provided by the second industrial revolution to create the economic conditions for some former masters to afford to become employers, and for many former slaves to afford to become independent 'peasants'. In Asante and Lagos, members of existing elites – especially chiefs and merchants, respectively – could take the lead in adopting cocoa cultivation, using the labour of slaves, pawns or subjects. When the farms were established as regular sources of substantial income, they could afford to hire wage-labourers. Where former slaves could obtain access to suitable land close to mechanized transport or ports, they could themselves become part-subsistence, part-export crop-producers, as was commonly the case in the groundnut basin of northern Nigeria. Denied such access, many former slaves evidently formed households, often in their original home areas, notably in the Sahel, but sent out male labour to earn money in the export-agriculture zones, especially in Senegambia and Ghana. Meanwhile, former slave-owners in areas unsuited by environment and location to growing the more profitable crops were vulnerable to economic and social decline: in the phrase Don Ohadike applied to south-east Nigeria, where presumably palm-oil – unlike cocoa in southwest Nigeria – did not support the employment of wage labour, 'when the slaves left, the owners wept'.[60]

Thus it was in this era – mostly from the 1890s to the 1920s, with major variations across the region – that the main unit of production for the market came to be, as since, the free independent 'peasant' household which, unswollen by slaves, relies fully on its members to produce their own crops for subsistence and sale, and/or selling their labour to outsiders, whether locally or at a (potentially great) distance.[61]

---

[59] As note 51.
[60] Don C. Ohadike, '"When the Slaves Left, the Owners wept": Entrepreneurs and Emancipation among the Igbo People', *Slavery and Colonial Rule in Africa*, ed. Suzanne Miers and Martin A. Klein (London, 1999), 189–207.
[61] Austin, 'Cash Crops and Freedom', 33–5.

A perspective: the long-term path(s) of African economic development

Let us place the two episodes discussed above in the context of very long-term patterns of interaction between West Africans and their environment. As with much of Sub-Saharan and especially tropical Africa, until well into the twentieth century most of West Africa was characterized by a relative scarcity of labour in relation to land, coupled with a natural environment that made intensive agriculture (the application of high proportions of labour and/or capital per unit of land) not only usually unnecessary, but relatively difficult to sustain. So, while intensive agriculture is ancient in African history, until the twentieth century it appears not to have spread.[62] This applied, not least, to the adoption of intensive methods by communities taking shelter from slave-raiding.[63] Rather, conditions favoured land-extensive agriculture. Where markets existed to encourage production for exchange, coercion was often the only feasible (as well as the cheapest) way to expand the workforce at household or even state level.[64] Meanwhile the main route to higher productivity lay through innovations that did not require much more labour or capital. Given also the relatively low number of indigenous cultivable plants in tropical Africa, this mainly meant the selective adoption of exotic crops and crop varieties.[65]

In this context, except where gold could be extracted with artisanal technology, tropical Africa was generally short of exportable commodities lucrative enough to pay for large quantities of imports.[66] Despite the scarcity of labour in Sub-Saharan Africa, the productivity of African labour, especially in producing for exchange, was often higher outside the continent than within it: the premise of the external slave trades.[67] Before and during the slave trade, West Africa exported

---

[62] Gareth Austin, 'Resources, Techniques and Strategies South of the Sahara: Revising the Factor Endowments Perspective on African Economic Development, 1500–2000', *EHR* 61/3 (2008), 587–624

[63] Walter Hawthorne, 'Nourishing a Stateless Society during the Slave Trade: the Rise of Balanta Paddy-rice Production in Guinea-Bissau', *JAH* 42/1 (2001), 1–24; Hawthorne, *Planting Rice and Harvesting Slaves: Transformations along the Guinea-Bissau Coast, 1400–1900* (Portsmouth NH, 2003).

[64] Gareth Austin, 'Factor Markets in Nieboer Conditions: Early Modern West Africa, c.1500–c.1900', *Continuity and Change* 24/1 (2009), 23–53.

[65] Austin, 'Resources, Techniques and Strategies'.

[66] David Eltis and Lawrence C. Jennings, 'Trade between Western Africa and the Atlantic World in the Pre-colonial era', *American Historical Review* 93/4 (1988), 936–59.

[67] Hopkins, *Economic History*, 105; Stefano Fenoaltea, 'Europe in the African Mirror: the Slave Trade and the Rise of Feudalism', *Rivista di Storia Economica* 15:/2 (1999), 123–65; Manning, *Slavery*

gold, gum, timber and other commodities, including small quantities of textiles.⁶⁸ But what stands out is the size of the export trade in humans.

Perhaps largely because of comparative proximity to the emerging plantation economies of the Atlantic, West Africa was drawn into the Atlantic slave trade on a scale matched only by West-Central Africa. A relatively low level of political centralization created strong incentives, indeed pressures, for rulers to participate in raiding and exporting captives,⁶⁹ despite the destruction and the aggravation of the labour shortage in the region as a whole. Participation in the Atlantic trade as a whole, as opposed to that in slaves as such, benefited tropical Africa generally by permitting the importation of a range of crops, including maize, whose adoption promised to ease labour scarcity in the long term⁷⁰ – even if the intensity of slave-raiding compromised or postponed this prospect. Meanwhile West African buyers used their particularly deep involvement in the Atlantic trade to import commodities that they used as currencies within their own region. Arguably, this contributed to the strengthening of commercial relations and the integration of intra-African markets. That, in turn, may have contributed to the growth of the demand for labour to produce for sale within West Africa, something which, if Searing's argument for Senegambia can be generalized, may help to explain why the numbers of slaves embarked each year in West Africa had been declining, on average, during the twenty years before the British abolition act.

The successive phases in the industrialization of Europe and North America provided successively larger markets for a small but widening range of silvan and agricultural commodities that could be produced cheaply using the land-extensive, labour- and capital-saving methods that were efficient under West African conditions.⁷¹ Not surprisingly in this setting, a large proportion of 'agricultural' commodities produced for sale in the mid-nineteenth century were tended but not deliberately planted. This was true of some palm-oil, and of the kola

and African Life, 33–4; Austin, 'Resources, Strategies and Techniques'; Gareth Austin, 'The Reversal of Fortune" Thesis and the Compression of History: Perspectives from African and Comparative Economic History', *Journal of International Development* 20/8 (2008), 1006.

[68] John Thornton, *Africa and Africans in the Making of the Atlantic World, 1400–1800*, 2nd edn (Cambridge: , 1998).

[69] Joseph E. Inikori, 'The Struggle against the Transatlantic Slave Trade: the Role of the State', *Fighting the Slave Trade: West African Strategies*, ed. Sylviane A. Diouf (Athens OH, 2003), 170–98.

[70] E.g. James C. McCann, *Maize and Grace: Africa's Encounter with a New World Crop 1500–2000* (Cambridge MA, 2005).

[71] E.g. Gareth Austin, 'Mode of Production or Mode of Cultivation: Explaining the Failure of European Cocoa Planters in Competition with African Farmers in Colonial Ghana', *Cocoa Pioneer Fronts: the Role of Smallholders, Planters and Merchants*, ed. W.G. Clarence-Smith (Basingstoke, 1996), 154–75.

that Asantes sold to Hausa merchants.[72] It was also to be true of the wild rubber boom later in the century. Increasingly, however, cultivation took over; and it was essential in the case of cocoa beans, another exotic crop whose rapid adoption by African farmers in Ghana and Nigeria created the most lucrative of export cultigens in colonial West Africa.

African labour, mainly in enslaved form in the Americas, helped create the conditions in which industrialization began in Britain.[73] By providing export markets for groundnuts and palm-oil the first industrial revolution raised the value-productivity of African labour in Africa. The demand for cocoa beans was hugely expanded by the invention of milk chocolate during the second industrial revolution, and the income from it enabled former masters of slaves or pawns to hire free labourers instead. To coin a long-disused word, we see something of a dialectical interaction between markets and labour coercion. Inside West Africa as well as in the Americas, the growth of market demand for agricultural products initially encouraged the acquisition and deployment of slaves. Later, the income generated by cash crops enabled some masters to become employers, and many slaves to become either wage labourers or free farmers.

Conclusions

This chapter has focused on two crucial episodes in West Africa's transition within the Atlantic market from an exporter mainly of captive people to an exporter of agricultural produce. It began by highlighting the fact that the numbers of slaves shipped from West Africa – unlike those from Angola – began to dwindle twenty years before the British abolition act began the long process of elimination. It is easy to assume that this resulted from disruptions to the trade elsewhere in the Atlantic world. But that would have reduced the prices that Europeans and Americans had to pay for slaves in West Africa. The observation that, on the contrary, slave prices were rising along that coast suggests that the cause of the pre-abolition decline in the volume of the trade lay in the interactions between the trade and West African economies and societies. Two hypotheses were explored here: the supply-side possibility that Atlantic slaving had begun to exhaust the capacity of West African elites to generate exportable captives at the same level, and the demand-side possibility that the slave-shippers were beginning to be outbid by a growing internal

[72] E.g. Austin, *Labour, Land and Capital*, 64–5, 474.
[73] Joseph I. Inikori, *Africans and the Industrial Revolution in England* (Cambridge, 2002).

demand for slaves. But further research is required before the explanation can be taken further.

The other decisive episode is the decline of slavery within West Africa itself, during the early colonial period. I argue that, while colonial legislation created necessary conditions for abolition, the timing and form of the decline of slave labour on the ground was determined primarily by the economic opportunities available to former slaves, and indeed to former masters. The geographically uneven spread of the 'cash crop revolution' of the same period enabled some former slaves to become export-crop producing peasants, while others formed farming households but relied on exporting male migrant labourers for a foothold in the agricultural export economy. Masters situated in what became the more lucrative cash-cropping areas were able to become employers instead; elsewhere, many former masters lacked that opportunity, and their wealth, status and households shrank. While the transition from slave exports to 'legitimate commerce' after 1807 may have marked the beginning of the modern economic structure of West Africa, the early-colonial decline of internal slavery inaugurated the modern social structure of the West African countryside, based on often smaller households composed of free members.

Both transitions are linked by the long history of West Africans' responses to the constraints and opportunities of their natural environments. The relative scarcity of labour within the region was combined with natural obstacles to higher productivity in both agriculture and manufacturing. It was industrialization elsewhere that created markets for crops that could be grown in appropriately land-extensive fashion on West African soils, ensuring that the closing of the Atlantic (and later, Saharan) slave markets was followed by mass entries of African farmers into the world market. The industrial revolution was by no means a purely exogenous event; enslaved (and some free) Africans were major participants in its origin.

Index

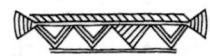

Abeokuta, 5, 21, 208, 210, 214, 215–23
abolitionism (also 'humanitarianism'), 1–8, 16–21, 24–7, 46, 56, 76, 116, 119, 125, 128, 138–48, 155, 158, 160–1, 174, 178–81, 184–7, 193–4, 198–9, 201–6, 211, 215, 225–6, 228, 238–40, 243–5, 248–9, 259–61, 263–5
de Abreu Castelo Branco, Nicolau, Governor of Angola, 227, 234
Accra, 31, 158, 166
Ada, 158, 166, 173
Adanson, Michael, 151
Afonso II, King of Kongo, 70
African Civilization Society, 208
African Improvement Society (AIS), 21, 208–15, 223–4
Agaja, King of Dahomey, 4 n.12, 119, 128
Agbebi, Mojola, 221–2
Agbowa, 221, 223
Agege, 221–3
Agostinians, 61
agricultural productivity, 11, 90–4, 136, 262, 264–5
agricultural techniques, in Africa, 10, 14, 20–3, 40, 46, 79–80, 83, 87, 91–2, 94–7, 102–3, 109, 113, 125–7, 185, 196, 208–13, 217, 224, 226, 238, 262
Ahanta, 121
Ake, 217
Akropong, 166, 167
Akuapem, 158, 166–79
alcohol (also 'liquor'), 41, 148
Alexandre, Valentim, 225–6, 230
Allen, William E., 194
d'Almada, André Alvares, 95, 104–5, 117 n.6
Alves, Catarina, 71
Amador, 65, 75
Amanopa ('Ammano-passo'), 166, 167
Ambaca, 231, 235
America/ Americas (also 'USA'), 2, 10, 11–12, 14, 17–23, 26, 30, 32, 33, 38, 39, 41–3, 45, 61, 76–78, 81, 83 n.15, 86, 87, 88, 90–2, 98, 100, 116–18, 121, 125, 128, 131–7, 139, 140–56, 162, 179, 180–4, 186, 192–202, 217, 219, 245–6, 249, 264; Spanish America, 67, 68; Latin America, 88; North America, 231, 263
American Colonization Society (ACS), 182, 192–4, 196–8, 201–2
Angola, colony, 13, 18, 21, 23, 24, 36, 41, 42, 44, 76–7, 225–42; region (also 'West-Central Africa'), 30–1, 43, 66–8, 80–1, 87, 244, 245, 246, 248, 250, 252, 258, 260, 264
Angolares, 65, 75, 77

Anomabu, 16, 44, 45
Antigua, 103, 107–8
Antilles, 68; French Antilles, 103; Lesser Antilles, 108
Anti-Slavery Society, 187
Antwerp, 72, 73
apprenticeship, 24, 195
Aquitamba, Bango, 230
Arguim, 58
Aro, 6
arrowroot, 185, 195, 209
Assiong, *okyeame*, 177
Atlantic Islands, 35 n.17, 80, 87, 105
Attiambo, Duke of Akuapem, see 'Obuobi Atiemo, Okuapenhene'
Awdaghust, 84
Axim, 120–2
Azores, 57, 77, 100 n.6

Badagry, 191
Badeau, Abbé, 148
Baga, 94
Bahia, 62, archdiocese, 68
Bainham, George, 106, 107 n.26
Balanta, 14, 94
Baltic and Guinea Company, 152, 161 n.10
Banana Islands, 204
Baptista, João, 62
Barbados, 45 n.50, 47, 100–1, 107–8
Barbosa, Baltasar, 89
Barboza, João Guilherme, 235–6
Barreira, Baltasar, 93
Basel Mission, 5, 207, 220 n.45
Batalha, Francisco Rodrigues, 229
Bayart, Jean-François, 97
bees-wax, 8, 22, 30, 82, 93, 114, 129, 153, 185, 191, 241
Bence Island, 100, 109, 114, 141
Benezet, Anthony, 2 n.2, 142–3, 146, 148, 156–7
Bengo, river, 44, 45, 227, 228 n.9, 232, 233
Benguela, 230, 241
Benin, kingdom, 8, 15, 66, 68, 80 n.5, 217; Republic of, 136
Bernardo I, King of Kongo, 71
Biafada, 83
Bight of Benin (also 'Slave Coast'), 15, 44, 118, 250–1, 252, 254–5
Bight of Biafra, 16, 27, 42, 44, 46, 57, 245, 252
Bioko: see 'Fernando Po'
Biørn, A. R., Governor of Danish Settlements on the West African Coast, 167, 172–7
Black Peter, 126–7

266

Bonny, 15, 44
Bosman, Willem, 121
Botelho, Sebastião Xavier, Governor of Cabo Verde, 227, 237, 238 n.10
Bowen, Thomas Jefferson, 218
Brazil, 10, 21, 36, 46, 56, 62–3, 67–8, 74–6, 77–8, 87, 116 n.1, 118, 128, 226–9, 233–6, 238–41
Bridgman, Richard, 102–3, 106, 107, 112
Bridgman, William, 102
Bristol, 192
Britain: see 'Great Britain'
Brown, Christopher Leslie, 3
Brown, Thomas, 195
Brüe, André, 109 n.33
Bubi, 56
Bugendo, 81, 89, 91 n.51, 93–6
Bulom, 191
Butri, 120–1
Buxton, Thomas Fowell, 3, 4, 20 n.70, 21, 22, 185, 189, 205, 206, 207

Cabinda, 44
Cabo Verde, islands, 13, 30, 42, 57, 58, 61, 77, 173, 81, 85–87, 88, 89–90, 91, 92–3, 94, 96, 97, 104–6, 117–8, 227
Cacheu, 9, 81, 96
Cadamosto, 84
Cadiz, 105
Calabar (also 'Old Calabar'), 68
Caldeira, Arlindo, 54, 68, 70
Calumbo, 235
Cambambe, 232, 235
Cameroon, 42
de Caminha, Álvaro, Captain of São Tomé, 58, 59, 60, 61, 63, 69, 70
Campbell, Major Henry Dundas, Lieutenant-Governor of Sierra Leone, 189
Campbell, Sir Neil, Governor of Sierra Leone, 187
camwood: see 'dyewoods'
Canaries, islands, 80, 85
candles, 9
cannon (also 'firearms'), 133, 171
Cape Coast Castle, 44 n.46–7, 45 n.50, 122, 125 n.64, 141, 145, 154
Cape Mount, 191
Cape of Good Hope, 62
Cape of Lopo Gonçalves, 68
Cape Palmas, 61, 186
Cape Verde, peninsula, 82, 105
caravans (also 'transport'), 13, 84–5
Carneiro, António, Captain of Princípé, 58, 66
Carney, Judith, 46, 83, 85, 91
Carolina (also 'South Carolina'), 116, 132, 139
Carr, John, Chief Justice, 212
Cartagena de las Indias, 67, 89, 91 n.44
Casamance/ Cassamansa, 14, 94, 96, 105
cassava (also 'manioc'), 13, 41–3, 44–6, 76, 185, 195
Castlereagh, Charles William Stewart, Secretary of State for Foreign Affairs, 205
Catholic Church (also 'Christianity', 'missions', 'missionaries'), 61–2, 69
Cavalunga, Cabanga, 234
Cazengo, 231, 234–8, 241
de Centinera, Martín, 86

de Chaves, Ana, 64, 71
children, 58, 70, 148, 185, 208, 211, 220, 238
Chilpern, Captain, 107
Choiseul, Étienne François, duc, 151
Christensen, B. M., 179
Christianity (also 'Catholic Church', 'missions', 'missionaries'), 5, 21, 162–3, 188, 193, 205, 215
Christiansborg Castle, 158, 164–7, 171–8
Christoph Welser and Brothers, trading company, 72
Christopher, Emma, 154
Church Missionary Society (CMS), 5, 20–1, 87, 206–8; Niger mission, 221; Yoruba Mission, 208, 210, 212, 215–24
Clarkson, Thomas, 139, 140, 145
Clegg, Thomas, 210
cloth (also 'textiles'), exported from Africa, 12, 15, 30, 35, 87, 263
Coates, Benjamin, 198
cocoa, 5, 12, 19, 21–2, 25, 54, 76, 136, 151–2, 160, 179, 183, 185, 221–4, 236, 261, 264
Cocolí, 94
Coelho, Francisco Lemos, 91, 113 n.41
coffee, 5, 11, 21, 36, 76, 136, 147, 151–2, 160, 162, 178, 185, 188, 193, 195, 197–9, 209, 224, 229–30, 234–7
Cohen, William, 147
Coker, Jacob Kehinde, 221–3, 224
colonialism, 6–8, 12, 18–19, 24–6, 36, 65, 81, 132–3, 136, 139, 141, 147, 153–4, 158–79, 181–6, 189, 201, 206, 214, 217, 219, 223, 225–7, 259–60, 264–5
commercial transition, 6–8, 13, 24, 27, 79, 225–42, 243, 258–61, 264–5
Compagnie du Sénégal, 109 n.33
Companhia de Agricultura e Indústria de Angola e Benguela, 227–8
Company of Merchants Trading to Africa, 124, 130, 131–5
convicts, 57, 58–9
Copenhagen, 158–60, 164–5, 166 n.33, 167–8, 171–4
Coromandel Coast, India, 100
corn: see 'grain'
Correia e Silva, António, 86
corsairs (also 'piracy'), 62, 65–6, 75
Côte d'Ivoire: see 'Ivory Coast'
cotton, 4, 5 11–2, 21, 87, 92, 109, 117–8, 120–7, 132–5, 141–2, 147, 151–4, 162, 166, 170, 173, 176–8, 184, 188, 193, 198, 202, 209–10, 212, 214, 217–20, 222, 224, 227, 229–32, 239, 241, 259–60
cowry shells (also 'currency'), 249, 253, 255–7
credit, 165
crisis of adaptation, 7, 13, 23, 237–42, 258–65
Cross, river, 45
Crowther, Samuel Ajayi, Bishop of the Niger Territory, 221
da Cunha, Pedro Alexandrino, Governor of Angola 230–1, 236, 238
da Cunha, Pedro José Paiva, 54, 73
Curaçao, 126–7
currency (also 'cowry shells'), 253–8, 263
Curtin, Philip, 30 n.3, 37, 43, 90, 140, 154, 181, 253

Daaku, K.Y., 119, 120 n.24, 130, 134–5
Dahomey, 4 n.12, 119, 128, 217, 256, 257,
Dande, river, 233
van Dantzig, Albert, 118–19, 130, 133
Davies, K. G., 98 n.3, 118, 129
Deans, Isaac, 196
Dembos, 234, 235
Denmark, 1, 2, 18, 21, 23, 138, 151 n.19, 152, 158–79, 219
Devonshire, 99
Dike, K.O., 6
Diogo I, King of Kongo, 70–1
disease, 57–8, 61–2, 77, 80, 183
Dodowa, 177, 178
Donelha, André, 93, 95
Drescher, Seymour, 20, 225
Duque de Bragança, *présidio*, 229
dyewoods, 8, 10, 29, 34–5, 129; camwood, 108, 114, 185, 191, 199

East Anglia, 99
East India Company, Dutch, 100; English, 100
East Indies (also 'India'), 117 n.6, 131, 144, 147
Egba, 214 n.36, 215–1, 222
Elmina, 56, 58, 61, 62, 66–7, 68, 80, 120–2
Eltis, David, 90, 133–4, 245
Encoje, 234, 235
Equiano, Olaudah, 3, 204
Erskine, Hopkins W., 197
Escobar, Pedro, 56
exports, from Africa (also 'produce trade, direct with Africa', 'provisions'), 3, 9–10, 12, 28–53, 69, 72–3, 75–7, 79, 82–8, 90–4, 96–8, 102, 104–6, 114, 116, 118, 121, 129–30, 185, 187–92, 199, 211–12, 214, 218, 220–1, 223, 228–9, 231–3, 237, 241, 244, 253, 258–62
Eyzaguirre, Pablo, 54, 71

Fante, 45
Fenton, Edward, 95
Fernandes, Valentim, 59, 63, 66, 83, 85, 86
Fernando Po, 22, 56, 58, 64, 66
Ferreira, Blas, 88–9, 93, 94
Fields-Black, Edda L., 83, 91
de Figueiredo Abreu e Castro, Bernardino Freire, 239–40
Findlay, Lieutenant-Colonel Alexander, Lieutenant-Governor of Sierra Leone, 188
firearms, 112, 133, 170, 251–2, 254
Flindt, Jens, 172, 174–6, 177, 178
foodstuffs: see 'provisions'
Fogo, island (Cabo Verde), 87, 92
Forcados River (Niger Delta), 67
Forros, 59, 61, 76, 77
France, 11, 13, 19, 24, 31, 62, 65–66, 68, 72, 78, 100, 101, 103, 112, 114, 131–2, 136, 138, 141, 147–9, 151–3, 154, 156, 211 n.22, 229 n.15, 245, 260
Francisco, D., 70
Fredensborg, fort, 158, 162 n.15, 166
free labour, 23–4, 26, 125, 140, 148, 181–3, 188, 191, 200, 237–8, 261, 264–5
Freetown, 21, 30, 180, 182, 184–6, 189, 191, 206, 208–10, 212–13
French Soudan (Mali), 136
Friederichsnopel, 158, 165–79

Friderichsstæd, 177
Funchal, diocese, 61
Fynberg, Ole, 176

Gabon, 4, 35, 68, 134–5
Galam, 142
Gambia, river, 37, 42, 45, 83, 84, 86, 91, 100, 103, 114, 126, 127, 141, 151–2, 153
Garfield, Robert, 54, 63, 71
gender relations, 58–9, 65, 95, 163, 185, 215, 220, 237–8, 253–4
Geraldini, Alessandro, 82–3, 86
Gezo, King of Dahomey, 217
Gibson, Jacob, 195 n.45
ginger, 109, 122–3, 185, 218, 237
Ghana ('also 'Gold Coast'), 22, 25, 261, 264
Godinho, Simoa, 71
gold, 2, 8, 10–11, 15, 31, 35–6, 40, 44–5, 58, 61, 67, 80, 129, 142, 151, 153, 185, 262–3
Gold Coast, 1, 5, 11–12, 15, 16, 18–19, 21, 24, 25, 34, 38, 42, 43, 44, 45, 56, 113, 118–37, 141, 149, 150, 152, 154, 158–79, 203, 220 n.45, 237, 250–1, 252
Golungo, 227, 230, 231, 234–6
Gomes, Diogo, 84
Gorée, 13, 141, 145, 151–2
da Graça, Joaquim Rodrigues, 235
Grace, Edward, 142
Graf, Rev. John Ulrich, 21, 207, 209–15, 224
grain (also 'corn'), trade in, 12–13, 16, 37, 42–5, 90–1, 96, 152–3, 185, 250, 256; maize, 13, 42–6, 86, 209, 263; millet, 13, 30, 42–6, 82–4, 86, 91, 96, 250
Grain Coast, 8, 15, 44
Grande, river, 83
Great Britain, 1–6, 13, 15, 17–21, 24–5, 29, 35, 40, 48, 64, 99–115, 118–9, 122–57, 180–225, 233, 236, 239–41, 243, 245–8, 254–6, 260, 263–4
Grotrian, Johan Herman, 176, 177
groundnuts, 17, 22, 26, 136, 185, 243, 259, 261, 264
Guadeloupe, 103–4, 111
Guatemala, 100
Guiana, Dutch colony, 131
Guinea-Bissau, 9, 14
Guinea entrepreneurs: see 'Pingel, Meyer, Prætorius & Co.'
Gujarat, 100
gum, 15, 17, 19, 22, 109, 129, 153, 241, 263; gum arabic, 9, 27, 35, 141–2, 250
guns: see 'firearms'
Gunn, Isabella, 107
Gurley, Ralph Randolph, ACS Secretary, 197
Gusmão, Antonio Gonçalves de, 88–90

Haïti, see Saint-Domingue
Hall, Trevor, 85
Hamburg, 199, 237
Hammond, R. J., 225
Hastings, 207, 210, 213, 215
Havana, 239
Hawthorne, Walter, 46, 83, 87, 91, 94–5
Henriques, Isabel de Castro, 63
Herrnhuters (Moravian Brethren), 162
Heywood, Linda M., 70
hides (also 'skins'), 8, 10, 30, 35, 82, 93, 147, 185

Hinderer, Rev. David, 212
Hogendorn, Jan, 249, 255–6
Hopkins, Tony, 6–7, 14–15, 22, 23, 25–7, 258–9
Houghton, Daniel Francis, 154
humanitarianism (also 'abolitionism'), 3, 128, 143–4, 156–7, 162, 164, 184, 186, 188, 198–9, 207, 214–15, 223

Ibadan (also 'Church Missionary Society, Yoruba Mission'), 210, 212, 214 n.36, 216, 219
Icolo, 233
Igboland, 204
Ijaye, 219
Ijebu, 219
Ikija, 218
illegal slave trade, 2, 18, 89–91, 204
imperialism, 7, 20–1, 149, 153–4, 200, 225–6,
India (also 'East Indies'), 56–7, 81, 100, 102, 103, 105, 106, 111, 115, 117, 126
indigo, 11–12, 98–115, 117–27, 129, 131–5, 139, 147–8, 151, 153–4, 162, 170, 188, 227, 259
industrialization (also 'manufacturing', 'raw materials'), 3, 34, 98–9, 210, 246, 261, 263–5
Inikori, Joseph, 9, 46, 119, 124, 128–9, 132, 133, 135 n.104, 154, 252, 254–5
iron, 83, 94, 148, 229
Irving, Dr Edward, 218
Isert, Carl Christopher, 172, 174, 176
Isert, Paul Erdmann, 152, 154, 158–79
ivory (also 'elephant's teeth'), 2, 8, 10, 15, 22, 30, 35, 37, 93, 114, 129, 142, 153, 185, 199, 241
Ivory Coast, 15

Jamaica, 37 n.23, 47, 100, 102–3, 110–11, 112, 131
James Fort (Gambia), 42
James Island (Gambia), 100–2, 114
du Jarric, Pierre, 89
Jeremie, Sir John, Governor of Sierra Leone, 189
Jesuits, 44, 89
João II, King of Portugal, 56, 69
João III, King of Portugal, 61, 67
Jobson, Captain James, 101
Johannes, 176, 177
Johnson, G.W., 219
Johnson, Henry, 21, 210, 213, 223–4
Johnson, Marion, 249, 253, 255–6
Jolof, 83, 84, 105

Kassanké, 14, 95, 96
Kea, Ray, 170, 237
Keta, 122, 158
Kilby, Randall, 197
Kipnasse, Governor of Danish settlements on the West African Coast, 166–7, 171–2, 174–5
Kirstein, Eric, 160
Klein, Martin A., 260
kola nuts, 9, 28, 84, 93–4, 260, 263–4
Komenda, 122–3
Kongensteen, fort, 158, 166, 173
Kongo, kingdom, 56, 57, 59, 61–2, 66, 67–8, 70–1, 80
Kru, 195
Kup, A. P., 98 n.3
Kwanza, river, 44, 45

Labat, Jean-Baptiste, 103, 109 n.33
Lagos, 25, 191–2, 200, 219–24, 261
Lains e Silva, Hélder, 73
land tenure, 11, 60, 81, 125, 150, 162, 164, 166, 168–72, 183–4, 186, 194–6, 201, 219, 229
Lang, 229
Law, Robin, 237, 256
de Leão, Duarte, 88–90, 93
Leeward Islands, 100, 102, 103, 107, 108, 114–15
legitimate trade, 1–6, 16, 20–1, 23–7, 46, 79, 129, 138–57, 180, 182, 184, 186, 189, 192–3, 198, 200, 203–6, 214, 216, 223, 227, 241, 243, 258–60, 265
Liberated Africans: see 'recaptives'
Liberia, 18, 20, 21, 24, 180–2, 186, 192–202
liquor, distilled in Africa, 76; rum, 107, 121, 128, 171; cane brandy, 227
Lisbon, 59, 61, 62, 66, 68, 70, 72, 74, 86–7, 227, 229, 231, 234
Liverpool, 35, 141, 192, 197 n.59
Liverpool, Robert Banks Jenkinson, Earl, 187
livestock, 92, 165, 256; cattle, 29, 60, 82, 197, 229, 232, 237; donkeys, 60; goats, 60; horses, 95, 237; mules, 35, pigs, 60, 72, 76, 197; poultry, 60, 76, 256; sheep 60
Livingstone, David, 5
Loango, 35, 44
Loko, 185, 189, 191
London, 1, 7, 21, 47–8, 99, 101, 104, 107–09, 112, 114, 119, 124, 128, 131, 135, 140, 141, 142, 183, 192, 208, 213
Lopes de Calheiros e Menezes, Sebastião, Governor of Angola, 237, 238, 239, 241
Lopes de Lima, José Joaquim, 227, 238
Louisiana, 195 n.25, 213
Lovejoy, Paul E., 245, 246, 248, 253, 259–60
Low Countries (also 'Netherlands'), 99
Luanda, 13, 29, 38 n.30, 42 n.38, 43, 44, 68, 227–37, 239–41
Lucala, river, 237
Lucas, Eliza, 103

Macaulay, Zachary, Governor of Sierra Leone, 183, 205
Macdonald, Norman William, Governor of Sierra Leone, 191, 212
Macqueen, James, 187–88
Madeira, 10, 56, 57, 69, 72, 74, 77, 80, 85, 99 n.5
Madagascar, 41 n.36, 45
de Magalhães Mesquita, Antonio, 232
Magbele, 189
Maghreb (also 'North Africa'), 87
maize: see 'grain'
malagueta: see 'pepper'
Malheiros, Manoel Eleutério, Governor of Angola, 232
Mali, empire, 85
Malphy, 166–7, 171
Manchester, 141, 210, 214
Mande, 83, 85
Mandinka, 83, 86, 94
Manicongo, 58, 67
Manicongo, Pêro de, 70
manioc: see 'cassava'
manufacturing (also 'industrialization', 'raw materials'), in Europe, 3, 9, 20, 22, 36, 98–9, 131,

140–3, 146, 156–7, 188–9, 210, 232, 254 ; in Africa, 12, 87, 94, 105–15, 121, 139, 156, 218–19, 265
Marques, João Pedro, 226
marronage (also 'slave resistance'), 55–6, 63–5, 74–5, 77, 112
Martinique ('Maritinico'), 157
Masatamba, Kassanké monarch, 95, 96
Massangano, 235
May, Captain, 107
McDonogh, John, 195 n.45
McKenzie, Captain Kenneth, 154
Mechlin, Joseph, 194
de Mello, João, Captain of São Tomé, 58
de Melo, Fernão, Captain of São Tomé, 66
Melvil Thomas, Governor of the Company of Merchants Trading to Africa on the Gold Coast, 124, 154
Mende, 185, 196,
Menino, João, 70
Meyer, Peder, 173–4
Middle East, 100
Miller, Joseph C., 258
Mina coast (also 'Bight of Benin', 'Slave Coast'), 68
Minchin, Susie, 38 n.27, 91 n.51, 96
mining, 206, 229, 254
Mintz, Sidney, 80
Miracle, Marvin, 83 n.15
Misevich, Phil, 30
missions (also 'Catholic Church', 'Christianity', 'missionaries'), Catholic, 61, 89; Protestant, 162, 207–8, 210, 212, 215–18, 221–2
missionaries (also 'Catholic Church', 'Christianity', 'missions'), 5, 20–21, 61, 89, 162, 171, 205–9, 215–18, 220–1, 224
Mlefi: see 'Malphy'
Mohora Suru, 191
molasses: see 'sugar'
Monrovia, 186
Monteiro, Joachim John, 241
Montserrat, 103, 106, 108
Morris, Edward, 196
Morse, Edward, 154
Mossamedes, 226, 233–4, 239–40, 242
Mughal Empire (also 'India'), 98
Münzer, Hieronymus, 69
Muslims, 102, 117

Naasamón, 82–3
de Narria, Juan, 89
de Nemours, Pierre Dupont, 148
Netherlands (also 'Low Countries'), 10, 11, 12, 17 n.61, 31, 36 n.20, 48 n.57, 56, 62, 68, 75, 78, 99, 100, 118–22, 125–8, 130–1, 133–7, 144, 149, 151, 152, 160 n.7, 167, 203, 212, 256
Nevis, 47
New Orleans, 196
Newson, Linda A., 38 n.27, 91 n.51, 96
New Spain, 100
New York, 194, 196
Nielsen, Peder, 176
Niger, river, 30 n.2, 45; delta, 6, 15, 57, 59, 66–7, 70; Niger Expedition (1841–42), 4, 19, 20, 21, 27, 180, 189, 191, 204, 205 n.3, 206; Niger Mission: see 'Church Missionary Society'

Nigeria, 6, 21, 22, 25–6, 136, 192, 204, 221 n.49, 222, 262, 264
Nile, river, 29
Ningo, 158
de Noronha, Antonio Manoel, Governor of Angola, 228–30
Northrup, David, 16
Nossa Senhora da Graça, church, 61
North Africa (also 'Maghreb'), 84, 85, 117
Nova Scotia, 183, 184
Nuñes, river, 16, 93, 94, 96, 105

Obuobi Atiemo, Okuapenhene, 167–70, 172
Ogunbona, 218
Ohadike, Don C., 261
O'Hara, Charles, Governor of Senegambia, 153
Old Calabar, 15, 44, 45, 239 n.109
ostrich feathers, 35
Oswald, Richard, 141
Oualata, 84
Ouidah, 13, 41–2, 43, 44, 118, 119, 121 n.29, 122, 123, 124 n.53, 126 n.61, 126
Our Lady of the Rosary, Catholic brotherhood, 61

Paiva, João de, 57
palm-oil, 9, 12, 13, 15–16, 22, 26–7, 29, 40, 42–3, 46, 76, 136, 179, 185–6, 189–92, 198–201, 243, 256, 259–61, 263–4; palm-kernels, 22, 199
de la Palma, Willem, Director-General of the Dutch WIC in West Africa, 120–1, 126, 129
Palmerston, Henry John Temple, Viscount, 192
Paquette, Gabriel, 226
Pará, 229
partition of Africa, 7
Paul III, Pope, 61
pawns, 219, 237, 261, 264
Pennsylvania, 196
Pepel, 81
pepper, 8, 10, 15, 35, 109, 123, 185, 198, 218; malagueta, 8, 13, 35, 43–4
Pereira, Duarte Pacheco, 83, 86
Pereira, Valentim José, 231 n.35
Pérez, Manuel Bautista, 91, 92, 96
Pernambuco, 232, 234
Peyton, Reverend Thomas, 209
Philadelphia, 29, Centennial Exhibition (1876), 197 n.59
Physiocrats, 147–8, 151
Pingel, Meyer, Prætorius & Co., 161 n.10, 164–4, 171
Pinheiro, Luís da Cunha, 54
piracy (also 'corsairs'), 62, 75
Pires, Mateu, 65
Pitt, William, 141
plantations, in Africa, 1, 10–12, 14, 17–18, 23, 26, 36, 45–6, 54, 56, 59–60, 63–5, 69–78, 80–1, 98, 100, 116, 118–25, 128–36, 140, 142, 151–2, 154, 160–1, 163, 166, 173, 176–9, 186, 189, 196, 203, 221–3, 234–5, 240–1; in Americas, 11, 17, 23–4, 26, 35, 45, 55, 63, 74–5, 77, 87, 100, 116–17, 130–1, 135–6, 152, 154, 159, 161, 178–9, 183, 186, 193, 206, 249, 263
Podore, 153
Ponny (Kpone), 166
Portendick, 141

Portugal, 13, 18, 21, 24, 30, 36, 42, 48 n.57, 54–5, 56–8, 60–4, 66–9, 72, 75–8, 81, 85, 87, 89, 93–4, 106, 117, 128, 135, 140, 203, 225–31, 233, 236, 240–2
Postlethwayt, Malachy, 17, 145–51, 156, 157
prices of slaves, 19, 32–4, 37–8, 47–53, 228, 239, 243, 245–50, 255–7, 259–60, 264
Prince Cain, 191
Príncipe, 54–78
Prindsensten, fort, 158
produce trade, direct with Africa (also 'exports from Africa'), 30–6, 47–53
provisions, trade in, 12–17, 19, 30, 37–46, 56, 60–1, 63, 67, 71, 76, 78, 82–97, 152, 250, 254, 257; see also 'cassava', 'grain', 'maize', 'manioc', 'millet', 'rice', 'yams'
Pungo Andongo, 229, 234, 235

Quifandongo, 238
Quitanda, market, 233

Ramsay, James, 139–40
Ratelband, Klaas, 75
raw materials (also 'industrialization', 'manufacturing'), 2–3, 9, 12–13, 98–101, 108, 141–2
recaptives 18, 20–21, 24, 184–7, 189, 191, 193, 201, 206–8, 211, 215–17, 220
Recife, 229
Regaire, Pedro, 232
Reynolds, Edward, 119, 130
Rhode Island, 41
Ribeira Grande, diocese, 61
rice, 12–13, 17, 29–30, 37, 42–3, 45–6, 82–3, 85–6, 89, 91, 93–4, 96, 116–17, 136, 139, 153, 209, 229
Richardson, David, 47, 245, 246–8, 249, 259–60
Ricketts, Rev. John, 221–2, 223
Rio de Janeiro, 41, 229
Rio Real (New Calabar river), 57, 58
Robbin, Henry, 210, 214, 218, 223–4
Rodney, Walter, 9, 46, 90, 98 n.3, 119, 129, 252
Rodrigues, Nicolao, 96
Røge, Pernille, 147–8, 153
von Rohr, Julius Benjamin, Lieutenant Colonel, 177
Rokelle, river, 189
Roubaud, Abbé Pierre-Joseph-André, 147
Royal African Company, English (RAC), 1 n.2, 11–12, 17, 35, 40, 47–8, 98–110, 112–15, 118–37, 148–50, 151
Royal Botanical Garden Kew (RBGK), 21, 208, 209, 210, 213
Royal Danish African Mission Institution, 162
Royal Navy (also 'suppression of the slave trade'), 2, 18, 204, 211, 244–6
rubber, 22, 264
rum: see 'liquor'

Saget, M., 151–2
Sahara, 41, 84, 117, 141, 250, 265
Sahel, 83, 85, 250 n.20, 261
St Croix, 177
Saint-Domingue, 19, 246; revolution (1791–1804), 245, 246
St Kitts, 139

Saint-Louis, 13, 109 n.33, British occupation, 141, 145
St Paul, river, 197
salt, 43, 105
Saluum, river, 82
Santa Casa de Misericórdia, charitable institution, 61, 69
Santa Maria, church, 61
de Santa Maria, Rodrigo, 70
de Santarém, João, 56
Santiago, island (Cabo Verde), 61, 91 n.51, 105
Santo André (Sassandra), river, 61
Santo Domingo, town, 30, 82
dos Santos Silva, Ana Joaquina, 232–3, 235
São Domingos, river, 88–9, 95, 96
São Francisco, church and monastery, 61
São Jorge da Mina, see 'Elmina'
São Salvador, diocese, 62
São Sebastião, fort, 62
São Tomé, 10, 14, 22, 30, 31, 35, 36, 42, 44, 54–78, 80, 81, 85, 86–7, 117–18, 128, 140, 239
Schimmelmann, Count Ernst, 160–4, 165, 168, 171–4, 177, 178
Schwartz, Stuart B., 71
Sea Island, 210
Searing, James, 16, 19, 250–1, 252, 254, 258 n.47, 260 n.56, 263
Seck, Ibrahim, 84
Sekondi, 120–1
Senegal, river, 9, 15, 16, 45, 83, 84, 142, 105, 153; region, 19, 27, 117, 145, 149, 151, 230 n.15
Senegambia, region, 12, 30, 34, 38, 41, 42, 43, 86–7, 90, 141–2, 245, 250–1, 252, 254, 258 n.47, 261; British colony, 17, 35, 145, 151, 153–3, 154
Serafim, Cristina Maria Seuanes, 54–5, 75
Sereer, 82, 252 n.25
van Sevenhuysen, Jan, Director-General of the Dutch WIC in West Africa, 120, 121, 126–7
Seville, 82, 105
Shama, 120, 122
Sharp, Granville, 183, 184
Sherbro, 34, 106, 107–9, 112, 126, 127
Shick, Tom, 201
Sibthorpe, A. B. C., 224
Sierra Leone, region, 9, 11, 12, 17 n.6, 30, 34, 84, 89, 93–4, 100, 102, 108–9, 112, 117–19, 122, 129, 131, 252; colony, 1, 5 n.17, 18, 20–1, 24, 25, 30, 34, 138, 140, 141, 160, 180–92, 193, 200–2, 204–24
Sierra Leone Company, 1, 2, 4, 5, 18, 21, 23, 40, 138, 183–4
Sijilmassa, 84
da Silva, Pedro, Governor of São Tomé, 76
da Silva Neves, José Augusto, 229
da Silveira Pinto, Adrião Acácio, Governor of Angola, 228, 233, 234
Slave Coast (also 'Bight of Benin'), 15, 43, 66, 118, 119, 122
slave resistance (also 'marronage'), 10, 19, 36, 54–6, 63–5, 67, 75, 77–8, 112–13, 115, 136, 153, 168, 175, 238–9, 245–6, 249
slave trade, trans-Atlantic, 1–4, 6, 8–11, 14–17, 19, 27–8, 30–53, 56–9, 61–3, 66–8, 76–8, 80, 82–3, 88–90, 93–5, 97, 100, 107, 114, 116–17, 119, 128–30, 134, 137, 140–51, 153–4, 156–8, 160–1, 174, 178, 181, 186, 188–9, 203–6, 214, 217,

225–8, 232–4, 237, 239, 242–61, 263–5
slavery, in Africa, 10–11, 21, 23–7, 55, 57, 59–61, 63, 70–1, 76–7, 80, 84–5, 87, 106, 112–13, 115–16, 118–19, 122, 125–6, 154, 162–4, 190, 194, 220, 237–44, 250, 253, 259–61, 264–5
Smeathman, Henry, 140, 152, 154, 183, 204, 214
soap, 9, 22
Society for Effecting the Abolition of the Slave Trade, 138, 141, 144
Sodeke, 216
soil conditions, 3–4, 10–12, 17, 45, 55–7, 69, 74, 77, 81, 85, 102–4, 136, 151, 166, 195, 197, 200, 210, 214, 218–19, 223, 227, 265
Sokoto Caliphate, 259
de Sousa e Almeida, João Maria, 76
South Carolina, 29, 100, 103, 116, 117
Soyo, 66, 67
Spain, 30, 31, 62, 64, 67, 68, 72, 78, 88, 100, 132
Stibbs, Bartholomew, 37
Straw, James K., 198
Sudan, Western, 250; Central, 258 (also 'French Soudan')
sugar, as export from Africa, 10–11, 14, 21, 30–1, 35–6, 54–7, 59–61, 63–5, 69–78, 80, 82, 85, 107–8, 117–18, 120–3, 126, 128, 131–4, 136, 139, 147–8, 152, 160, 162, 188, 197–9, 227, 229–30, 232–4, 241–2; molasses, 198–9
sugar, as export from the Americas, 2, 29, 36, 56, 74, 76, 100, 104, 108, 115, 117–18, 131–2
suppression of the slave trade (also 'Royal Navy'), 2, 6, 20, 189, 191–3, 201, 211, 244, 246
Suriname: see 'Guiana, Dutch colony'
Swan, James, 146
Swedenbourg, Emanuel, 150

Talleyrand, Charles, 153
Tambaka, 191
Tasso, island, 102, 108, 109, 112
de Tavora, Lourenço Pires, Governor of São Tomé, 75
tea, 36
Temne, 185, 191
Tenreiro, Francisco José, 63
textiles: see 'cloth'
Thomas, Sir Dalby, 35, 122, 131
Thompson, Thomas Perronet, Governor of Sierra Leone, 205
Thornton, John, 70, 81
timber (also 'wood', 'forest resources'), as export from Africa, 22, 185, 191, 205–6, 263
Tlemcen, 84
tobacco, as export from Africa, 5, 21, 118, 120, 123, 128, 131–2, 139, 153, 170, 217, 230–2, 233, 236
Tobago, 152
Torrão, Maria Manuel Ferraz, 86, 89 n.35, 91 n.51
Townsend, Rev. Henry, 217, 220
traders, 2, 7–8, 11, 14–16, 26, 36, 38, 41, 48, 67–8, 72, 76, 81, 84–5, 88–9, 91, 93–6, 99, 101, 103–4, 107, 115–16, 121, 124, 127, 129–33, 139, 141–3, 146–50, 152–4, 156, 164–5, 185, 187, 191, 194, 196, 201, 205, 217, 228–30, 250, 258, 261, 264
Trafford, Henry, 141
transport (also 'caravans'), 28–30, 45, 93, 137, 166, 173, 218, 223, 229, 230, 237, 252, 257, 261
trans-Saharan trade, 8, 14, 16, 41, 82, 84, 87–88, 117, 250, 253
Tubreku (Togbloku), 173
Turkey, 132
Turner, Major-General Charles, Governor of Sierra Leone, 185

Upper Guinea, 12, 13, 16, 17, 30–1, 34, 42, 79–115, 119, 141
USA, 18, 24, 198, 199, 239

Vai, 195
Venn, Rev. Henry, CMS Secretary, 5, 21, 206, 207–8, 212 n.26, 213, 218, 221, 222, 223
Vera Cruz, 67
Vili, 44
Virgina, 231,
Volta, river, 158, 160 n.7, 165–7, 173–4

war, 2, 11, 19, 34, 41, 64–65, 78, 101, 114, 122, 129, 139, 141–2, 144–5, 147, 151–3, 167, 196, 200, 202, 215–16, 218–20, 222, 245–6, 251
Warri, 68
wax: see 'bees-wax'
Webster, J. B., 218
Weeks, John, Bishop of Sierra Leone, 220
Welwitsch, Frederico, 231, 238
West-Central Africa (also 'Angola'), 19, 29, 47, 244–5, 249–50, 252, 258, 263
West India Company, Dutch (WIC), 11–12, 15, 40, 62, 68, 75, 118, 120–2, 125–9, 130–1, 133, 134–5, 151, 152
West Indies, 11, 101, 103–7, 111–12, 115, 117, 123, 125–7, 130–2, 134, 139–40, 147, 153, 154, 160–1, 163, 173, 176–9, 181, 183, 187, 194, 206, 211, 215, 221
Whalley, Andrew, 119
Wilberforce, William, 2
Wilhelm, Andrew, 217
Willis, John, 43
Wiltshire, 99
Winneba, 122
Wolf, Eric, 94–5
women, 58, 64–5, 71, 87, 95, 103, 106–7, 112, 165, 175, 185, 215, 232–5, 237–8, 253–4
Woodard, Gilbert, 179

yams, trade in, 13, 16, 42–5, 63
York Island, 100–1, 102, 104, 107–8, 109, 114–15
Yorkshire, 99
Yoruba, 186, 215; recaptives, 186, 191; Yorubaland, 23, 186, 210, 215–24; Yoruba mission: see 'Church Missionary Society'

Zenza, 235
Zimmermann, Johannes, 220 n.45

www.ingramcontent.com/pod-product-compliance
Lightning Source LLC
Chambersburg PA
CBHW051604230426
43668CB00013B/1981